普通高等教育“十二五”规划教材

Modern Control Theory

（现代控制理论）

胡 健 刘丽娜 编著

国防工业出版社
·北京·

内 容 简 介

本书内容共分为6章，主要涉及系统建模、系统分析和系统优化设计。第1章通过引入控制系统的一些基本概念，给出了系统的数学描述方式，例如：状态空间模型，传递函数矩阵。第2章在时域内对系统进行了定量分析。第3章和第4章主要进行了系统的定性分析。其中，第3章讨论了系统的稳定性问题，第4章讨论了系统的能控性和能观性问题。第5章研究了系统综合设计的方法，例如：状态反馈，利用状态观测器进行状态重构。在第6章中，研究了离散系统的建模、分析与综合设计。

本书可作为自动化、电气工程及其自动化等专业高年级本科生以及控制科学与工程、电气工程等学科研究生学习现代控制理论双语课程的教材，也可作为学习高级宏、微观经济学的经济、管理学科研究生的辅助教材。

图书在版编目(CIP)数据

现代控制理论 = Modern Control Theory：英文/胡健，刘丽娜编著．—北京：国防工业出版社，2012.3(2024.1重印)

普通高等教育“十二五”规划教材

ISBN 978-7-118-07997-5

Ⅰ.①现… Ⅱ.①胡… ②刘… Ⅲ.①现代控制理论—高等学校—教材—英文 Ⅳ.①O231

中国版本图书馆CIP数据核字(2012)第029179号

※

国防工业出版社出版发行
(北京市海淀区紫竹院南路23号 邮政编码 100048)
北京虎彩文化传播有限公司印刷
新华书店经售

*

开本 787×1092 1/16 **印张** 12¼ **字数** 276千字
2024年1月第1版第2次印刷 **印数** 4001—4500册 **定价** 28.00元

国防书店：(010)88540777　　发行邮购：(010)88540776
发行传真：(010)88540755　　发行业务：(010)88540717

Preface

Modern control theory is greatly developed after 1960s. Based on the state space method and the mathematical tools such as matrix theory and differential equation, modern control theory plays an important role in control science and engineering and other research areas.

The contents of the book were organized into six chapters and mainly dealt with system modeling, system analysis and system optimization design. In chapter 1, the basic concepts of control systems were introduced, and the mathematical descriptions for the systems were discussed. In chapter 2, the quantitative analysis in time domain for the linear dynamic systems was dealt with. The qualitative analysis problems were discussed in chapter 3 for stability, in chapter 4 for controllability and observability, respectively. Chapter 5 covered some important approaches for system synthesis, such as state feedback and state reconstruction with observer. In chapter 6, a study of modeling, analysis and synthesis design for the discrete-time systems was carried on.

The book focuses on the basic concepts, theorems and methods in modern control theory. Many of the concepts are very abstract. Furthermore, learning and understanding them need to use a lot of mathematical knowledge. So, we try to use the straightforward narrative approach to explain that obscure knowledge and provide a wealth of examples to help students deepen their understanding of these important concepts.

Bilingual teaching is the inevitable trend of education reform, also the important action to improve undergraduate education quality, and the requirement of cultivating compound-type talent in the 21 century. This book was written based on the practice of bilingual education for modern control theory course, and is a result of the construction for the fine course. This book can be used as a bilingual material of modern control theory for the senior undergraduates, such as automation, electrical engineering and automation specialties, as well as for the postgraduates, such as control science and engineering, electrical engineering, and so on. In addition, for the economics and management postgraduates, which want to study advanced macroeconomics and microeconomics, the book can also serve as reference.

Chapter 1～5 of the book were written by Prof. Hu Jian, and chapter 6 was written by Liu Lina. The authors thank Prof. Li Suling, Prof. Du Qinjun and Prof. Zhang Cunshan for their support and many useful comments. We also appreciate the help of Mr. Zhai Xingbo, Mr. Du Jun and Ms. Mu Lili in the edition and publication of the book. Furthermore, we would especially like to thank the students, who give us many valuable comments and suggestions over the bilingual teaching of modern control theory.

Hu Jian Liu Lina

Dec 10, 2011

前 言

20 世纪 60 年代以来，现代控制理论取得了巨大的发展。通过引入状态空间方法和利用矩阵理论、微分方程等数学工具，现代控制理论在控制科学与工程以及一些其他的研究领域都发挥了非常重要的作用。

本书内容共分为 6 章，主要涉及系统建模、系统分析和系统优化设计。第 1 章通过引入控制系统的一些基本概念，给出了系统的数学描述方式，例如：状态空间模型，传递函数矩阵。第 2 章在时域内对系统进行了定量分析。第 3 章和第 4 章主要进行了系统的定性分析。其中，第 3 章讨论了系统的稳定性问题，第 4 章讨论了系统的能控性和能观性问题。第 5 章研究了系统综合设计的方法，例如：状态反馈，利用状态观测器进行状态重构。第 6 章研究了离散系统的建模、分析与综合设计。

本书介绍了现代控制理论的基本概念、理论和方法。其中，很多概念是非常抽象的，对它们的学习和理解需要用到较多的数学知识。因此，我们在阐述这些概念时尽量做到深入浅出，并提供了丰富的算例以帮助学生加深对这些重要概念的理解。

双语教学是高等教育改革的一个重要趋势，也是改进高等教育教学质量、培养 21 世纪复合型人才的重要举措。本书的编写得益于在现代控制理论教学中进行的双语教学实践，同时也是现代控制理论精品课程建设的成果之一。本书可作为自动化、电气工程及其自动化等专业高年级本科生以及控制科学与工程、电气工程等学科研究生学习现代控制理论双语课程的教材，也可作为学习高级宏、微观经济学的经济、管理学科研究生的辅助教材。

本书的第 1～5 章由胡健副教授编写，第 6 章由刘丽娜老师编写。我们要感谢李素玲教授、杜钦君教授和张存山教授为本书的编写所提供的支持和提出的很多宝贵意见。我们也要感谢翟兴波先生、杜钧先生和穆丽丽女士为本书的编辑和出版提供的帮助。我们要特别感谢选修现代控制理论双语课程的学生们，在教学过程中，他们提出了很多非常有价值的意见和建议。

由于作者水平所限，书中不妥之处在所难免，恳请指正。

胡 健　刘丽娜

2011 年 12 月

Contents

Rudolf Emil Kalman (1930—) is a Hungarian-American electrical engineer, mathematical system theorist, and college professor. He is most noted for his co-invention and development of the *Kalman filter*, a mathematical formulation that is widely used in control systems, avionics, and outer space manned and unmanned vehicles. For this work, U.S. President Barack Obama awarded Kalman with the *National Medal of Science* on October 7, 2009.

During the 1960s, Kalman was the leader in the development of a rigorous theory of control systems. Among his many outstanding contributions were the formulation and study of most fundamental state-space notions (including controllability, observability, minimality, realizability from input/output data, matrix Riccati equations, linear-quadratic control, and the separation principle) that are today ubiquitous in control. While some of these concepts were also encountered in other contexts, such as optimal control theory, it was Kalman who recognized the central role that they play in systems analysis. The paradigms formulated by Kalman and the basic results he established have become an intrinsic part of the foundations of control and systems theory and are standard tools in research as well as in every exposition of the area, from undergraduate engineering textbooks to graduate-level mathematics research monographs.

During the 1970s Kalman played a major role in the introduction of algebraic and geometric techniques in the study of linear and nonlinear control systems. His work since the 1980s has focused on a system-theoretic approach to the foundations of statistics, econometric modeling, and identification, as a natural complement to his earlier studies of minimality and realizability.

Chapter1　State Space Description

1.1　Definition of State Space

1.1.1　Example

Example 1.1　A very simple RLC network shown in Figure 1.1 is considered.

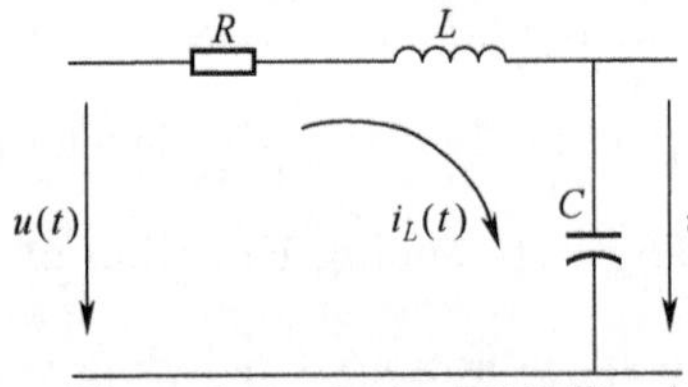

Figure 1.1　Series RLC Circuit

Suppose that the voltage $u(t)$ is the input to the RLC network. This circuit contains two energy-storage elements: the inductor and the capacitor. Applying Kirchhoff's laws, the voltage $u_c(t)$ across the capacitor C and the current $i_L(t)$ through the inductor L satisfy the following differential equations.

$$C\frac{\mathrm{d}u_c(t)}{\mathrm{d}t}=i_L(t) \tag{1.1}$$

$$L\frac{\mathrm{d}i_L(t)}{\mathrm{d}t}+R\cdot i_L(t)+u_c(t)=u(t) \tag{1.2}$$

The voltage $u_c(t)$ across the capacitor C is considered to be the output $y(t)$, then

$$y(t)=u_c(t) \tag{1.3}$$

Differentiating (1.1) and substituting it into (1.2), we get a second-order differential equation.

$$LC\frac{\mathrm{d}^2u_c(t)}{\mathrm{d}t^2}+RC\frac{\mathrm{d}u_c(t)}{\mathrm{d}t}+u_c(t)=u(t) \tag{1.4}$$

The second-order differential equation (1.4) is called the differential equation description of the system.

The differential equation description can be directly converted to the transfer function description by **Laplace transform.** By taking the Laplace transform of (1.4) and assuming the **zero initial conditions** hold true, the transfer function description of the RLC network is obtained as

$$G(s)=\frac{U_c(s)}{U(s)}=\frac{1}{LCs^2+RCs+1} \tag{1.5}$$

From the description (1.4) and (1.5), it can be seen that the differential equation description and the transfer function description are all the external descriptions of a system.

If we make the definitions, $x_1(t)=u_c(t)$ and $x_2(t)=i_L(t)$, for $t\in(0,t]$, the following differential equations can be obtained from (1.1) and (1.2).

$$\dot{x}_1(t)=\frac{1}{C}x_2(t) \tag{1.6}$$

$$\dot{x}_2(t)=-\frac{1}{L}x_1(t)-\frac{R}{L}x_2(t)+\frac{1}{L}u(t) \tag{1.7}$$

$x_1(t)$ and $x_2(t)$ are called state variables and are the components of the state vector $\boldsymbol{X}(t)=[x_1(t) \quad x_2(t)]^{\mathrm{T}}$.

Moreover, the differential equations above are expressed in matrix form notation as

$$\begin{bmatrix}\dot{x}_1(t)\\ \dot{x}_2(t)\end{bmatrix}=\begin{bmatrix}0 & \frac{1}{C}\\ -\frac{1}{L} & -\frac{R}{L}\end{bmatrix}\begin{bmatrix}x_1(t)\\ x_2(t)\end{bmatrix}+\begin{bmatrix}0\\ \frac{1}{L}\end{bmatrix}[u(t)] \tag{1.8}$$

It can also be expressed in a more compact form as

$$\dot{\boldsymbol{X}}(t)=\boldsymbol{AX}(t)+\boldsymbol{bu}(t) \tag{1.9}$$

Where

$$\boldsymbol{X}(t)=\begin{bmatrix}x_1(t)\\ x_2(t)\end{bmatrix};\dot{\boldsymbol{X}}(t)=\begin{bmatrix}\dot{x}_1(t)\\ \dot{x}_2(t)\end{bmatrix};$$

$$\boldsymbol{A}=\begin{bmatrix}0 & \frac{1}{C}\\ -\frac{1}{L} & -\frac{R}{L}\end{bmatrix};\boldsymbol{b}=\begin{bmatrix}0\\ \frac{1}{L}\end{bmatrix};\boldsymbol{u}(t)=[u(t)].$$

The set of the differential equations in matrix form (1.8) or (1.9) is called the state equation of the system.

Rewrite (1.3) in matrix form as

$$\boldsymbol{y}(t)=[1 \quad 0]\begin{bmatrix}u_c(t)\\ i_L(t)\end{bmatrix} \tag{1.10}$$

It can be expressed in a more compact form also as

$$\boldsymbol{y}(t)=\boldsymbol{cX}(t)+\boldsymbol{du}(t) \tag{1.11}$$

Where

$$\boldsymbol{y}(t)=[u_c(t)];$$
$$\boldsymbol{c}=[1 \quad 0];\boldsymbol{d}=[0].$$

The set of the algebraic equations in the matrix form (1.11) is called the output equation of the system.

Both the state equation and the output equation are called the state space description of asystem. The state space description is an internal description of system.

1.1.2 Definitions

Definition 1.1 The **state** of a system is a minimum set of variables x_i, called the

state variables, to describe the system's behavior. The initial values of this set and the inputs of system for $t \in (t_0, \infty)$ are sufficient to describe uniquely the system's response for all $t \geqslant t_0$.

There are several remarks about state variables or state must be emphasized that,

(1) the choice of state variables or state is not unique;

(2) the state variables of a system may be purely mathematical quantities, so they may or may not be easily interpretable in physical terms;

(3) the number of state variables is called the **order** or **dimension** of the system;

(4) the components of the state is linearly independent at any instant, in other words, x_i can not be expressed by x_j linearly, $j \neq i, i, j = 1, 2, \cdots, n$;

(5) the state of a system at any fixed time can be represented by the values of a set of state variables x_i, and the state can be represented by an n dimension **state vector** $\boldsymbol{X}(t) = [x_1(t) \quad x_2(t) \quad \cdots \quad x_n(t)]^{\mathrm{T}}$.

Definition 1.2 **State space** is defined as the n dimension space in which the components of the state vector represent its coordinate axes.

Definition 1.3 **State trajectory** is defined as the path produced in the state space by the state vector as it changes with the passage of time.

1.1.3 State Space Description

Suppose the system that we are considering has r output signals denoted by $y_1(t)$, $y_2(t), \cdots, y_r(t)$ and m input signals denoted by $u_1(t), u_2(t), \cdots, u_m(t)$. The **MIMO** (multiple-input multiple-output) system can be illustrated by Figure 1.2.

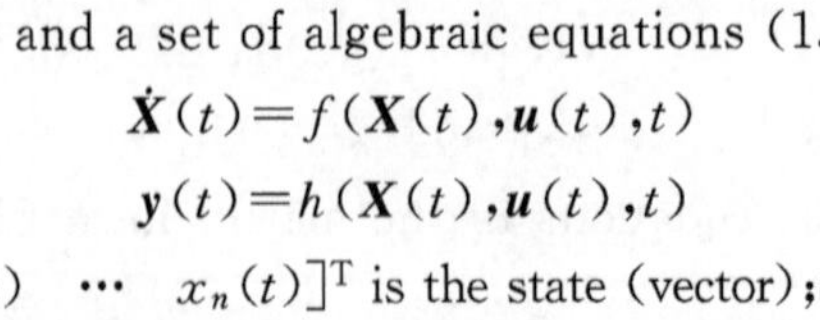

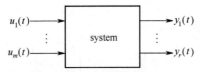

Figure 1.2 MIMO System

The continuous-time dynamic system can be represented by a set of first-order differential equations (1.12) and a set of algebraic equations (1.13).

$$\dot{\boldsymbol{X}}(t) = f(\boldsymbol{X}(t), \boldsymbol{u}(t), t) \tag{1.12}$$

$$\boldsymbol{y}(t) = h(\boldsymbol{X}(t), \boldsymbol{u}(t), t) \tag{1.13}$$

Where $\boldsymbol{X}(t) = [x_1(t) \quad x_2(t) \quad \cdots \quad x_n(t)]^{\mathrm{T}}$ is the state (vector); $\boldsymbol{u}(t) = [u_1(t) \quad u_2(t) \quad \cdots \quad u_m(t)]^{\mathrm{T}}$ is the input (vector); $\boldsymbol{y}(t) = [y_1(t) \quad y_2(t) \quad \cdots \quad y_r(t)]^{\mathrm{T}}$ is the output (vector).

Equation (1.12) governs the behavior of the state and is called a **state equation.** Equation (1.13) gives the output and is called an **output equation.** Both the state equation and the output equation are called the **state space description** of a system.

When the function f and h are nonlinear functions, then the corresponding system is called a **nonlinear system.** However, the linear case is of major importance and also lends itself to more detailed analysis. The most general state space description of a linear continuous-time dynamic system is given by

$$\dot{\boldsymbol{X}}(t)=\boldsymbol{A}(t)\boldsymbol{X}(t)+\boldsymbol{B}(t)\boldsymbol{u}(t) \tag{1.14}$$

$$\boldsymbol{y}(t)=\boldsymbol{C}(t)\boldsymbol{X}(t)+\boldsymbol{D}(t)\boldsymbol{u}(t) \tag{1.15}$$

Where $\boldsymbol{A}(t)$, $\boldsymbol{B}(t)$, $\boldsymbol{C}(t)$ and $\boldsymbol{D}(t)$ are $n\times n$, $n\times m$, $r\times n$, $r\times m$ matrixes.

Since $\boldsymbol{A}(t)$, $\boldsymbol{B}(t)$, $\boldsymbol{C}(t)$ and $\boldsymbol{D}(t)$ change with time, the system described by the equation(1.14) and (1.15) is called a **linear time-varying (LTV) system.**

If $\boldsymbol{A}(t)$, $\boldsymbol{B}(t)$, $\boldsymbol{C}(t)$ and $\boldsymbol{D}(t)$ are independent of time, that is, they are constant matrixes, then the equation (1.14) and (1.15) reduce to

$$\dot{\boldsymbol{X}}(t)=\boldsymbol{A}\boldsymbol{X}(t)+\boldsymbol{B}\boldsymbol{u}(t) \tag{1.16}$$

$$\boldsymbol{y}(t)=\boldsymbol{C}\boldsymbol{X}(t)+\boldsymbol{D}\boldsymbol{u}(t) \tag{1.17}$$

In the state space description (1.16) and (1.17), $\boldsymbol{A}$ is called the **system matrix** or **coefficient matrix**, and is a $n\times n$ real constant matrix; $\boldsymbol{B}$ is called the **input matrix**, and is a $n\times m$ real constant matrix; $\boldsymbol{C}$ is called the **output matrix**, and is a $r\times n$ real constant matrix; $\boldsymbol{D}$ is called the **forward matrix**, and is a $r\times m$ real constant matrix.

The system described by (1.16) and (1.17) is called a **linear time-invariant (LTI) system.**

The state space description (1.16) and (1.17) are shown in **block diagram** form in Figure 1.3. The block diagram is sometimes called the **simulation diagram** of a state space description.

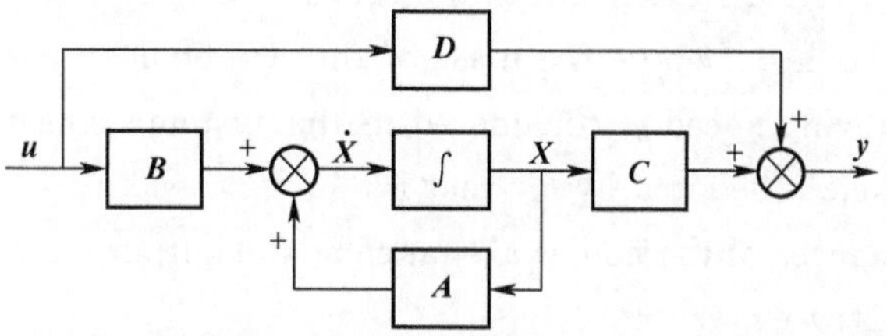

Figure 1.3 Block Diagram of the Linear Time-Invariant System

1.1.4 Transfer Function Matrix

A MIMO system illustrated by Figure 1.2 is considered.

Definition 1.4 If the Laplace transforms of $\boldsymbol{y}(t)$ and $\boldsymbol{u}(t)$ are $\boldsymbol{Y}(s)$ and $\boldsymbol{U}(s)$, respectively, the $r\times m$ matrix

$$\boldsymbol{G}(s)=\begin{bmatrix} G_{11}(s) & G_{12}(s) & \cdots & G_{1m}(s) \\ G_{21}(s) & G_{22}(s) & \cdots & G_{2m}(s) \\ \vdots & \vdots & & \vdots \\ G_{r1}(s) & G_{r2}(s) & \cdots & G_{rm}(s) \end{bmatrix} \tag{1.18}$$

is called the **transfer function matrix** of the system.

Where

$$G_{ij}(s)=\frac{Y_i(s)}{U_j(s)},i=1,2,\cdots,r,j=1,2,\cdots,m \tag{1.19}$$

and $Y_i(s)$ and $U_j(s)$ are the Laplace transforms of $y_i(t)$ and $u_j(t)$.

According to the concept of transfer function matrix, the following equation can be obtained as

$$\boldsymbol{Y}(s)=\boldsymbol{G}(s)\boldsymbol{U}(s) \tag{1.20}$$

Definition 1.5 $\boldsymbol{G}(s)$ is **proper rational function matrix** iff (if and only if) $G_{ij}(s)$ is proper rational function.

Definition 1.6 $\boldsymbol{G}(s)$ is **strictly proper rational function matrix** iff $G_{ij}(s)$ is strictly proper rational function.

Thus, if $\boldsymbol{G}(s)$ is strictly proper rational matrix, $\lim\limits_{s\to\infty}\boldsymbol{G}(s)=\boldsymbol{0}$

Example 1.2 A moving body system, which is shown in Figure 1.4, will be considered.

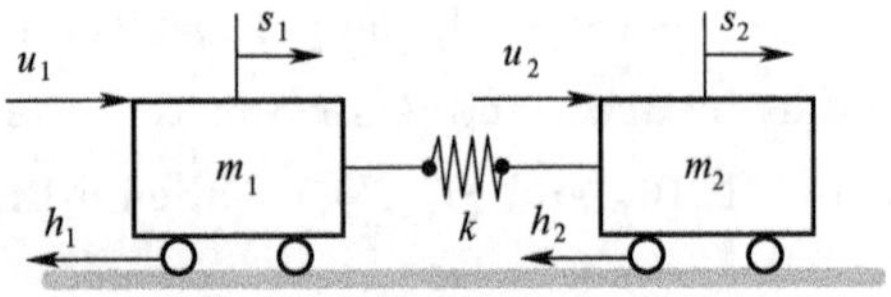

Figure 1.4 Moving Body System

Figure 1.4 shows a system of two bodies connected by a spring with elastic coefficient k. The moving distances of the bodies are denoted by $s_1(t)$ and $s_2(t)$, and supposed zero when the system stays in the still state, respectively. $u_1(t)$ and $u_2(t)$ are the forces onto the two bodies. Denote the mass of the two bodies by m_1 and m_2. The friction raised by the moving speed is considered in the system. The two bodies have the same frictional coefficient denoted by h, that is, $h_1=h_2=h$.

The moving distances of the bodies are taken as the output $y(t)$ of the system, that are $y_1(t)=s_1(t), y_2(t)=s_2(t)$.

From the 2nd Newton Law, the dynamic equations of the system can be obtained as

$$m_1\frac{\mathrm{d}^2 s_1(t)}{\mathrm{d}t^2}=u_1-h\frac{\mathrm{d}s_1(t)}{\mathrm{d}t}-k[s_1(t)-s_2(t)] \tag{1.21}$$

$$m_2\frac{\mathrm{d}^2 s_2(t)}{\mathrm{d}t^2}=u_2-h\frac{\mathrm{d}s_2(t)}{\mathrm{d}t}-k[s_2(t)-s_1(t)] \tag{1.22}$$

The state variables are selected as $x_1(t)=s_1(t), x_2(t)=s_2(t), x_3(t)=\dot{s}_1(t)$ and $x_4(t)=\dot{s}_2(t)$, the following first-order differential equations can be obtained as

$$\dot{x}_1(t)=x_3(t) \tag{1.23}$$

$$\dot{x}_2(t)=x_4(t) \tag{1.24}$$

$$\dot{x}_3(t)=-\frac{k}{m_1}x_1(t)+\frac{k}{m_1}x_2(t)-\frac{h}{m_1}x_3(t)+\frac{1}{m_1}u_1(t) \tag{1.25}$$

$$\dot{x}_4(t)=\frac{k}{m_2}x_1(t)-\frac{k}{m_2}x_2(t)-\frac{h}{m_2}x_4(t)+\frac{1}{m_2}u_2(t) \tag{1.26}$$

The differential equations are expressed in matrix form as

$$\begin{bmatrix}\dot{x}_1(t)\\ \dot{x}_2(t)\\ \dot{x}_3(t)\\ \dot{x}_4(t)\end{bmatrix}=\begin{bmatrix}0 & 0 & 1 & 0\\ 0 & 0 & 0 & 1\\ -\frac{k}{m_1} & \frac{k}{m_1} & -\frac{h}{m_1} & 0\\ \frac{k}{m_2} & -\frac{k}{m_2} & 0 & -\frac{h}{m_2}\end{bmatrix}\begin{bmatrix}x_1(t)\\ x_2(t)\\ x_3(t)\\ x_4(t)\end{bmatrix}+\begin{bmatrix}0 & 0\\ 0 & 0\\ \frac{1}{m_1} & 0\\ 0 & \frac{1}{m_2}\end{bmatrix}\begin{bmatrix}u_1(t)\\ u_2(t)\end{bmatrix} \tag{1.27}$$

It can be expressed in a more compact form as

$$\dot{\boldsymbol{X}}(t)=\boldsymbol{AX}(t)+\boldsymbol{Bu}(t) \tag{1.28}$$

Where $\boldsymbol{X}(t)=\begin{bmatrix}x_1(t)\\ x_2(t)\\ x_3(t)\\ x_4(t)\end{bmatrix}$ is the state (vector) of system; $\boldsymbol{u}(t)=\begin{bmatrix}u_1(t)\\ u_2(t)\end{bmatrix}$ is the input (vector) of system; $\dot{\boldsymbol{X}}(t)=\begin{bmatrix}\dot{x}_1(t)\\ \dot{x}_2(t)\\ \dot{x}_3(t)\\ \dot{x}_4(t)\end{bmatrix}$; $\boldsymbol{A}=\begin{bmatrix}0 & 0 & 1 & 0\\ 0 & 0 & 0 & 1\\ -\frac{k}{m_1} & \frac{k}{m_1} & -\frac{h}{m_1} & 0\\ \frac{k}{m_2} & -\frac{k}{m_2} & 0 & -\frac{h}{m_2}\end{bmatrix}$; $\boldsymbol{B}=\begin{bmatrix}0 & 0\\ 0 & 0\\ \frac{1}{m_1} & 0\\ 0 & \frac{1}{m_2}\end{bmatrix}$.

Eq. (1.27) or (1.28) is the state equation of the moving body system.

The output equation is given as

$$\boldsymbol{y}(t)=\boldsymbol{CX}(t) \tag{1.29}$$

Where $\boldsymbol{y}(t)=\begin{bmatrix}y_1(t)\\ y_2(t)\end{bmatrix}$ is the output (vector) of the system;

$\boldsymbol{C}=\begin{bmatrix}1 & 0 & 0 & 0\\ 0 & 1 & 0 & 0\end{bmatrix}$.

Since $y_1(t)=s_1(t)$, $y_2(t)=s_2(t)$, the following differential equations can be obtained from (1.21) and (1.22).

$$\ddot{y}_1(t)+\frac{h}{m_1}\dot{y}_1(t)+\frac{k}{m_1}y_1(t)-\frac{k}{m_1}y_2(t)=\frac{1}{m_1}u_1(t) \tag{1.30}$$

$$\ddot{y}_2(t)+\frac{h}{m_2}\dot{y}_2(t)+\frac{k}{m_2}y_2(t)-\frac{k}{m_2}y_1(t)=\frac{1}{m_2}u_2(t) \tag{1.31}$$

By taking the Laplace transforms of (1.30) and (1.31) and assuming the zero initial conditions hold true, the following equations in frequency domain are obtained as

$$\left(s^2+\frac{h}{m_1}s+\frac{k}{m_1}\right)Y_1(s)-\frac{k}{m_1}Y_2(s)=\frac{1}{m_1}U_1(s) \tag{1.32}$$

$$\left(s^2+\frac{h}{m_2}s+\frac{k}{m_2}\right)Y_2(s)-\frac{k}{m_2}Y_1(s)=\frac{1}{m_2}U_2(s) \tag{1.33}$$

The equations (1.32) and (1.32) are expressed in matrix form as

$$\begin{bmatrix}s^2+\frac{h}{m_1}s+\frac{k}{m_1} & -\frac{k}{m_1}\\ -\frac{k}{m_2} & s^2+\frac{h}{m_2}s+\frac{k}{m_2}\end{bmatrix}\begin{bmatrix}Y_1(s)\\ Y_2(s)\end{bmatrix}=\begin{bmatrix}\frac{1}{m_1} & 0\\ 0 & \frac{1}{m_2}\end{bmatrix}\begin{bmatrix}U_1(s)\\ U_2(s)\end{bmatrix}$$

By taking $\boldsymbol{Y}(s)=[Y_1(s) \quad Y_2(s)]^{\mathrm{T}}$ and $\boldsymbol{U}(s)=[U_1(s) \quad U_2(s)]^{\mathrm{T}}$

$$\boldsymbol{Y}(s)=\begin{bmatrix} s^2+\dfrac{h}{m_1}s+\dfrac{k}{m_1} & -\dfrac{k}{m_1} \\ -\dfrac{k}{m_2} & s^2+\dfrac{h}{m_2}s+\dfrac{k}{m_2} \end{bmatrix}^{-1}\begin{bmatrix} \dfrac{1}{m_1} & 0 \\ 0 & \dfrac{1}{m_2} \end{bmatrix}\boldsymbol{U}(s)$$

$$=\boldsymbol{G}(s)\boldsymbol{U}(s) \tag{1.34}$$

Comparing (1.34) with (1.20), $\boldsymbol{G}(s)$ is the transfer function matrix of the moving body system and

$$\boldsymbol{G}(s)=\begin{bmatrix} s^2+\dfrac{h}{m_1}s+\dfrac{k}{m_1} & -\dfrac{k}{m_1} \\ -\dfrac{k}{m_2} & s^2+\dfrac{h}{m_2}s+\dfrac{k}{m_2} \end{bmatrix}^{-1}\begin{bmatrix} \dfrac{1}{m_1} & 0 \\ 0 & \dfrac{1}{m_2} \end{bmatrix}$$

1.2 Obtaining State Space Description from I/O Description

State space description can be developed directly from a physical system, such as electrical network and mechanical system. Besides this direct method which is developed in section 1.1, we can also build up the state description from differential equation and transfer function. In other words, we can build up the internal description of system from its external description. It is, in fact, a realization problem. In this section, the discussion is restricted to **SISO** (single input single output) constant coefficient system.

1.2.1 Obtaining State Space Description from Differential Equation

The general differential equation description of a SISO LTI n-order system is shown as

$$y^{(n)}+a_1y^{(n-1)}+\cdots+a_{n-1}\dot{y}+a_ny=b_0u^{(m)}+b_1u^{(m-1)}+\cdots+b_{m-1}\dot{u}+b_mu\ (m\leqslant n) \tag{1.35}$$

In this subsection, we will mainly discuss how to obtain the state space description of a system from its differential equation description, such as (1.35). Two case will be considered, that is, the case $m=0$ and the case $m\leqslant n$.

Case 1: $m=0$

In this case, the differential equation (1.35) reduces to

$$y^{(n)}+a_1y^{(n-1)}+\cdots+a_{n-1}\dot{y}+a_ny=b_0u \tag{1.36}$$

The state variables can be selected as the phase variables, which are defined by

$$\begin{cases} x_1=y \\ x_2=\dot{y} \\ \quad\vdots \\ x_{n-1}=y^{(n-2)} \\ x_n=y^{(n-1)} \end{cases} \tag{1.37}$$

Differentiating (1.37), we get a set of first-order differential equations shown as

$$\begin{cases}\dot{x}_1=\dot{y}=x_2\\ \dot{x}_2=\ddot{y}=x_3\\ \vdots\\ \dot{x}_{n-1}=y^{(n-1)}=x_n\\ \dot{x}_n=y^{(n)}=-a_n y-a_{n-1}\dot{y}-\cdots-a_1 y^{(n-1)}+b_0 u\\ \quad =-a_n x_1-a_{n-1}x_2-\cdots-a_1 x_n+b_0 u\end{cases} \tag{1.38}$$

The equation (1.38) may be written in matrix form as

$$\begin{bmatrix}\dot{x}_1\\ \dot{x}_2\\ \vdots\\ \dot{x}_{n-1}\\ \dot{x}_n\end{bmatrix}=\begin{bmatrix}0 & 1 & 0 & \cdots & 0\\ 0 & 0 & 1 & \cdots & 0\\ \vdots & \vdots & \vdots & \ddots & \vdots\\ 0 & 0 & 0 & \cdots & 1\\ -a_n & -a_{n-1} & -a_{n-2} & \cdots & -a_1\end{bmatrix}\begin{bmatrix}x_1\\ x_2\\ \vdots\\ x_{n-1}\\ x_n\end{bmatrix}+\begin{bmatrix}0\\ 0\\ \vdots\\ 0\\ b_0\end{bmatrix}u \tag{1.39}$$

The output equation of the system is given by

$$y=\begin{bmatrix}1 & 0 & 0 & \cdots & 0\end{bmatrix}\begin{bmatrix}x_1\\ x_2\\ x_3\\ \vdots\\ x_n\end{bmatrix} \tag{1.40}$$

The system matrix $\boldsymbol{A}$ contains the number 1 in the super diagonal, and the negative of the coefficients, which belong to those left elements of differential equation (1.36), in the nth row. In this simple form, the matrix $\boldsymbol{A}$ is called a **companion matrix.**

Obviously, Eq. (1.39) and Eq. (1.40) can be written directly from the original differential equation (1.36) and shown in Figure 1.5.

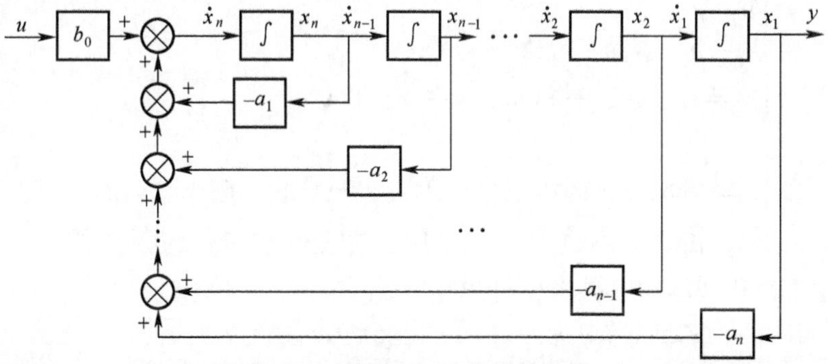

Figure 1.5 Simulation Diagram

Example 1.3 Find a state space description for the system described by the following differential equation.

$$\dddot{y}+2\ddot{y}+11\dot{y}+6y=3u$$

Solution For the system

$$n=3; a_1=2, a_2=11, a_3=6; b_0=3$$

The state variables can be chosen as

$$x_1=y, x_2=\dot{y}=\dot{x}_1, x_3=\ddot{y}=\dot{x}_2$$

Thus, the state equation of the system is

$$\begin{bmatrix}\dot{x}_1\\ \dot{x}_2\\ \dot{x}_3\end{bmatrix}=\begin{bmatrix}0 & 1 & 0\\ 0 & 0 & 1\\ -a_3 & -a_2 & -a_1\end{bmatrix}\begin{bmatrix}x_1\\ x_2\\ x_3\end{bmatrix}+\begin{bmatrix}0\\ 0\\ b_0\end{bmatrix}u$$

$$=\begin{bmatrix}0 & 1 & 0\\ 0 & 0 & 1\\ -6 & -11 & -2\end{bmatrix}\begin{bmatrix}x_1\\ x_2\\ x_3\end{bmatrix}+\begin{bmatrix}0\\ 0\\ 3\end{bmatrix}u$$

and the output equation is

$$y=[1 \quad 0 \quad 0]\begin{bmatrix}x_1\\ x_2\\ x_3\end{bmatrix}$$

Case 2: $m\leqslant n$

In this case, the differential equation description of a SISO LTI n-order system is the general form shown as (1.35).

Firstly, the parameters β_i $(i=0,\cdots,n)$ are constructed with the coefficients of (1.35) as

$$\begin{cases}\beta_0=b_0\\ \beta_1=b_1-a_1\beta_0\\ \beta_2=b_2-a_1\beta_1-a_2\beta_0\\ \vdots\\ \beta_{n-1}=b_{n-1}-a_1\beta_{n-2}-a_2\beta_{n-3}-\cdots-a_{n-2}\beta_1-a_{n-1}\beta_0\\ \beta_n=b_n-a_1\beta_{n-1}-a_2\beta_{n-2}-\cdots-a_{n-2}\beta_2-a_{n-1}\beta_1-a_n\beta_0\end{cases} \tag{1.41}$$

i. e.

$$\begin{cases}b_0=\beta_0\\ b_1=\beta_1+a_1\beta_0\\ b_2=\beta_2+a_1\beta_1+a_2\beta_0\\ \vdots\\ b_{n-1}=\beta_{n-1}+a_1\beta_{n-2}+a_2\beta_{n-3}+\cdots+a_{n-2}\beta_1+a_{n-1}\beta_0\\ b_n=\beta_n+a_1\beta_{n-1}+a_2\beta_{n-2}+\cdots+a_{n-2}\beta_2+a_{n-1}\beta_1+a_n\beta_0\end{cases} \tag{1.42}$$

The state variables can be defined by

$$\begin{cases}x_1=y-\beta_0 u\\ x_2=\dot{y}-\beta_0\dot{u}-\beta_1 u\\ x_3=\ddot{y}-\beta_0\ddot{u}-\beta_1\dot{u}-\beta_2 u\\ \vdots\\ x_{n-1}=y^{(n-2)}-\beta_0 u^{(n-2)}-\cdots-\beta_{n-3}\dot{u}-\beta_{n-2}u\\ x_n=y^{(n-1)}-\beta_0 u^{(n-1)}-\cdots-\beta_{n-2}\dot{u}-\beta_{n-1}u\end{cases} \tag{1.43}$$

So, the phase variables $y, \dot{y}, \cdots, y^{(n-1)}$ can be taken from (1.43) as

$$\begin{cases} y = x_1 + \beta_0 u \\ \dot{y} = x_2 + \beta_0 \dot{u} + \beta_1 u \\ \ddot{y} = x_3 + \beta_0 \ddot{u} + \beta_1 \dot{u} + \beta_2 u \\ \vdots \\ y^{(n-2)} = x_{n-1} + \beta_0 u^{(n-2)} + \cdots + \beta_{n-3} \dot{u} + \beta_{n-2} u \\ y^{(n-1)} = x_n + \beta_0 u^{(n-1)} + \cdots + \beta_{n-2} \dot{u} + \beta_{n-1} u \end{cases} \tag{1.44}$$

Substituting (1.44) into (1.35), we may deduce that

$$\begin{aligned} y^{(n)} &= -a_1 y^{(n-1)} - \cdots - a_{n-1} \dot{y} - a_n y + b_0 u^{(n)} + b_1 u^{(n-1)} + \cdots + b_{n-1} \dot{u} + b_n u \\ &= -a_1 \{x_n + \beta_0 u^{(n-1)} + \cdots + \beta_{n-2} \dot{u} + \beta_{n-1} u\} - a_2 \{x_{n-1} + \beta_0 u^{(n-2)} + \cdots \\ &\quad + \beta_{n-3} \dot{u} + \beta_{n-2} u\} - \cdots - a_{n-1} \{x_2 + \beta_0 \dot{u} + \beta_1 u\} - a_n \{x_1 + \beta_0 u\} \\ &\quad + b_0 u^{(n)} + b_1 u^{(n-1)} + \cdots + b_{n-1} \dot{u} + b_n u \\ &= -\{a_1 x_n + a_2 x_{n-1} + \cdots + a_{n-1} x_2 + a_n x_1\} - \{a_1 \beta_0\} u^{(n-1)} \\ &\quad - \{a_1 \beta_1 + a_2 \beta_0\} u^{(n-2)} - \cdots - \{a_1 \beta_{n-2} + a_2 \beta_{n-3} + \cdots + a_{n-2} \beta_1 + a_{n-1} \beta_0\} \dot{u} \\ &\quad - \{a_1 \beta_{n-1} + a_2 \beta_{n-2} + \cdots + a_{n-2} \beta_2 + a_{n-1} \beta_1 + a_n \beta_0\} u + \beta_0 u^{(n)} \\ &\quad + \{\beta_1 + a_1 \beta_0\} u^{(n-1)} + \cdots + \{\beta_{n-1} + a_1 \beta_{n-2} + a_2 \beta_{n-3} + \cdots + a_{n-2} \beta_1 + a_{n-1} \beta_0\} \dot{u} \\ &\quad + \{\beta_n + a_1 \beta_{n-1} + a_2 \beta_{n-2} + \cdots + a_{n-2} \beta_2 + a_{n-1} \beta_1 + a_n \beta_0\} u \\ &= -\{a_1 x_n + a_2 x_{n-1} + \cdots + a_{n-1} x_2 + a_n x_1\} + \beta_0 u^{(n)} + \beta_1 u^{(n-1)} + \cdots + \beta_{n-1} \dot{u} + \beta_n u \end{aligned} \tag{1.45}$$

Differentiating x_i for $i=1,\cdots,n$, the set of first-order differential equations are obtained as

$$\begin{cases} \dot{x}_1 = \dot{y} - \beta_0 \dot{u} = (x_2 + \beta_0 \dot{u} + \beta_1 u) - \beta_0 \dot{u} = x_2 + \beta_1 u \\ \dot{x}_2 = \ddot{y} - \beta_0 \ddot{u} - \beta_1 u = x_3 + \beta_2 u \\ \dot{x}_3 = x_4 + \beta_3 u \\ \vdots \\ \dot{x}_{n-1} = x_n + \beta_{n-1} u \\ \dot{x}_n = y^{(n)} - \beta_0 u^{(n)} - \cdots - \beta_{n-2} \ddot{u} - \beta_{n-1} \dot{u} \\ \quad = -\{a_1 x_n + a_2 x_{n-1} + \cdots + a_{n-1} x_2 + a_n x_1\} + \beta_0 u^{(n)} + \beta_1 u^{(n-1)} \\ \qquad + \cdots + \beta_{n-1} \dot{u} + \beta_n u - \beta_0 u^{(n)} - \cdots - \beta_{n-2} \ddot{u} - \beta_{n-1} \dot{u} \\ \quad = (-a_n x_1 - a_{n-1} x_2 - \cdots - a_2 x_{n-1} - a_1 x_n) + \beta_n u \end{cases} \tag{1.46}$$

and

$$y = x_1 + \beta_0 u \tag{1.47}$$

By expressing (1.46) and (1.47) in matrix form, we can obtain the state space description of the system as

$$\begin{bmatrix} \dot{x}_1 \\ \dot{x}_2 \\ \dot{x}_3 \\ \vdots \\ \dot{x}_{n-1} \\ \dot{x}_n \end{bmatrix} = \begin{bmatrix} 0 & 1 & 0 & \cdots & 0 & 0 \\ 0 & 0 & 1 & \cdots & 0 & 0 \\ 0 & 0 & 0 & \ddots & 0 & 0 \\ \vdots & \vdots & \vdots & \ddots & \ddots & \vdots \\ 0 & 0 & 0 & \cdots & 0 & 1 \\ -a_n & -a_{n-1} & -a_{n-2} & \cdots & -a_2 & -a_1 \end{bmatrix} \begin{bmatrix} x_1 \\ x_2 \\ x_3 \\ \vdots \\ x_{n-1} \\ x_n \end{bmatrix} + \begin{bmatrix} \beta_1 \\ \beta_2 \\ \beta_3 \\ \vdots \\ \beta_{n-1} \\ \beta_n \end{bmatrix} u \tag{1.48}$$

$$y=\begin{bmatrix}1 & 0 & 0 & \cdots & 0 & 0\end{bmatrix}\begin{bmatrix}x_1\\ x_2\\ x_3\\ \vdots\\ x_{n-1}\\ x_n\end{bmatrix}+[\beta_0]u \tag{1.49}$$

Example 1.4 Find a state space description for the system described by the following differential equation

$$\dddot{y}+18\ddot{y}+192\dot{y}+640y=160\dot{u}+640u$$

Solution For the system

$$a_1=18, a_2=192, a_3=640; b_0=b_1=0, b_2=160, b_3=640$$

The parameters $\beta_i\ (i=0,\cdots,3)$ can be chosen as

$$\begin{cases}\beta_0=b_0=0\\ \beta_1=b_1-a_1\beta_0=0\\ \beta_2=b_2-a_1\beta_1-a_2\beta_0=b_2=160\\ \beta_3=b_3-a_1\beta_2-a_2\beta_1-a_3\beta_0=640-18\times160=-2240\end{cases}$$

Furthermore, the state variables can be defined by

$$\begin{cases}x_1=y-\beta_0 u\\ x_2=\dot{y}-\beta_0\dot{u}-\beta_1 u\\ x_3=\ddot{y}-\beta_0\ddot{u}-\beta_1\dot{u}-\beta_2 u\end{cases}$$

Then, the state space description of system can be obtained as

$$\begin{aligned}\dot{\boldsymbol{X}}&=\begin{bmatrix}0 & 1 & 0\\ 0 & 0 & 1\\ -a_3 & -a_2 & -a_1\end{bmatrix}\boldsymbol{X}+\begin{bmatrix}\beta_1\\ \beta_2\\ \beta_3\end{bmatrix}u\\ &=\begin{bmatrix}0 & 1 & 0\\ 0 & 0 & 1\\ -640 & -192 & -18\end{bmatrix}\boldsymbol{X}+\begin{bmatrix}0\\ 160\\ -2240\end{bmatrix}u\\ y&=\begin{bmatrix}1 & 0 & 0\end{bmatrix}\boldsymbol{X}+[\beta_0]u=\begin{bmatrix}1 & 0 & 0\end{bmatrix}\boldsymbol{X}\end{aligned}$$

Where $\boldsymbol{X}=\begin{bmatrix}x_1 & x_2 & x_3\end{bmatrix}^{\mathrm{T}}$.

Example 1.5 Find a state space description for the system described by the following differential equation

$$\dddot{y}+16\ddot{y}+194\dot{y}+640y=4\dddot{u}+160\dot{u}+720u$$

Solution For the system

$$a_1=16, a_2=194, a_3=640; b_0=4, b_1=0, b_2=160, b_3=720$$

The parameters $\beta_i\ (i=0,\cdots,3)$ can be chosen as

$$\begin{cases}\beta_0=b_0=4\\ \beta_1=b_1-a_1\beta_0=0-16\times4=-64\\ \beta_2=b_2-a_1\beta_1-a_2\beta_0=160-16\times(-64)-194\times4=408\\ \beta_3=b_3-a_1\beta_2-a_2\beta_1-a_3\beta_0=720-16\times408-194\times(-64)-640\times4=4048\end{cases}$$

The state variables can be defined by

$$\begin{cases} x_1 = y - \beta_0 u \\ x_2 = \dot{y} - \beta_0 \dot{u} - \beta_1 u \\ x_3 = \ddot{y} - \beta_0 \ddot{u} - \beta_1 \dot{u} - \beta_2 u \end{cases}$$

So, the state space description of system can be obtained as

$$\dot{\boldsymbol{X}} = \begin{bmatrix} 0 & 1 & 0 \\ 0 & 0 & 1 \\ -a_3 & -a_2 & -a_1 \end{bmatrix} \boldsymbol{X} + \begin{bmatrix} \beta_1 \\ \beta_2 \\ \beta_3 \end{bmatrix} u$$

$$= \begin{bmatrix} 0 & 1 & 0 \\ 0 & 0 & 1 \\ -640 & -194 & -16 \end{bmatrix} \boldsymbol{X} + \begin{bmatrix} -64 \\ 408 \\ 4048 \end{bmatrix} u$$

$$y = [1 \quad 0 \quad 0]\boldsymbol{X} + [\beta_0]u = [1 \quad 0 \quad 0]\boldsymbol{X} + 4u$$

Where $\boldsymbol{X} = [x_1 \quad x_2 \quad x_3]^{\mathrm{T}}$.

1.2.2 Obtaining State Space Description from Transfer Function

In this section, two methods for finding the state space description from transfer function of a system will be discussed. The discussion in the section is restricted to SISO constant coefficient system described by the following transfer function.

$$G(s) = \frac{Y(s)}{U(s)} = \frac{b_0 s^m + b_1 s^{m-1} + b_2 s^{m-2} + \cdots + b_{m-1} s + b_m}{s^n + a_1 s^{n-1} + \cdots + a_{n-1} s + a_n} \tag{1.50}$$

Method 1: Direct Decomposition

A system described by (1.50) is called proper rational system if $m \leqslant n$. By the way, it is called strictly proper system if $m < n$. Firstly, the strictly proper system described as (1.51) will be discussed.

$$G(s) = \frac{Y(s)}{U(s)} = \frac{b_1 s^{n-1} + b_2 s^{n-2} + \cdots + b_{n-1} s + b_n}{s^n + a_1 s^{n-1} + \cdots + a_{n-1} s + a_n} \tag{1.51}$$

Let us introduce an intermediate variable

$$Z(s) = \frac{1}{s^n + a_1 s^{n-1} + \cdots + a_{n-1} s + a_n} \cdot U(s) \tag{1.52}$$

Thus, the system can be decomposed as

$$Y(s) = (b_1 s^{n-1} + b_2 s^{n-2} + \cdots + b_{n-1} s + b_n) Z(s) \tag{1.53}$$

and can be shown in Figure 1.6.

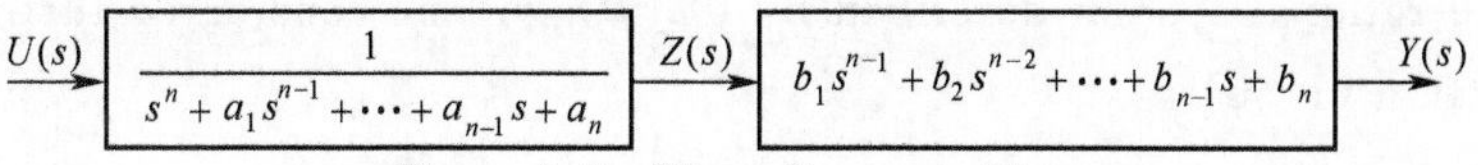

Figure 1.6 Direct Decomposition

The equation (1.54) in frequency domain can be derived from (1.52) as

$$s^n Z(s) = -a_1 s^{n-1} Z(s) - \cdots - a_{n-1} s Z(s) - a_n Z(s) + U(s) \tag{1.54}$$

By taking the inverse Laplace transform of (1.54) and assuming the zero initial conditions hold true, the following equation is obtained as

$$z^{(n)}=-a_1z^{(n-1)}-\cdots-a_{n-1}\dot{z}-a_nz+u \tag{1.55}$$

The state variables can be selected as the phase variables, which are defined by

$$\begin{cases}x_1=z\\x_2=\dot{z}\\\vdots\\x_{n-1}=z^{(n-2)}\\x_n=z^{(n-1)}\end{cases} \tag{1.56}$$

Differentiating (1.56) and substituting (1.55) into its last equation, we get a set of first-order differential equations shown as

$$\begin{cases}\dot{x}_1=\dot{z}=x_2\\\dot{x}_2=\ddot{z}=x_3\\\vdots\\\dot{x}_{n-1}=z^{(n-1)}=x_n\\\dot{x}_n=z^{(n)}=-a_1z^{(n-1)}-\cdots-a_{n-1}\dot{z}-a_nz+u\\\quad=-a_nx_1-a_{n-1}x_2-\cdots-a_1x_n+u\end{cases} \tag{1.57}$$

The state equation (1.57) may be written in matrix form as

$$\begin{bmatrix}\dot{x}_1\\\dot{x}_2\\\vdots\\\dot{x}_{n-1}\\\dot{x}_n\end{bmatrix}=\begin{bmatrix}0&1&0&\cdots&0\\0&0&1&\cdots&0\\\vdots&\vdots&\vdots&\ddots&\vdots\\0&0&0&\cdots&1\\-a_n&-a_{n-1}&-a_{n-2}&\cdots&-a_1\end{bmatrix}\begin{bmatrix}x_1\\x_2\\\vdots\\x_{n-1}\\x_n\end{bmatrix}+\begin{bmatrix}0\\0\\\vdots\\0\\1\end{bmatrix}u \tag{1.58}$$

By taking the inverse Laplace transform of (1.53) and assuming the zero initial conditions hold true, the output equation can be obtained as

$$y=b_nx_1+b_{n-1}x_2+\cdots+b_2x_{n-1}+b_1x_n \tag{1.59}$$

The output equation (1.59) may be written in matrix form as

$$y=[b_n\quad b_{n-1}\quad\cdots\quad b_2\quad b_1]\begin{bmatrix}x_1\\x_2\\\vdots\\x_{n-1}\\x_n\end{bmatrix} \tag{1.60}$$

Obviously, the system matrix is the companion matrix. The state space description, such as (1.58) and (1.60), is well known as the **controllable canonical form.**

A proper rational system described by (1.50) will be considered further. Rewrite transfer function (1.50) as

$$\begin{aligned}G(s)&=\frac{Y(s)}{U(s)}\\&=b_0+\frac{(b_1-b_0a_1)s^{n-1}+(b_2-b_0a_2)s^{n-2}+\cdots+(b_{n-1}-b_0a_{n-1})s+(b_n-b_0a_n)}{s^n+a_1s^{n-1}+\cdots+a_{n-1}s+a_n}\end{aligned} \tag{1.61}$$

Then

$$Y(s)=b_0U(s)+\frac{(b_1-b_0a_1)s^{n-1}+(b_2-b_0a_2)s^{n-2}+\cdots+(b_{n-1}-b_0a_{n-1})s+(b_n-b_0a_n)}{s^n+a_1s^{n-1}+\cdots+a_{n-1}s+a_n}\cdot U(s) \tag{1.62}$$

Let us introduce an intermediate variable

$$\tilde{Y}(s)=\frac{(b_1-b_0a_1)s^{n-1}+(b_2-b_0a_2)s^{n-2}+\cdots+(b_{n-1}-b_0a_{n-1})s+(b_n-b_0a_n)}{s^n+a_1s^{n-1}+\cdots+a_{n-1}s+a_n}U(s) \tag{1.63}$$

Substituting (1.63) into (1.62), we get the output equation (1.64) and the system is shown in Figure 1.7.

$$Y(s)=b_0U(s)+\tilde{Y}(s)=b_0U(s)+\tilde{G}(s)U(s) \tag{1.64}$$

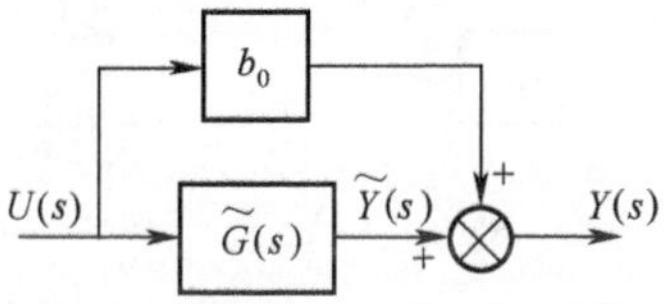

Figure 1.7 The Proper Rational System

Where

$$\tilde{G}(s)=\frac{\tilde{Y}(s)}{U(s)}=\frac{(b_1-b_0a_1)s^{n-1}+(b_2-b_0a_2)s^{n-2}+\cdots+(b_{n-1}-b_0a_{n-1})s+(b_n-b_0a_n)}{s^n+a_1s^{n-1}+\cdots+a_{n-1}s+a_n} \tag{1.65}$$

is a strictly proper transfer function.

According to (1.58) and (1.60), the state space description of the transfer function $\tilde{G}(s)$ is obtained as

$$\begin{cases}\dot{\boldsymbol{X}}=\boldsymbol{AX}+\boldsymbol{b}u\\ \tilde{y}=\boldsymbol{cX}\end{cases} \tag{1.66}$$

Where $\boldsymbol{A}=\begin{bmatrix}0 & 1 & 0 & \cdots & 0\\ 0 & 0 & 1 & \cdots & 0\\ \vdots & \vdots & \vdots & \ddots & \vdots\\ 0 & 0 & 0 & \cdots & 1\\ -a_n & -a_{n-1} & -a_{n-2} & \cdots & -a_1\end{bmatrix}$; $\boldsymbol{b}=\begin{bmatrix}0\\0\\ \vdots\\0\\1\end{bmatrix}$;

$$\boldsymbol{c}=[b_n-b_0a_n \quad b_{n-1}-b_0a_{n-1} \quad \cdots \quad b_2-b_0a_2 \quad b_1-b_0a_1]. \tag{1.67}$$

By taking the inverse Laplace transform of (1.64) and assuming the zero initial conditions hold true, the output equation of proper rational system can be obtained

$$y=\tilde{y}+b_0u=\boldsymbol{cX}+\boldsymbol{d}u \tag{1.68}$$

Where $\boldsymbol{d}=[b_0]$.

It is clear that the state space description of proper rational system represented as

$$\begin{cases}\dot{\boldsymbol{X}}=\boldsymbol{AX}+\boldsymbol{b}u \\ y=\boldsymbol{cX}+\boldsymbol{d}u\end{cases} \tag{1.69}$$

may be a controllable canonical form, and can be shown in Figure 1.8.

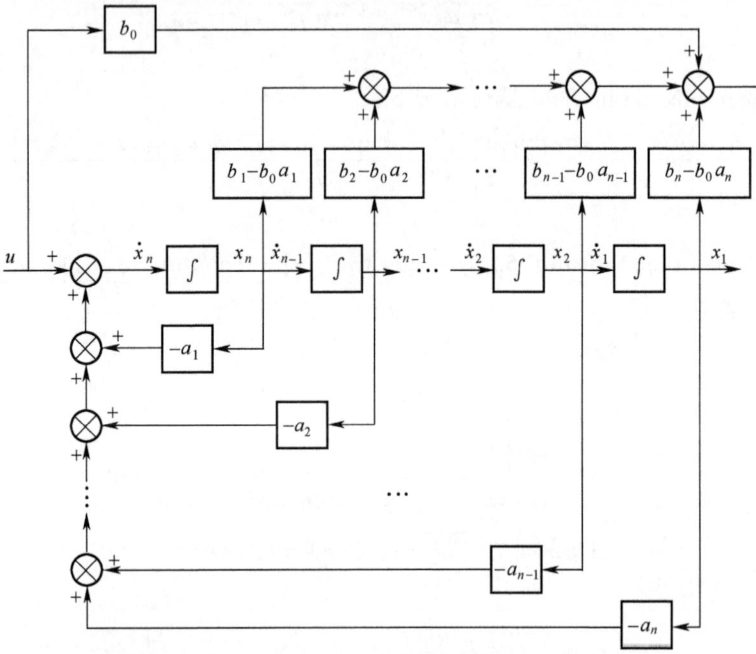

Figure 1.8 The Block Diagram of Proper Rational System

Particularly, $\boldsymbol{d}=\boldsymbol{0}$ when $b_0=0$, in other words, when the system is a strictly proper system.

Example 1.6 Consider a system described by the transfer function

$$G(s)=\frac{s+3}{s^3+9s^2+24s+20}$$

Find a state space representation by the method of direct decomposition.

Solution The numerator and denominator of the transfer function can be separated by method of direct decomposition as shown in Figure 1.9 and the intermediate variable $Z(s)$ is labeled in the figure.

The state variables can be selected as the phase variables, which are defined by

$$\begin{cases}x_1=z \\ x_2=\dot{z} \\ x_3=\ddot{z}\end{cases}$$

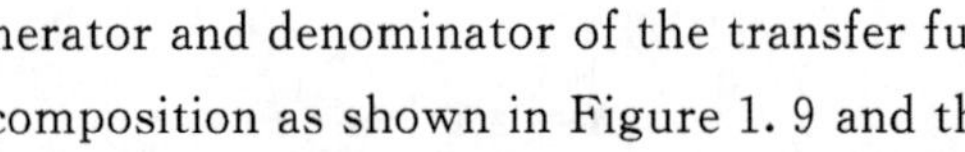

Figure 1.9 The Direct Decomposition of $G(s)$

For the strictly proper system

$$a_1=9, a_2=24, a_3=20; b_0=0, b_1=0, b_2=1, b_3=3$$

The state space description of system can be obtained as

$$\dot{\boldsymbol{X}}=\begin{bmatrix}0&1&0\\0&0&1\\-a_3&-a_2&-a_1\end{bmatrix}\boldsymbol{X}+\begin{bmatrix}0\\0\\1\end{bmatrix}u=\begin{bmatrix}0&1&0\\0&0&1\\-20&-24&-9\end{bmatrix}\boldsymbol{X}+\begin{bmatrix}0\\0\\1\end{bmatrix}u$$

$$y=[b_3\quad b_2\quad b_1]\boldsymbol{X}=[3\quad 1\quad 0]\boldsymbol{X}$$

Where $\boldsymbol{X}=[x_1\quad x_2\quad x_3]^{\mathrm{T}}$.

The state space description above is shown in Figure 1.10.

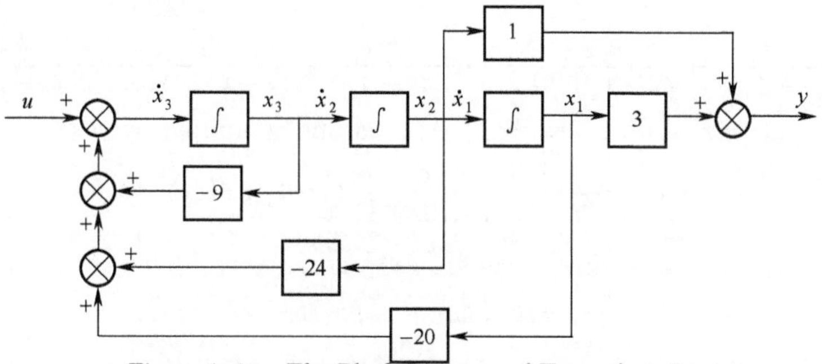

Figure 1.10 The Block Diagram of Example 1.6

Method 2: Parallel Decomposition

Firstly, let us recognize some important matrixes before introduce the method of parallel decomposition.

We have recognized the **diagonal matrix**(1.70) in the Linear Algebra course.

$$\boldsymbol{\Lambda}=\begin{bmatrix}\lambda_1&0&\cdots&0\\0&\lambda_2&\ddots&\vdots\\\vdots&\ddots&\ddots&0\\0&\cdots&0&\lambda_n\end{bmatrix}\tag{1.70}$$

The state space description (1.16) is called the **diagonal canonical form** if its system matrix $\boldsymbol{A}$ is a diagonal matrix such as (1.70).

A matrix such as (1.71) is the called **Jordan block**. A **Jordan matrix** is a special diagonal matrix because the main diagonal of it is composed of several Jordan blocks $\boldsymbol{J}_i$, for $i=1,\cdots,m$, such as (1.72). The state space description (1.16) is called the **Jordan canonical form** if its system matrix $\boldsymbol{A}$ is a Jordan matrix.

$$\boldsymbol{J}_i=\begin{bmatrix}\lambda_i&1&0&\cdots&0\\0&\lambda_i&1&\ddots&\vdots\\0&0&\lambda_i&\ddots&0\\\vdots&\ddots&\ddots&\ddots&1\\0&\cdots&0&0&\lambda_i\end{bmatrix}\tag{1.71}$$

$$\boldsymbol{J}=\begin{bmatrix}\boldsymbol{J}_1&\boldsymbol{0}&\cdots&\boldsymbol{0}\\\boldsymbol{0}&\boldsymbol{J}_2&\ddots&\vdots\\\vdots&\ddots&\ddots&\boldsymbol{0}\\\boldsymbol{0}&\cdots&\boldsymbol{0}&\boldsymbol{J}_m\end{bmatrix}\tag{1.72}$$

If the denominator $D(s)$ of the transfer function (1.50), which is also called the **characteristic polynomial**, can be decomposed as

$$D(s)=(s-\lambda_1)^r(s-\lambda_{r+1})\cdots(s-\lambda_n)$$

Where λ_1 is the characteristic root of multiplicity r and $\lambda_{r+1},\cdots,\lambda_n$ are the other distinct characteristic roots of $D(s)$.

By the partial fraction expansion, when $m=n$ the transfer function (1.50) admits the following decomposition

$$G(s)=\frac{Y(s)}{U(s)}=k_0+G(s)$$

$$=k_0+\frac{k_{11}}{(s-\lambda_1)^r}+\frac{k_{12}}{(s-\lambda_1)^{r-1}}+\cdots+\frac{k_{1r}}{(s-\lambda_1)}+\frac{k_{r+1}}{(s-\lambda_{r+1})}+\cdots+\frac{k_n}{(s-\lambda_n)} \quad (1.73)$$

Where $k_{1j}, j=1,\cdots,r$, and $k_i, i=r+1,\cdots,n$, can be computed by

$$k_{1j}=\frac{1}{(j-1)!}\lim_{s\to\lambda_1}\frac{\mathrm{d}^{j-1}}{\mathrm{d}s^{j-1}}[(s-\lambda_1)^r\widetilde{G}(s)]$$

$$k_i=\lim_{s\to\lambda_i}[(s-\lambda_i)\widetilde{G}(s)]$$

$$k_0=\lim_{s\to\infty}G(s)=b_0$$

Note that $k_0=0$ if $m<n$ or $b_0=0$.

The state variables $x_1,x_2,\cdots,x_n$ are selected and their Laplace transforms take the form of

$$\begin{cases} x_1(s)=\dfrac{1}{(s-\lambda_1)^r}U(s) \\ x_2(s)=\dfrac{1}{(s-\lambda_1)^{r-1}}U(s) \\ \vdots \\ x_r(s)=\dfrac{1}{(s-\lambda_1)}U(s) \\ x_{r+1}(s)=\dfrac{1}{(s-\lambda_{r+1})}U(s) \\ \vdots \\ x_n(s)=\dfrac{1}{(s-\lambda_n)}U(s) \end{cases} \quad (1.74)$$

From (1.74), we can get

$$\begin{cases} x_1(s)=\dfrac{1}{(s-\lambda_1)}x_2(s) \\ x_2(s)=\dfrac{1}{(s-\lambda_1)}x_3(s) \\ \vdots \\ x_{r-1}(s)=\dfrac{1}{(s-\lambda_1)}x_r(s) \\ x_r(s)=\dfrac{1}{(s-\lambda_1)}U(s) \\ x_{r+1}(s)=\dfrac{1}{(s-\lambda_{r+1})}U(s) \\ \vdots \\ x_n(s)=\dfrac{1}{(s-\lambda_n)}U(s) \end{cases} \quad (1.75)$$

It is not difficult to deduce the state equations from (1.75) as

$$\begin{cases}\dot{x}_1=\lambda_1 x_1+x_2\\ \dot{x}_2=\lambda_1 x_2+x_3\\ \vdots\\ \dot{x}_{r-1}=\lambda_1 x_{r-1}+x_r\\ \dot{x}_r=\lambda_1 x_r+u\\ \dot{x}_{r+1}=\lambda_{r+1}x_{r+1}+u\\ \vdots\\ \dot{x}_n=\lambda_n x_n+u\end{cases} \tag{1.76}$$

From (1.73) and (1.74), the equation (1.77) can be obtained as

$$Y(s)=k_{11}x_1(s)+k_{12}x_2(s)+\cdots+k_{1r}x_r(s)+k_{r+1}x_{r+1}(s)+\cdots+k_n x_n(s)+k_0 U(s) \tag{1.77}$$

Applying the inverse Laplace transform to (1.77), the output equation is obtained as

$$y(t)=k_{11}x_1(t)+k_{12}x_2(t)+\cdots+k_{1r}x_r(t)+k_{r+1}x_{r+1}(t)+\cdots+k_n x_n(t)+k_0 u(t) \tag{1.78}$$

Consequently, (1.76) and (1.78) can be rewritten in the matrix form as

$$\begin{bmatrix}\dot{x}_1\\ \dot{x}_2\\ \vdots\\ \dot{x}_r\\ \dot{x}_{r+1}\\ \vdots\\ \dot{x}_n\end{bmatrix}=\begin{bmatrix}\lambda_1 & 1 & 0 & \cdots & 0 & 0 & \cdots & \cdots & 0\\ 0 & \lambda_1 & \ddots & \ddots & \vdots & & & & \\ 0 & \ddots & \ddots & \ddots & 0 & \vdots & & & \vdots\\ \vdots & & \ddots & \ddots & 1 & & & & \\ 0 & \cdots & \cdots & 0 & \lambda_1 & 0 & \cdots & \cdots & 0\\ 0 & & \cdots & & 0 & \lambda_{r+1} & 0 & \cdots & 0\\ \vdots & & & & \vdots & 0 & \ddots & \ddots & \vdots\\ \vdots & & & & \vdots & \vdots & \ddots & \ddots & 0\\ 0 & & \cdots & & 0 & 0 & \cdots & 0 & \lambda_n\end{bmatrix}\begin{bmatrix}x_1\\ x_2\\ \vdots\\ x_r\\ x_{r+1}\\ \vdots\\ x_n\end{bmatrix}+\begin{bmatrix}0\\ \vdots\\ 0\\ 1\\ 1\\ \vdots\\ 1\end{bmatrix}u \tag{1.79}$$

$$y=\begin{bmatrix}k_{11} & k_{12} & \cdots & k_{1r} & k_{r+1} & \cdots & k_n\end{bmatrix}\begin{bmatrix}x_1\\ x_2\\ \vdots\\ x_r\\ x_{r+1}\\ \vdots\\ x_n\end{bmatrix}+k_0 u \tag{1.80}$$

Obviously, the state space description above is a Jordan canonical form.

Particularly, if λ_1 is an eigenvalue of multiplicity 1, in other words, the characteristic roots of system $\lambda_i, i=1,\cdots,n$, are all distinct, the state space description (1.79) and (1.80) can be reduced to a diagonal canonical form such as

$$\begin{bmatrix} \dot{x}_1 \\ \dot{x}_2 \\ \vdots \\ \dot{x}_n \end{bmatrix} = \begin{bmatrix} \lambda_1 & 0 & \cdots & 0 \\ 0 & \lambda_2 & \ddots & \vdots \\ \vdots & \ddots & \ddots & 0 \\ 0 & \cdots & 0 & \lambda_n \end{bmatrix} \begin{bmatrix} x_1 \\ x_2 \\ \vdots \\ x_n \end{bmatrix} + \begin{bmatrix} 1 \\ 1 \\ \vdots \\ 1 \end{bmatrix} u \tag{1.81}$$

$$y = [k_1 \quad k_2 \quad \cdots \quad k_n] \begin{bmatrix} x_1 \\ x_2 \\ \vdots \\ x_n \end{bmatrix} + k_0 u \tag{1.82}$$

Example 1.7 Consider a system described by the transfer function

$$G(s) = \frac{s+3}{s^3+9s^2+24s+20}$$

Find its state space representation by the method of parallel decomposition.

Solution The characteristic polynomial can be decomposed as

$$D(s) = s^3+9s^2+24s+20 = (s+2)^2(s+5)$$

Then, $\lambda_1 = -2$ is the 2 multiplicity characteristic root of $D(s)$, and $\lambda_3 = -5$ is the distinct one.

By the partial fraction expansion, the transfer function can be decomposed as

$$G(s) = \frac{Y(s)}{U(s)} = \frac{k_{11}}{(s+2)^2} + \frac{k_{12}}{(s+2)} + \frac{k_3}{(s+5)}$$

Where

$$k_{11} = \frac{1}{(1-1)!} \lim_{s \to -2} [(s+2)^2 G(s)] = \frac{1}{3}$$

$$k_{12} = \frac{1}{(2-1)!} \lim_{s \to -2} \frac{\mathrm{d}}{\mathrm{d}s} [(s+2)^2 G(s)] = \frac{2}{9}$$

$$k_3 = \lim_{s \to -5} [(s+5) G(s)] = -\frac{2}{9}$$

The state variables x_1, x_2, x_3 are selected and their Laplace transforms take the form of

$$\begin{cases} x_1(s) = \dfrac{1}{(s-\lambda_1)^2} U(s) \\ x_2(s) = \dfrac{1}{(s-\lambda_1)} U(s) \\ x_3(s) = \dfrac{1}{(s-\lambda_3)} U(s) \end{cases}$$

Consequently, the state space description of the system can be deduced as

$$\dot{\boldsymbol{X}} = \begin{bmatrix} \lambda_1 & 1 & 0 \\ 0 & \lambda_1 & 0 \\ 0 & 0 & \lambda_3 \end{bmatrix} \boldsymbol{X} + \begin{bmatrix} 0 \\ 1 \\ 1 \end{bmatrix} u = \begin{bmatrix} -2 & 1 & 0 \\ 0 & -2 & 0 \\ 0 & 0 & -5 \end{bmatrix} \boldsymbol{X} + \begin{bmatrix} 0 \\ 1 \\ 1 \end{bmatrix} u$$

$$y = [k_{11} \quad k_{12} \quad k_3] \boldsymbol{X} = \begin{bmatrix} \frac{1}{3} & \frac{2}{9} & -\frac{2}{9} \end{bmatrix} \boldsymbol{X}$$

The state space description above is shown in Figure 1.11.

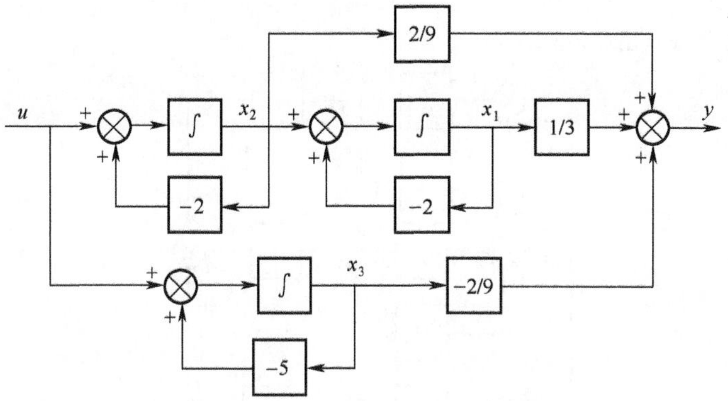

Figure 1.11 The Block Diagram of Example 1.7

Comparing Example 1.7 with Example 1.6, it is clear that state space description of a system or the realization of a transfer function $G(s)$ is not unique because the choice of state variables is different.

Example 1.8 Consider a system described by the transfer function

$$G(s)=\frac{4s+10}{s^3+8s^2+19s+12}$$

Find a state space representation by the method of parallel decomposition.

Solution The characteristic polynomial can be decomposed as

$$D(s)=s^3+8s^2+19s+12=(s+1)(s+3)(s+4)$$

The characteristic roots $\lambda_1=-1, \lambda_2=-3, \lambda_3=-4$ are distinct.

By the partial fraction expansion, the transfer function can be decomposed as

$$G(s)=\frac{Y(s)}{U(s)}=\frac{k_1}{(s+1)}+\frac{k_2}{(s+3)}+\frac{k_3}{(s+4)}$$

Where

$$k_1=\lim_{s\to -1}[(s+1)G(s)]=1$$

$$k_2=\lim_{s\to -3}[(s+3)G(s)]=1$$

$$k_3=\lim_{s\to -4}[(s+4)G(s)]=-2$$

The state variables x_1, x_2, x_3 are selected and their Laplace transforms take the form of

$$\begin{cases} x_1(s)=\dfrac{1}{(s-\lambda_1)}U(s) \\ x_2(s)=\dfrac{1}{(s-\lambda_2)}U(s) \\ x_3(s)=\dfrac{1}{(s-\lambda_3)}U(s) \end{cases}$$

Then, the state space description of the system can be deduced as

$$\dot{\boldsymbol{X}}=\begin{bmatrix} \lambda_1 & 0 & 0 \\ 0 & \lambda_2 & 0 \\ 0 & 0 & \lambda_3 \end{bmatrix}\boldsymbol{X}+\begin{bmatrix} 1 \\ 1 \\ 1 \end{bmatrix}u=\begin{bmatrix} -1 & 0 & 0 \\ 0 & -3 & 0 \\ 0 & 0 & -4 \end{bmatrix}\boldsymbol{X}+\begin{bmatrix} 1 \\ 1 \\ 1 \end{bmatrix}u$$

$$y=[k_1 \quad k_2 \quad k_3]\boldsymbol{X}=[1 \quad 1 \quad -2]\boldsymbol{X}$$

and it is shown in Figure 1.12.

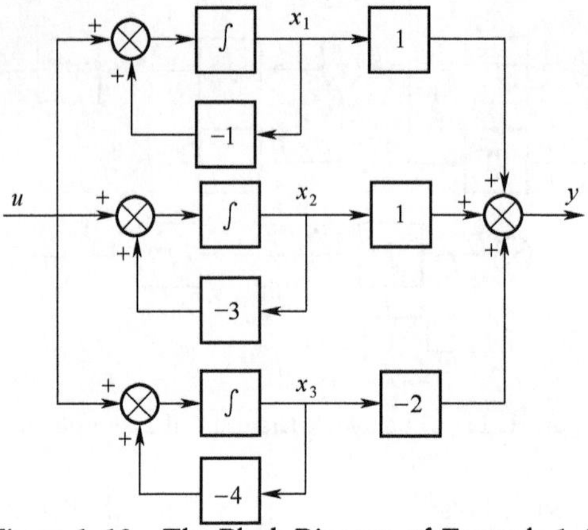

Figure 1.12 The Block Diagram of Example 1.8

1.2.3 Obtaining State Space Description from Block Diagram

In general, a control system can be represented by the block diagram intuitively. In this section, we will discuss how to obtain the state space description from the block diagram.

The outputs of every integrator will be selected as the state variables of a system usually. But for the majority of control system which is represented by a block diagram, the integrator is not made out directly. Table 1.1 shows how to obtain the equivalent system, in which the integrators are made out easily and the state variables are selected directly, from the block diagram.

Table 1.1 The Equivalent System of Block Diagram

Block Diagram	Equivalent System
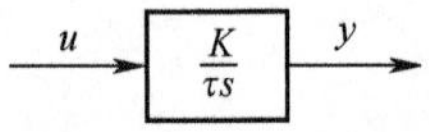	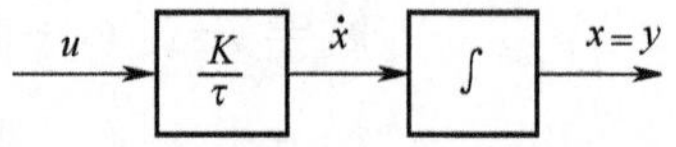
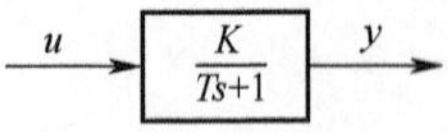	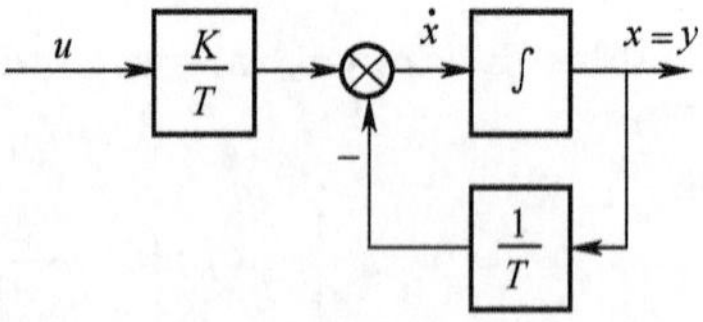
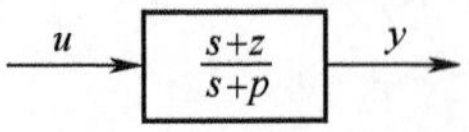	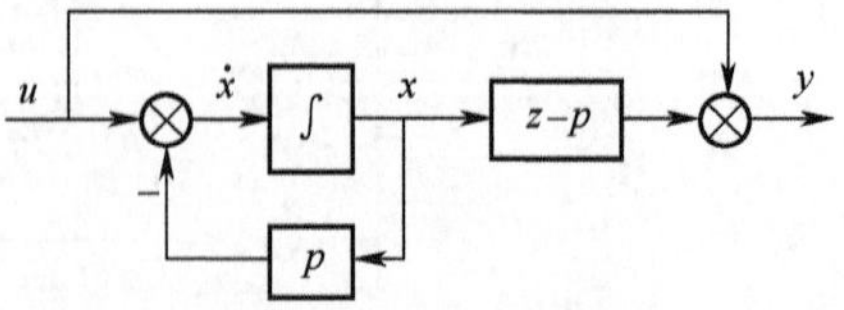
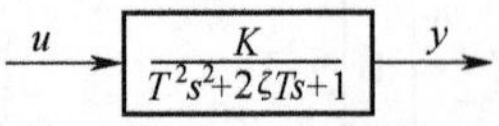	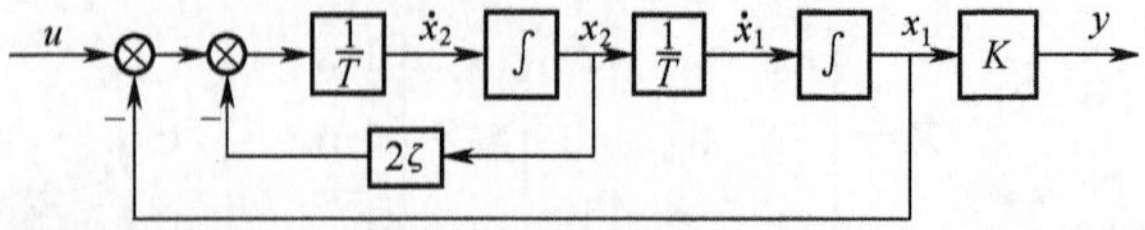

Example 1.9 A closed-loop control system is represented by the block diagram shown in Figure 1.13. Obtain the state space description from the block diagram.

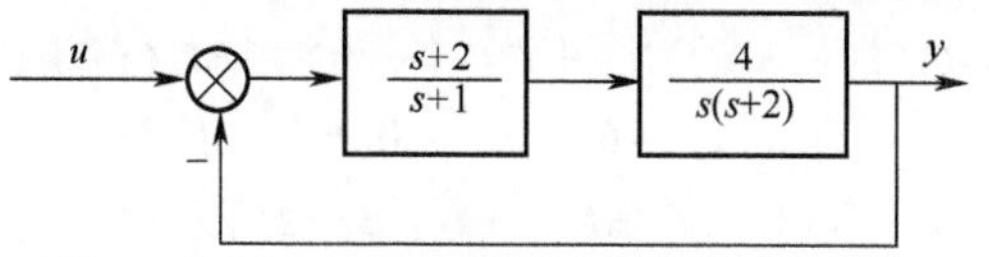

Figure 1.13 A Closed-Loop Control System

Solution According to Table 1.1, the equivalent system is developed in Figure 1.14, and the state variables x_1, x_2 and x_3 are identified on the model.

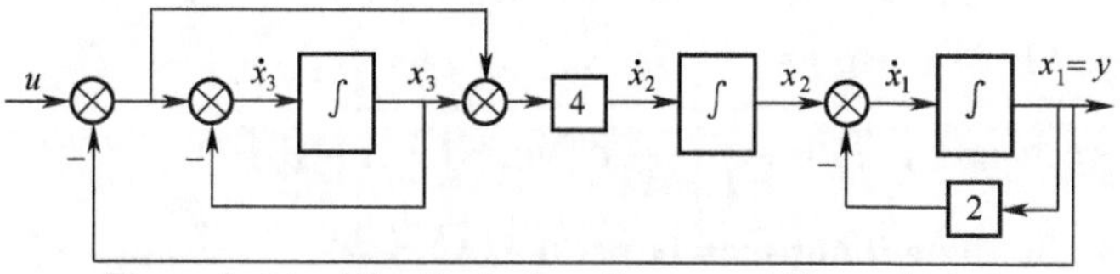

Figure 1.14 The Equivalent System of Example 1.9

The differential equation can be written from Figure 1.14 as

$$\begin{cases}\dot{x}_1=-2x_1+x_2\\ \dot{x}_2=-4x_1+4x_3+4u\\ \dot{x}_3=-x_1-x_3+u\end{cases}$$

and the corresponding output equation is

$$y=x_1$$

So, the state space description of the closed-loop system is

$$\dot{\boldsymbol{X}}=\begin{bmatrix}-2 & 1 & 0\\ -4 & 0 & 4\\ -1 & 0 & -1\end{bmatrix}\boldsymbol{X}+\begin{bmatrix}0\\ 4\\ 1\end{bmatrix}u$$

$$y=[1\quad 0\quad 0]\boldsymbol{X}$$

1.3 Obtaining Transfer Function Matrix from State Space Description

In the study of the LTI system, the transfer function matrix is often used. This section will be devoted to obtain the transfer function matrix from the state space description.

Consider the LTI system such as (1.16)and(1.17). Taking the Laplace transform and assuming $\boldsymbol{X}(0)=\boldsymbol{X}_0$, we will obtain

$$s\boldsymbol{X}(s)-\boldsymbol{X}_0=\boldsymbol{AX}(s)+\boldsymbol{BU}(s) \tag{1.83}$$

$$\boldsymbol{Y}(s)=\boldsymbol{CX}(s)+\boldsymbol{DU}(s) \tag{1.84}$$

Where $\boldsymbol{X}(s)$, $\boldsymbol{U}(s)$, and $\boldsymbol{Y}(s)$ denote the Laplace transforms of the state, input, and output.

Rewriting (1.83), we have

$$\boldsymbol{X}(s)=(s\boldsymbol{I}-\boldsymbol{A})^{-1}\boldsymbol{X}_0+(s\boldsymbol{I}-\boldsymbol{A})^{-1}\boldsymbol{B}\boldsymbol{U}(s) \tag{1.85}$$

Substituting (1.85) into (1.84), the equation (1.86) is obtained as

$$\boldsymbol{Y}(s)=\boldsymbol{C}(s\boldsymbol{I}-\boldsymbol{A})^{-1}\boldsymbol{X}_0+\boldsymbol{C}(s\boldsymbol{I}-\boldsymbol{A})^{-1}\boldsymbol{B}\boldsymbol{U}(s)+\boldsymbol{D}\boldsymbol{U}(s) \tag{1.86}$$

If the initial state of system $\boldsymbol{X}_0=\boldsymbol{0}$, the equation (1.86) can be reduced to

$$\boldsymbol{Y}(s)=[\boldsymbol{C}(s\boldsymbol{I}-\boldsymbol{A})^{-1}\boldsymbol{B}+\boldsymbol{D}]\boldsymbol{U}(s) \tag{1.87}$$

According to the concept of transfer function matrix, we have

$$\boldsymbol{G}(s)=\boldsymbol{C}(s\boldsymbol{I}-\boldsymbol{A})^{-1}\boldsymbol{B}+\boldsymbol{D} \tag{1.88}$$

Hence, if the LTI system is described by transfer function matrix $G(s)$ and the state space description $\{\boldsymbol{A},\boldsymbol{B},\boldsymbol{C},\boldsymbol{D}\}$, the two descriptions must be related by (1.88).

We can write (1.88) also as

$$\boldsymbol{G}(s)=\frac{1}{|s\boldsymbol{I}-\boldsymbol{A}|}\boldsymbol{C}[\mathrm{adj}(s\boldsymbol{I}-\boldsymbol{A})]\boldsymbol{B}+\boldsymbol{D}$$

Where $\mathrm{adj}(s\boldsymbol{I}-\boldsymbol{A})$ is the **adjoint matrix** of $(s\boldsymbol{I}-\boldsymbol{A})$.

Since the degree of every entry polynomial of $\mathrm{adj}(s\boldsymbol{I}-\boldsymbol{A})$ is strictly less than the degree of the determinant of $(s\boldsymbol{I}-\boldsymbol{A})$, $\boldsymbol{C}(s\boldsymbol{I}-\boldsymbol{A})^{-1}\boldsymbol{B}$ is a strictly proper rational function matrix. So, if $\boldsymbol{D}$ is a nonzero matrix, $\boldsymbol{C}(s\boldsymbol{I}-\boldsymbol{A})^{-1}\boldsymbol{B}+\boldsymbol{D}$ is a proper rational matrix, and

$$\boldsymbol{D}=\lim_{s\to\infty}\boldsymbol{G}(s)$$

Note that the system is realizable iff $\boldsymbol{G}(s)$ is a proper rational matrix.

Example 1.10 A system is described by

$$\dot{\boldsymbol{X}}=\begin{bmatrix}-5 & -1\\ 3 & -1\end{bmatrix}\boldsymbol{X}+\begin{bmatrix}2\\ 5\end{bmatrix}u$$

$$y=[1 \quad 2]\boldsymbol{X}$$

Find the transfer function of it.

Solution From the state description $\{\boldsymbol{A},\boldsymbol{b},\boldsymbol{c}\}$ of the system, we can obtain

$$(s\boldsymbol{I}-\boldsymbol{A})=\begin{bmatrix}s+5 & 1\\ -3 & s+1\end{bmatrix}$$

Its inverse matrix is

$$(s\boldsymbol{I}-\boldsymbol{A})^{-1}=\frac{1}{(s+1)(s+5)+3}\begin{bmatrix}s+1 & -1\\ 3 & s+5\end{bmatrix}$$

$$=\frac{1}{s^2+6s+8}\begin{bmatrix}s+1 & -1\\ 3 & s+5\end{bmatrix}$$

Then, the transfer function of the system can be calculated as

$$G(s)=\boldsymbol{c}(s\boldsymbol{I}-\boldsymbol{A})^{-1}\boldsymbol{b}+\boldsymbol{d}$$

$$=[1 \quad 2](\frac{1}{s^2+6s+8})\begin{bmatrix}s+1 & -1\\ 3 & s+5\end{bmatrix}\begin{bmatrix}2\\ 5\end{bmatrix}=\frac{12s+59}{s^2+6s+8}$$

Example 1.11 A system is described by

$$\dot{\boldsymbol{X}}=\begin{bmatrix}0 & 1\\ -2 & -3\end{bmatrix}\boldsymbol{X}+\begin{bmatrix}1 & 0\\ 1 & 1\end{bmatrix}\boldsymbol{u}$$

$$y=\begin{bmatrix}1 & 0\\ 1 & 1\end{bmatrix}X$$

Find the transfer function matrix of it.

Solution From the state description $\{A, B, C\}$ of the system, we can obtain

$$(sI-A)=\begin{bmatrix}s & -1\\ 2 & s+3\end{bmatrix}$$

Its inverse matrix is

$$(sI-A)^{-1}=\frac{1}{s(s+3)+2}\begin{bmatrix}s+3 & 1\\ -2 & s\end{bmatrix}=\frac{1}{s^2+3s+2}\begin{bmatrix}s+3 & 1\\ -2 & s\end{bmatrix}$$

Then, the transfer function matrix of the system can be obtained as

$$\begin{aligned}G(s)&=C(sI-A)^{-1}B+D\\ &=\begin{bmatrix}1 & 0\\ 1 & 1\end{bmatrix}\left(\frac{1}{s^2+3s+2}\right)\begin{bmatrix}s+3 & 1\\ -2 & s\end{bmatrix}\begin{bmatrix}1 & 0\\ 1 & 1\end{bmatrix}\\ &=\begin{bmatrix}\dfrac{s+4}{(s+1)(s+2)} & \dfrac{1}{(s+1)(s+2)}\\ \dfrac{2}{s+2} & \dfrac{1}{s+2}\end{bmatrix}\end{aligned}$$

1.4 Description of Composite Systems

1.4.1 Basic Connection of Composite Systems

In engineering, a system is often built by interconnecting a number of subsystems. Such a system is called a **composite system.** Composite systems are mostly built from three basic connection forms: the parallel, the series, and the feedback connection.

Suppose one subsystem $\sum_1(A_1, B_1, C_1, D_1)$ is

$$\begin{cases}\dot{X}_1(t)=A_1X_1(t)+B_1u_1(t)\\ y_1(t)=C_1X_1(t)+D_1u_1(t)\end{cases}\tag{1.89}$$

and its transfer function matrix is

$$G_1(s)=\frac{Y_1(s)}{U_1(s)}=C_1(sI-A_1)^{-1}B_1+D_1\tag{1.90}$$

Another subsystem $\sum_2(A_2, B_2, C_2, D_2)$ is

$$\begin{cases}\dot{X}_2(t)=A_2X_2(t)+B_2u_2(t)\\ y_2(t)=C_2X_2(t)+D_2u_2(t)\end{cases}\tag{1.91}$$

and its transfer function matrix is

$$G_2(s)=\frac{Y_2(s)}{U_2(s)}=C_2(sI-A_2)^{-1}B_2+D_2\tag{1.92}$$

So, three basic connection forms of the composite system are shown in Figure 1.15~Figure 1.17.

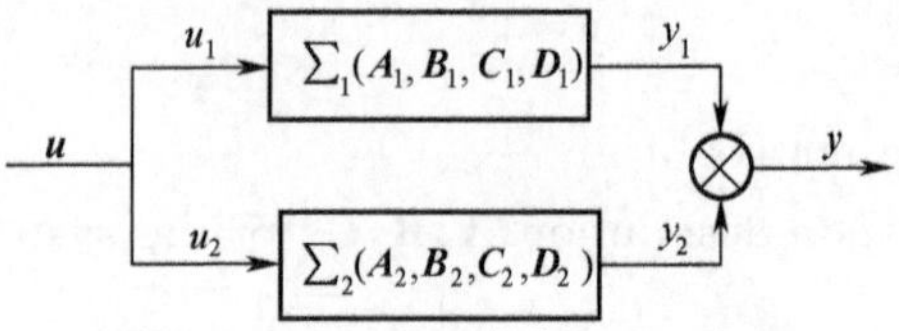

Figure 1. 15 Parallel Connection

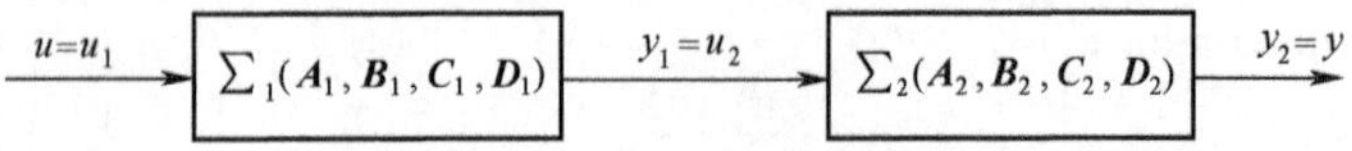

Figure 1. 16 Series Connection

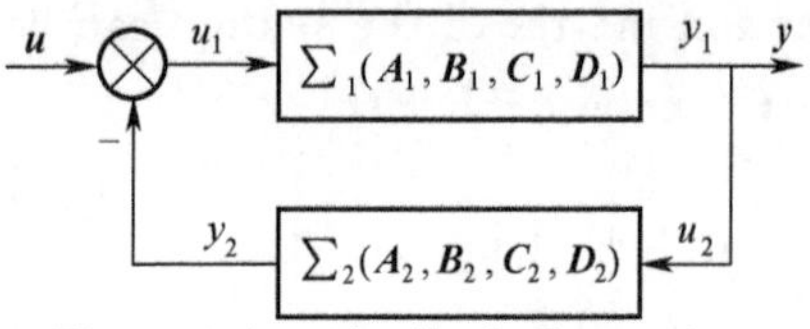

Figure 1. 17 Feedback Connection

1. 4. 2 Description of the Series Composite Systems

The state space description of the series composite system, which is connected with subsystem $\sum_1(\boldsymbol{A}_1, \boldsymbol{B}_1, \boldsymbol{C}_1, \boldsymbol{D}_1)$ and $\sum_2(\boldsymbol{A}_2, \boldsymbol{B}_2, \boldsymbol{C}_2, \boldsymbol{D}_2)$, is given by

$$\begin{bmatrix} \dot{\boldsymbol{X}}_1 \\ \dot{\boldsymbol{X}}_2 \end{bmatrix} = \begin{bmatrix} \boldsymbol{A}_1 & \boldsymbol{0} \\ \boldsymbol{B}_2\boldsymbol{C}_1 & \boldsymbol{A}_2 \end{bmatrix} \begin{bmatrix} \boldsymbol{X}_1 \\ \boldsymbol{X}_2 \end{bmatrix} + \begin{bmatrix} \boldsymbol{B}_1 \\ \boldsymbol{B}_2\boldsymbol{D}_1 \end{bmatrix} \boldsymbol{u}$$

$$\boldsymbol{y} = [\boldsymbol{D}_2\boldsymbol{C}_1 \quad \boldsymbol{C}_2] \begin{bmatrix} \boldsymbol{X}_1 \\ \boldsymbol{X}_2 \end{bmatrix} + [\boldsymbol{D}_2\boldsymbol{D}_1]\boldsymbol{u} \tag{1.93}$$

In order to calculate the transfer function matrix for the series composite system, the following matrix inversion formula is needed.

$$\begin{bmatrix} \boldsymbol{A}_{11} & \boldsymbol{0} \\ \boldsymbol{A}_{21} & \boldsymbol{A}_{22} \end{bmatrix}^{-1} = \begin{bmatrix} \boldsymbol{A}_{11}^{-1} & \boldsymbol{0} \\ -\boldsymbol{A}_{22}^{-1}\boldsymbol{A}_{21}\boldsymbol{A}_{11}^{-1} & \boldsymbol{A}_{22}^{-1} \end{bmatrix}$$

So, the transfer function matrix for the series composite system is

$$\begin{aligned} \boldsymbol{G}(s) &= [\boldsymbol{D}_2\boldsymbol{C}_1 \quad \boldsymbol{C}_2] \begin{bmatrix} s\boldsymbol{I}_1 - \boldsymbol{A}_1 & \boldsymbol{0} \\ -\boldsymbol{B}_2\boldsymbol{C}_1 & s\boldsymbol{I}_2 - \boldsymbol{A}_2 \end{bmatrix}^{-1} \begin{bmatrix} \boldsymbol{B}_1 \\ \boldsymbol{B}_2\boldsymbol{D}_1 \end{bmatrix} + [\boldsymbol{D}_2\boldsymbol{D}_1] \\ &= [\boldsymbol{D}_2\boldsymbol{C}_1 \quad \boldsymbol{C}_2] \begin{bmatrix} (s\boldsymbol{I}_1 - \boldsymbol{A}_1)^{-1} & \boldsymbol{0} \\ (s\boldsymbol{I}_2 - \boldsymbol{A}_2)^{-1}\boldsymbol{B}_2\boldsymbol{C}_1(s\boldsymbol{I}_1 - \boldsymbol{A}_1)^{-1} & (s\boldsymbol{I}_2 - \boldsymbol{A}_2)^{-1} \end{bmatrix} \begin{bmatrix} \boldsymbol{B}_1 \\ \boldsymbol{B}_2\boldsymbol{D}_1 \end{bmatrix} + [\boldsymbol{D}_2\boldsymbol{D}_1] \\ &= [\boldsymbol{C}_2(s\boldsymbol{I}_2 - \boldsymbol{A}_2)^{-1}\boldsymbol{B}_2 + \boldsymbol{D}_2][\boldsymbol{C}_1(s\boldsymbol{I}_1 - \boldsymbol{A}_1)^{-1}\boldsymbol{B}_1 + \boldsymbol{D}_1] \\ &= \boldsymbol{G}_2(s)\boldsymbol{G}_1(s) \end{aligned} \tag{1.94}$$

1. 4. 3 Description of the Parallel Composite Systems

The state space description of the parallel composite system, which is connected with subsystem $\sum_1(\boldsymbol{A}_1,\boldsymbol{B}_1,\boldsymbol{C}_1,\boldsymbol{D}_1)$ and $\sum_2(\boldsymbol{A}_2,\boldsymbol{B}_2,\boldsymbol{C}_2,\boldsymbol{D}_2)$, is given by

$$\begin{bmatrix}\dot{\boldsymbol{X}}_1\\ \dot{\boldsymbol{X}}_2\end{bmatrix}=\begin{bmatrix}\boldsymbol{A}_1 & \boldsymbol{0}\\ \boldsymbol{0} & \boldsymbol{A}_2\end{bmatrix}\begin{bmatrix}\boldsymbol{X}_1\\ \boldsymbol{X}_2\end{bmatrix}+\begin{bmatrix}\boldsymbol{B}_1\\ \boldsymbol{B}_2\end{bmatrix}\boldsymbol{u}$$
$$\boldsymbol{y}=[\boldsymbol{C}_1\quad \boldsymbol{C}_2]\begin{bmatrix}\boldsymbol{X}_1\\ \boldsymbol{X}_2\end{bmatrix}+[\boldsymbol{D}_1+\boldsymbol{D}_2]\boldsymbol{u} \tag{1.95}$$

So, the transfer function matrix for the parallel composite system is

$$\begin{aligned}\boldsymbol{G}(s)&=[\boldsymbol{C}_1\quad \boldsymbol{C}_2]\begin{bmatrix}s\boldsymbol{I}_1-\boldsymbol{A}_1 & \boldsymbol{0}\\ \boldsymbol{0} & s\boldsymbol{I}_2-\boldsymbol{A}_2\end{bmatrix}^{-1}\begin{bmatrix}\boldsymbol{B}_1\\ \boldsymbol{B}_2\end{bmatrix}+[\boldsymbol{D}_1+\boldsymbol{D}_2]\\
&=[\boldsymbol{C}_1\quad \boldsymbol{C}_2]\begin{bmatrix}(s\boldsymbol{I}_1-\boldsymbol{A}_1)^{-1} & \boldsymbol{0}\\ \boldsymbol{0} & (s\boldsymbol{I}_2-\boldsymbol{A}_2)^{-1}\end{bmatrix}\begin{bmatrix}\boldsymbol{B}_1\\ \boldsymbol{B}_2\end{bmatrix}+[\boldsymbol{D}_1+\boldsymbol{D}_2]\\
&=[\boldsymbol{C}_1(s\boldsymbol{I}_1-\boldsymbol{A}_1)^{-1}\boldsymbol{B}_1+\boldsymbol{D}_1+\boldsymbol{C}_2(s\boldsymbol{I}_2-\boldsymbol{A}_2)^{-1}\boldsymbol{B}_2+\boldsymbol{D}_2]\\
&=\boldsymbol{G}_1(s)+\boldsymbol{G}_2(s)\end{aligned} \tag{1.96}$$

1.4.4 Description of the Feedback Composite Systems

The state space description and the transfer function matrix of the feedback composite system, which is connected with subsystem $\sum_1(\boldsymbol{A}_1,\boldsymbol{B}_1,\boldsymbol{C}_1,\boldsymbol{D}_1)$ and $\sum_2(\boldsymbol{A}_2,\boldsymbol{B}_2,\boldsymbol{C}_2,\boldsymbol{D}_2)$, is given by (1.97) and (1.98) without proofs.

$$\begin{bmatrix}\dot{\boldsymbol{X}}_1\\ \dot{\boldsymbol{X}}_2\end{bmatrix}=\begin{bmatrix}\boldsymbol{A}_1-\boldsymbol{B}_1(\boldsymbol{I}+\boldsymbol{D}_2\boldsymbol{D}_1)^{-1}\boldsymbol{D}_2\boldsymbol{C}_1 & -\boldsymbol{B}_1(\boldsymbol{I}+\boldsymbol{D}_2\boldsymbol{D}_1)^{-1}\boldsymbol{C}_2\\ \boldsymbol{B}_2(\boldsymbol{I}+\boldsymbol{D}_1\boldsymbol{D}_2)^{-1}\boldsymbol{C}_1 & \boldsymbol{A}_2-\boldsymbol{B}_2(\boldsymbol{I}+\boldsymbol{D}_1\boldsymbol{D}_2)^{-1}\boldsymbol{D}_1\boldsymbol{C}_2\end{bmatrix}\begin{bmatrix}\boldsymbol{X}_1\\ \boldsymbol{X}_2\end{bmatrix}$$
$$+\begin{bmatrix}\boldsymbol{B}_1(\boldsymbol{I}+\boldsymbol{D}_2\boldsymbol{D}_1)^{-1}\\ \boldsymbol{B}_2(\boldsymbol{I}+\boldsymbol{D}_1\boldsymbol{D}_2)^{-1}\boldsymbol{D}_1\end{bmatrix}\boldsymbol{u} \tag{1.97}$$
$$\boldsymbol{y}=[(\boldsymbol{I}+\boldsymbol{D}_1\boldsymbol{D}_2)^{-1}\boldsymbol{C}_1\quad -(\boldsymbol{I}+\boldsymbol{D}_1\boldsymbol{D}_2)^{-1}\boldsymbol{D}_1\boldsymbol{C}_2]\begin{bmatrix}\boldsymbol{X}_1\\ \boldsymbol{X}_2\end{bmatrix}+[(\boldsymbol{I}+\boldsymbol{D}_1\boldsymbol{D}_2)^{-1}\boldsymbol{D}_1]\boldsymbol{u}$$
$$\boldsymbol{G}(s)=\boldsymbol{G}_1(s)(\boldsymbol{I}+\boldsymbol{G}_2(s)\boldsymbol{G}_1(s))^{-1}=(\boldsymbol{I}+\boldsymbol{G}_1(s)\boldsymbol{G}_2(s))^{-1}\boldsymbol{G}_1(s) \tag{1.98}$$

1.5 State Transformation of the LTI system

1.5.1 Eigenvalue and Eigenvector

Consider the LTI system described by the state equation

$$\dot{\boldsymbol{X}}=\boldsymbol{A}\boldsymbol{X}+\boldsymbol{B}\boldsymbol{u}$$

Where $\boldsymbol{X}\in\mathbf{R}^n$ is the state vector; $\boldsymbol{A}\in\mathbf{R}^{n\times n}$ is the system matrix and plays an important role in system properties. When $\boldsymbol{u}=\boldsymbol{0}$, the system given by (1.99) is called a free system.

$$\dot{\boldsymbol{X}}=\boldsymbol{A}\boldsymbol{X} \tag{1.99}$$

One case is $\boldsymbol{X}$ and $\dot{\boldsymbol{X}}$ have the same directions in the state space but may differ in magnitude by a scalar proportionality factor λ. That means

$$\boldsymbol{A}\boldsymbol{X}=\lambda\boldsymbol{X}$$

and the Eq. (1.100)

$$(\lambda\boldsymbol{I}-\boldsymbol{A})\boldsymbol{X}=\boldsymbol{0} \tag{1.100}$$

has non zero solution.

In this case, the matrix $\lambda\boldsymbol{I}-\boldsymbol{A}$ must not be full rank, or $\text{rank}(\lambda\boldsymbol{I}-\boldsymbol{A})<n$, and the determinant $|\lambda\boldsymbol{I}-\boldsymbol{A}|$ must be zero.

The polynomial about λ

$$Q(\lambda)=|\lambda\boldsymbol{I}-\boldsymbol{A}|=\lambda^n+\sum_{i=0}^{n-1}\alpha_i\lambda^i \tag{1.101}$$

is called the **characteristic polynomial**, and $Q(\lambda)=0$ is called the **characteristic equation** of system .

If the polynomial $Q(\lambda)$ can be written in factored form as

$$Q(\lambda)=\det(\lambda\boldsymbol{I}-\boldsymbol{A})=\prod_{i=1}^{n}(\lambda-\lambda_i) \tag{1.102}$$

the roots $\lambda_i(i=1,2,\cdots,n)$ of the characteristic equation are called the **eigenvalues** of $\boldsymbol{A}$.

Some important properties of eigenvalues are given as follows.

(1) If the elements of $\boldsymbol{A}$ are real, then its eigenvalues are either real or in complex conjugate pairs.

(2) If $\lambda_i(i=1,2,\cdots,n)$ are the eigenvalues of $\boldsymbol{A}$, then

$$\text{tr}(\boldsymbol{A})=\sum_{i=1}^{n}\lambda_i \tag{1.103}$$

That is, the trace of $\boldsymbol{A}$ is the sum of all eigenvalues of $\boldsymbol{A}$.

(3) If $\lambda_i(i=1,2,\cdots,n)$ are eigenvalues of $\boldsymbol{A}$, then they are the eigenvalues of $\boldsymbol{A}^{\text{T}}$.

(4) If $\boldsymbol{A}$ is nonsingular, with eigenvalues $\lambda_i(i=1,2,\cdots,n)$, then $\frac{1}{\lambda_i}$ are the eigenvalues of $\boldsymbol{A}^{-1}$.

Any nonzero vector $\boldsymbol{V}_i$ which satisfies the matrix equation

$$(\lambda_i\boldsymbol{I}-\boldsymbol{A})\boldsymbol{V}_i=\boldsymbol{0} \tag{1.104}$$

is called the **eigenvector** of $\boldsymbol{A}$ associated with eigenvalue $\lambda_i(i=1,2,\cdots,n)$. If $\boldsymbol{A}$ has distinct eigenvalues, the eigenvectors can be solved directly from Eq. (1.104). Eigenvectors are useful in modern control methods, one of which is state transformation.

It should be pointed out that if $\boldsymbol{A}$ has multiplicity eigenvalues and is nonsymmetrical, not all eigenvectors can be found using Eq. (1.104). Let us assume that λ_1 is the m multiplicity eigenvalue of $\boldsymbol{A}$ and the remaining $n-m$ distinct eigenvalues are λ_{m+1}, $\lambda_{m+2},\cdots,\lambda_n$. Based on $\lambda_{m+1},\lambda_{m+2},\cdots,\lambda_n$, we can find $n-m$ linearly independent eigenvectors $\boldsymbol{V}_{m+1},\boldsymbol{V}_{m+2},\cdots,\boldsymbol{V}_n$ from Eq. (1.104). But if $\text{rank}(\lambda_1\boldsymbol{I}-\boldsymbol{A})=n-1$, then from the equation $(\lambda_1\boldsymbol{I}-\boldsymbol{A})\boldsymbol{V}=\boldsymbol{0}$ we can only find one linearly independent eigenvectors

V_{11}. We can also obtain $m-1$ linearly independent **generalized eigenvectors** $V_{12}, \cdots, V_{1m}$ from the following $m-1$ vector equations (1.105).

$$\begin{aligned}(\lambda_1 I - A)V_{12} &= -V_{11}\\ (\lambda_1 I - A)V_{13} &= -V_{12}\\ &\vdots\\ (\lambda_1 I - A)V_{1m} &= -V_{1(m-1)}\end{aligned} \tag{1.105}$$

Example 1.12 Determine the eigenvalues of A and the corresponding eigenvectors.

$$A = \begin{bmatrix} 0 & 6 & -5 \\ 1 & 0 & 2 \\ 3 & 2 & 4 \end{bmatrix}$$

Solution The characteristic equation is

$$|\lambda I - A| = (\lambda - 2)(\lambda - 1)^2 = 0$$

The eigenvalues of A are $\lambda_1 = 2, \lambda_{2,3} = 1$. Thus, A has a 2 multiplicity eigenvalue at 1.

The eigenvector associated with $\lambda_1 = 2$ is determined by Eq. (1.104). Thus,

$$(\lambda_1 I - A)V_1 = \begin{bmatrix} 2 & -6 & 5 \\ -1 & 2 & -2 \\ -3 & -2 & -2 \end{bmatrix} \begin{bmatrix} v_{11} \\ v_{12} \\ v_{13} \end{bmatrix} = \mathbf{0} \tag{1.106}$$

Since there are only two independent equations in Eq. (1.106), we arbitrarily set $v_{11} = 2$, and have

$$V_1 = \begin{bmatrix} 2 \\ -1 \\ -2 \end{bmatrix}$$

For the 2 multiplicity eigenvalue, we substitute $\lambda_2 = 1$ into the equation(1.104).

$$(\lambda_2 I - A)V_{21} = \begin{bmatrix} 1 & -6 & 5 \\ -1 & 1 & -2 \\ -3 & -2 & -3 \end{bmatrix} \begin{bmatrix} v_{211} \\ v_{212} \\ v_{213} \end{bmatrix} = \mathbf{0}$$

Setting $v_{211} = 1$ arbitrarily, we have the eigenvector associated with eigenvalue λ_2.

$$V_{21} = \begin{bmatrix} 1 \\ -3/7 \\ -5/7 \end{bmatrix}$$

Substituting $\lambda_2 = 1$ into the first equation of(1.105)

$$(\lambda_2 I - A)V_{22} = \begin{bmatrix} 1 & -6 & 5 \\ -1 & 1 & -2 \\ -3 & -2 & -3 \end{bmatrix} \begin{bmatrix} v_{221} \\ v_{222} \\ v_{223} \end{bmatrix} = -V_{21} = \begin{bmatrix} -1 \\ 3/7 \\ 5/7 \end{bmatrix}$$

and setting $v_{221} = 1$ arbitrarily, we have the generalized eigenvector associated with eigenvalue λ_2.

$$V_{22} = \begin{bmatrix} 1 \\ -22/49 \\ -46/49 \end{bmatrix}$$

1.5.2 State Transformation

Consider the LTI system described by

$$\begin{aligned}\dot{\boldsymbol{X}}(t)&=\boldsymbol{AX}(t)+\boldsymbol{Bu}(t)\\ y(t)&=\boldsymbol{CX}(t)+\boldsymbol{Du}(t)\end{aligned} \tag{1.107}$$

As we have seen, this is not a unique description of the system and the selection of state variables for a dynamic system is not unique. We consider a change of state from $\boldsymbol{X}(t)$ to $\overline{\boldsymbol{X}}(t)$, that is the **linear state transformation** of $\boldsymbol{X}(t)$. For a $n\times n$ nonsingular matrix $\boldsymbol{P}$, we let

$$\boldsymbol{X}(t)=\boldsymbol{P}\overline{\boldsymbol{X}}(t) \tag{1.108}$$

or

$$\overline{\boldsymbol{X}}(t)=\boldsymbol{P}^{-1}\boldsymbol{X}(t) \tag{1.109}$$

Eq. (1.108)or Eq. (1.109) is called the linear nonsingular state transformation.

By substituting (1.108) and (1.109) into (1.107), we have the new state equation in terms of the new state $\overline{\boldsymbol{X}}(t)$ as

$$\dot{\overline{\boldsymbol{X}}}(t)=\boldsymbol{P}^{-1}\boldsymbol{AP}\overline{\boldsymbol{X}}(t)+\boldsymbol{P}^{-1}\boldsymbol{Bu}(t)=\overline{\boldsymbol{A}}\overline{\boldsymbol{X}}(t)+\overline{\boldsymbol{B}}\boldsymbol{u}(t) \tag{1.110}$$

Where

$$\overline{\boldsymbol{A}}=\boldsymbol{P}^{-1}\boldsymbol{AP} \tag{1.111}$$

$$\overline{\boldsymbol{B}}=\boldsymbol{P}^{-1}\boldsymbol{B} \tag{1.112}$$

The transformation (1.111) is called the **similarity transformation** of A.

By substituting (1.108) into (1.107), we have the output equation in terms of the new state $\overline{\boldsymbol{X}}(t)$

$$y(t)=\boldsymbol{CP}\overline{\boldsymbol{X}}(t)+\boldsymbol{Du}(t)=\overline{\boldsymbol{C}}\overline{\boldsymbol{X}}(t)+\overline{\boldsymbol{D}}\boldsymbol{u}(t) \tag{1.113}$$

Where

$$\overline{\boldsymbol{C}}=\boldsymbol{CP} \tag{1.114}$$

$$\overline{\boldsymbol{D}}=\boldsymbol{D} \tag{1.115}$$

Given the general matrixes $\boldsymbol{A}$, $\boldsymbol{B}$, $\boldsymbol{C}$ and $\boldsymbol{D}$, we should like to find the nonsingular transformation $\boldsymbol{X}(t)=\boldsymbol{P}\overline{\boldsymbol{X}}(t)$ such that $\overline{\boldsymbol{A}}$, $\overline{\boldsymbol{B}}$, $\overline{\boldsymbol{C}}$ and $\overline{\boldsymbol{D}}$ are in a particular form, for example, Jordan canonical form or controllable canonical form. Moreover, **the characteristic equation, eigenvalues and transfer function are all preserved by the nonsingular state transformation.**

1.5.3 Invariance Properties of the State Transformation

The characteristic equation of the system described by (1.110) and (1.113) is

$$|s\boldsymbol{I}-\overline{\boldsymbol{A}}|=|s\boldsymbol{I}-\boldsymbol{P}^{-1}\boldsymbol{AP}|=|s\boldsymbol{P}^{-1}\boldsymbol{P}-\boldsymbol{P}^{-1}\boldsymbol{AP}|=|\boldsymbol{P}^{-1}(s\boldsymbol{I}-\boldsymbol{A})\boldsymbol{P}|=0 \tag{1.116}$$

Since the determinant of a product matrix is equal to the product of the determinants of the matrices, Eq. (1.116) becomes

$$|sI-\overline{A}|=|P^{-1}||sI-A||P|=|sI-A|=0$$

Thus, the characteristic equation is preserved by the nonsingular state transformation, which naturally leads to the same eigenvalues and eigenvectors.

The transfer function of the system described by (1.110) and (1.113) can be calculated as

$$\begin{aligned}\overline{G}(s)&=\overline{C}(sI-\overline{A})^{-1}\overline{B}+\overline{D}=CP(sI-P^{-1}AP)^{-1}P^{-1}B+D\\&=C[P(sI-P^{-1}AP)P^{-1}]^{-1}B+D=C[P(sP^{-1}P-P^{-1}AP)P^{-1}]^{-1}B+D\\&=C[PP^{-1}(sI-A)PP^{-1}]^{-1}B+D=C(sI-A)^{-1}B+D\\&=G(s)\end{aligned}$$

It means that the transfer function is preserved by the nonsingular state transformation also.

In fact, controllability and observability are all preserved by the nonsingular state transformation. It will be discussed later.

1.5.4 Obtaining the Diagonal Canonical Form by State Transformation

Consider a n dimension system, described by the state space description

$$\begin{aligned}\dot{X}&=AX+Bu\\y&=CX+Du\end{aligned}\qquad(1.117)$$

Let $\lambda_1,\lambda_2,\cdots,\lambda_n$ be the distinct eigenvalues of A, and let V_i be the eigenvector of A associated with the eigenvalue $\lambda_i(i=1,2,\cdots,n)$.

Then, the matrix

$$P=[V_1\quad V_2\quad\cdots\quad V_n]\qquad(1.118)$$

is a nonsingular matrix. Since $AV_i=\lambda_iV_i$, we have

$$\begin{aligned}AP&=[AV_1\quad\cdots\quad AV_n]=[\lambda_1V_1\quad\cdots\quad\lambda_nV_n]\\&=[V_1\quad\cdots\quad V_n]\begin{bmatrix}\lambda_1&0&0\\0&\ddots&0\\0&0&\lambda_n\end{bmatrix}=P\begin{bmatrix}\lambda_1&0&0\\0&\ddots&0\\0&0&\lambda_n\end{bmatrix}\end{aligned}$$

Hence

$$P^{-1}AP=\begin{bmatrix}\lambda_1&0&0\\0&\ddots&0\\0&0&\lambda_n\end{bmatrix}\qquad(1.119)$$

So, if a system described by (1.117) and A has distinct eigenvalues $\lambda_1,\lambda_2,\cdots,\lambda_n$, there is a nonsingular transformation

$$X(t)=P\overline{X}(t)$$

which transforms the general state description, such as (1.117), into the diagonal canonical form, such as (1.120).

$$\begin{aligned}\dot{\overline{X}}&=\overline{A}\ \overline{X}+\overline{B}u\\y&=\overline{C}\ \overline{X}+\overline{D}u\end{aligned}\qquad(1.120)$$

Where $\overline{A}=P^{-1}AP=\begin{bmatrix}\lambda_1 & & 0\\ & \ddots & \\ 0 & & \lambda_n\end{bmatrix}$ is a diagonal matrix; $\overline{B}=P^{-1}B$; $\overline{C}=CP$; $\overline{D}=D$.

Example 1.13 The state space description of a system is

$$\dot{X}=\begin{bmatrix}2 & -1 & -1\\ 0 & -1 & 0\\ 0 & 2 & 1\end{bmatrix}X+\begin{bmatrix}7\\ 2\\ 3\end{bmatrix}u \quad y=[1 \quad 0 \quad 1]X$$

Determine the transformation matrix P and transform the state space description into the diagonal canonical form by the state transformation $X(t)=P\overline{X}(t)$.

Solution

(1) The characteristic equation

$$|\lambda I-A|=\begin{vmatrix}\lambda-2 & 1 & 1\\ 0 & \lambda+1 & 0\\ 0 & -2 & \lambda-1\end{vmatrix}=(\lambda-2)(\lambda-1)(\lambda+1)=0$$

yields the eigenvalues $\lambda_1=2, \lambda_2=1, \lambda_3=-1$.

(2) Since the eigenvalues are distinct, the state equation can be converted into a diagonal canonical form by the means of a state transformation $X(t)=P\overline{X}(t)$.

When $\lambda_1=2$ the equation $\lambda_1 V_1=AV_1$ yields the characteristic vector $V_1=\begin{bmatrix}C\\ 0\\ 0\end{bmatrix}$ of A corresponding to λ_1, and $V_1=\begin{bmatrix}1\\ 0\\ 0\end{bmatrix}$ is selected here.

When $\lambda_2=1$, the equation $\lambda_2 V_2=AV_2$ yields the characteristic vector $V_2=\begin{bmatrix}1\\ 0\\ 1\end{bmatrix}$ of A corresponding to λ_2.

When $\lambda_3=-1$ the equation $\lambda_3 V_3=AV_3$ yields the characteristic vector $V_3=\begin{bmatrix}0\\ 1\\ -1\end{bmatrix}$ of A corresponding to λ_3.

Then, the nonsingular matrix is constructed as

$$P=[V_1 \quad V_2 \quad V_3]=\begin{bmatrix}1 & 1 & 0\\ 0 & 0 & 1\\ 0 & 1 & -1\end{bmatrix}$$

and

$$P^{-1}=\begin{bmatrix}1 & -1 & -1\\ 0 & 1 & 1\\ 0 & 1 & 0\end{bmatrix}$$

So the diagonal canonical form can be obtained by the state transformation $\boldsymbol{X}(t)=\boldsymbol{P}\overline{\boldsymbol{X}}(t)$ as

$$\dot{\overline{\boldsymbol{X}}}=\overline{\boldsymbol{A}}\,\overline{\boldsymbol{X}}+\overline{\boldsymbol{b}}u=\boldsymbol{P}^{-1}\boldsymbol{A}\boldsymbol{P}\,\overline{\boldsymbol{X}}+\boldsymbol{P}^{-1}\boldsymbol{b}u=\begin{bmatrix}2&0&0\\0&1&0\\0&0&-1\end{bmatrix}\overline{\boldsymbol{X}}+\begin{bmatrix}2\\5\\2\end{bmatrix}u$$

$$y=\overline{\boldsymbol{c}}\,\overline{\boldsymbol{X}}=\boldsymbol{c}\boldsymbol{P}\,\overline{\boldsymbol{X}}=[1\quad 2\quad -1]\overline{\boldsymbol{X}}$$

The state space description above can be shown in Figure 1.18. It is apparent that one of the advantages of a diagonal canonical form is that the transformed states are uncoupled from each other.

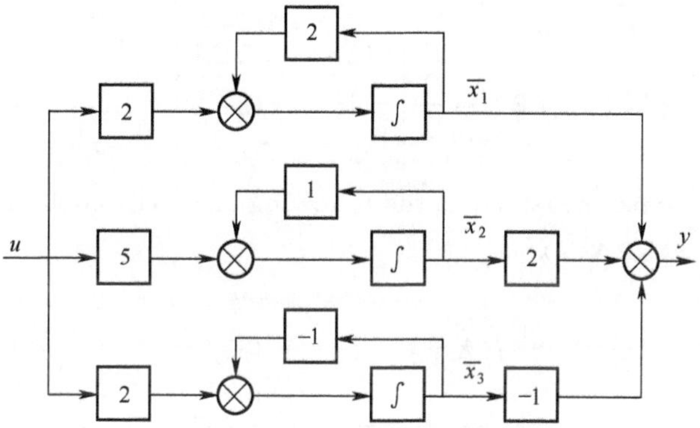

Figure 1.18 The Block Diagram of Example 1.13

Specially, if the system matrix $\boldsymbol{A}$ of the state space description (1.117) is a companion matrix, such as

$$\boldsymbol{A}=\begin{bmatrix}0&1&0&\cdots&0\\0&0&1&\cdots&0\\\vdots&\vdots&\vdots&\ddots&\vdots\\0&0&0&\cdots&1\\-a_n&-a_{n-1}&-a_{n-2}&\cdots&-a_1\end{bmatrix}\tag{1.121}$$

and $\boldsymbol{A}$ has distinct eigenvalues, there is a nonsingular transformation $\boldsymbol{X}(t)=\boldsymbol{P}\overline{\boldsymbol{X}}(t)$ transforms the state description to diagonal canonical form such as (1.120), where the nonsingular transformation matrix is the **Vandermonde matrix** shown as (1.122) and λ_1, $\lambda_2,\cdots,\lambda_n$ are those distinct eigenvalues of $\boldsymbol{A}$.

$$\boldsymbol{P}=\begin{bmatrix}1&1&\cdots&1\\\lambda_1&\lambda_2&\cdots&\lambda_n\\\vdots&\vdots&&\vdots\\\lambda_1^{n-1}&\lambda_2^{n-1}&\cdots&\lambda_n^{n-1}\end{bmatrix}\tag{1.122}$$

Example 1.14 The state space description of a system is

$$\dot{\boldsymbol{X}}=\begin{bmatrix}0&1&0\\0&0&1\\-24&-26&-9\end{bmatrix}\boldsymbol{X}+\begin{bmatrix}1\\0\\2\end{bmatrix}u\quad y=[3\quad 3\quad 1]\boldsymbol{X}$$

Determine the transformation matrix $\boldsymbol{P}$ and transform the state space description to the diagonal canonical form by the state transformation $\boldsymbol{X}(t)=\boldsymbol{P}\overline{\boldsymbol{X}}(t)$.

Solution The characteristic equation $|\lambda\boldsymbol{I}-\boldsymbol{A}|=0$ yields the roots

$$\lambda_1=-2,\lambda_2=-3,\lambda_3=-4$$

Since these eigenvalues are distinct and the matrix $\boldsymbol{A}$ is a companion matrix, the Vandermonde matrix can be constructed as the transformation matrix

$$\boldsymbol{P}=\begin{bmatrix}1 & 1 & 1\\ \lambda_1 & \lambda_2 & \lambda_3\\ \lambda_1^2 & \lambda_2^2 & \lambda_3^2\end{bmatrix}=\begin{bmatrix}1 & 1 & 1\\ -2 & -3 & -4\\ 4 & 9 & 16\end{bmatrix}$$

and

$$\boldsymbol{P}^{-1}=\frac{1}{2}\begin{bmatrix}12 & 7 & 1\\ -16 & -12 & 2\\ 6 & 5 & 1\end{bmatrix}$$

So the states can be uncoupled as the following diagonal canonical form by the state transformation $\boldsymbol{X}(t)=\boldsymbol{P}\overline{\boldsymbol{X}}(t)$

$$\dot{\overline{\boldsymbol{X}}}=\overline{\boldsymbol{A}}\,\overline{\boldsymbol{X}}+\overline{\boldsymbol{b}}u=\boldsymbol{P}^{-1}\boldsymbol{A}\boldsymbol{P}\,\overline{\boldsymbol{X}}+\boldsymbol{P}^{-1}\boldsymbol{b}u=\begin{bmatrix}-2 & 0 & 0\\ 0 & -3 & 0\\ 0 & 0 & -4\end{bmatrix}\overline{\boldsymbol{X}}+\begin{bmatrix}7\\ 10\\ 4\end{bmatrix}u$$

$$y=\overline{\boldsymbol{c}}\,\overline{\boldsymbol{X}}=\boldsymbol{c}\boldsymbol{P}\,\overline{\boldsymbol{X}}=[1\quad 3\quad 7]\overline{\boldsymbol{X}}$$

1.5.5 Obtaining the Jordan Canonical Form by State Transformation

According to the discussion above, if $\boldsymbol{A}$ has multiplicity eigenvalues, it is not always possible to find n linearly independent eigenvectors. Consequently, the matrix $\boldsymbol{A}$ can not be always transformed into a diagonal form.

Let us assume that $\boldsymbol{\lambda}_1$ is the m multiplicity eigenvalue of $\boldsymbol{A}$ and the remaining $n-m$ distinct eigenvalues are $\lambda_{m+1},\lambda_{m+2},\cdots,\lambda_n$.

Case 1

If $\text{rank}(\lambda_1\boldsymbol{I}-\boldsymbol{A})=n-m$, then from the equation $(\lambda_1\boldsymbol{I}-\boldsymbol{A})\boldsymbol{V}_1=0$ we can find m linearly independent eigenvectors $\boldsymbol{V}_{11},\boldsymbol{V}_{12},\cdots,\boldsymbol{V}_{1m}$. Based on the remaining $n-m$ distinct eigenvalues, $\lambda_{m+1},\lambda_{m+2},\cdots,\lambda_n$, we can find $n-m$ linearly independent eigenvectors $\boldsymbol{V}_{m+1},\boldsymbol{V}_{m+2},\cdots,\boldsymbol{V}_n$ from the equation $(\lambda_j\boldsymbol{I}-\boldsymbol{A})\boldsymbol{V}_j=0, j=m+1,\cdots,n$. Then, a nonsingular matrix is constructed as

$$\boldsymbol{P}=[\boldsymbol{V}_{11}\quad \boldsymbol{V}_{12}\quad \cdots\quad \boldsymbol{V}_{1m}\quad \boldsymbol{V}_{m+1}\quad \cdots\quad \boldsymbol{V}_n] \tag{1.123}$$

and the general state space description also can be transformed into the diagonal canonical form such as (1.124) by the state transformation $\boldsymbol{X}(t)=\boldsymbol{P}\overline{\boldsymbol{X}}(t)$.

$$\begin{aligned}\dot{\overline{\boldsymbol{X}}}&=\overline{\boldsymbol{A}}\,\overline{\boldsymbol{X}}+\overline{\boldsymbol{B}}\,\boldsymbol{u}\\ \boldsymbol{y}&=\overline{\boldsymbol{C}}\,\overline{\boldsymbol{X}}+\boldsymbol{D}\boldsymbol{u}\end{aligned} \tag{1.124}$$

Where

$$\overline{A}=P^{-1}AP=\begin{bmatrix}\lambda_1 & & & & & \mathbf{0}\\ & \ddots & & & & \\ & & \lambda_1 & & & \\ & & & \lambda_{m+1} & & \\ & & & & \ddots & \\ \mathbf{0} & & & & & \lambda_n\end{bmatrix} \tag{1.125}$$

is a diagonal matrix and $\overline{B}=P^{-1}B, \overline{C}=CP, \overline{D}=D$.

Example 1.15 The state equation of a system is

$$\dot{X}=\begin{bmatrix}1 & 0 & -1\\ 0 & 1 & 0\\ 0 & 0 & 2\end{bmatrix}X+\begin{bmatrix}1\\ 0\\ 2\end{bmatrix}u$$

Determine the transformation matrix P and transform the state space description into the diagonal canonical form by the state transformation $X(t)=P\overline{X}(t)$.

Solution The characteristic equation $|\lambda I-A|=0$ yields the roots

$$\lambda_{1,2}=1, \lambda_3=2$$

When $\lambda_{1,2}=1$ since $\mathrm{rank}(\lambda_1 I-A)=1$, then from the equation

$$(\lambda_1 I-A)V_1=\begin{bmatrix}0 & 0 & 1\\ 0 & 0 & 0\\ 0 & 0 & -1\end{bmatrix}V_1=\mathbf{0}$$

we can find $n-\mathrm{rank}(\lambda_1 I-A)=2$ linearly independent eigenvectors of A corresponding to $\lambda_{1,2}$. Obviously

$$V_{11}=\begin{bmatrix}1\\ 0\\ 0\end{bmatrix}, V_{12}=\begin{bmatrix}0\\ 1\\ 0\end{bmatrix}$$

can be selected.

When $\lambda_3=2$ the equation $\lambda_3 V_3=AV_3$ yields the eigenvector

$$V_3=\begin{bmatrix}-1\\ 0\\ 1\end{bmatrix}$$

of A corresponding to λ_3.

Then the nonsingular matrix is constructed as

$$P=[V_{11} \quad V_{12} \quad V_3]=\begin{bmatrix}1 & 0 & -1\\ 0 & 1 & 0\\ 0 & 0 & 1\end{bmatrix}$$

and

$$P^{-1}=\begin{bmatrix}1 & 0 & 1\\ 0 & 1 & 0\\ 0 & 0 & 1\end{bmatrix}$$

So the diagonal canonical form can be obtained by the state transformation $\boldsymbol{X}(t)=\boldsymbol{P}\overline{\boldsymbol{X}}(t)$ as

$$\dot{\overline{\boldsymbol{X}}}=\boldsymbol{P}^{-1}\boldsymbol{A}\boldsymbol{P}\overline{\boldsymbol{X}}+\boldsymbol{P}^{-1}\boldsymbol{b}u=\begin{bmatrix}1&0&0\\0&1&0\\0&0&2\end{bmatrix}\overline{\boldsymbol{X}}+\begin{bmatrix}3\\0\\2\end{bmatrix}u$$

Case 2

If $\text{rank}(\lambda_1\boldsymbol{I}-\boldsymbol{A})=n-1$, then from the equation $(\lambda_1\boldsymbol{I}-\boldsymbol{A})\boldsymbol{V}_1=\boldsymbol{0}$ we can find only one linearly independent eigenvector $\boldsymbol{V}_1$. The $m-1$ linearly independent generalized eigenvectors $\boldsymbol{V}_{12},\cdots,\boldsymbol{V}_{1m}$ can be determined from Eq. (1.105). Based on the remaining $n-m$ distinct eigenvalues, $\lambda_{m+1},\lambda_{m+2},\cdots,\lambda_n$, we can find $n-m$ linearly independent eigenvectors $\boldsymbol{V}_{m+1},\boldsymbol{V}_{m+2},\cdots,\boldsymbol{V}_n$ from the equation $(\lambda_j\boldsymbol{I}-\boldsymbol{A})\boldsymbol{V}_j=0, j=m+1,\cdots,n$. Then, a nonsingular matrix $\boldsymbol{P}$ can be constructed as (1.123) and the general state space description can be transformed into the Jordan canonical form such as (1.124) by the state transformation $\boldsymbol{X}(t)=\boldsymbol{P}\overline{\boldsymbol{X}}(t)$.

Where

$$\overline{\boldsymbol{A}}=\boldsymbol{P}^{-1}\boldsymbol{A}\boldsymbol{P}=\begin{bmatrix}\lambda_1&1&0&\cdots&0&&&&\boldsymbol{0}\\0&\lambda_1&\ddots&\ddots&\vdots&&&&\\\vdots&\ddots&\ddots&\ddots&0&&&&\\\vdots&&\ddots&\lambda_1&1&&&&\\0&\cdots&\cdots&0&\lambda_1&&&&\\&&&&&\lambda_{m+1}&0&\cdots&0\\&&&&&0&\ddots&\ddots&\vdots\\&&&&&\vdots&\ddots&\ddots&0\\\boldsymbol{0}&&&&&0&\cdots&0&\lambda_n\end{bmatrix}\tag{1.126}$$

is a Jordan matrix and $\overline{\boldsymbol{B}}=\boldsymbol{P}^{-1}\boldsymbol{B}, \overline{\boldsymbol{C}}=\boldsymbol{C}\boldsymbol{P}, \overline{\boldsymbol{D}}=\boldsymbol{D}$.

Example 1.16 The state equation of a system is

$$\dot{\boldsymbol{X}}=\begin{bmatrix}0&1&0\\0&0&1\\2&3&0\end{bmatrix}\boldsymbol{X}+\begin{bmatrix}0\\0\\1\end{bmatrix}u$$

Determine the transformation matrix $\boldsymbol{P}$ and transform the state space description into the Jordan canonical form by the state transformation $\boldsymbol{X}(t)=\boldsymbol{P}\overline{\boldsymbol{X}}(t)$.

Solution The characteristic equation

$$|\lambda\boldsymbol{I}-\boldsymbol{A}|=\lambda^3-3\lambda-2=(\lambda+1)^2(\lambda-2)=0$$

yields the roots $\lambda_{1,2}=-1, \lambda_3=2$

When $\lambda_{1,2}=-1$, since $\text{rank}(\lambda_1\boldsymbol{I}-\boldsymbol{A})=2$, then from the equation

$$(\lambda_1\boldsymbol{I}-\boldsymbol{A})\boldsymbol{V}_1=\begin{bmatrix}-1&-1&0\\0&-1&-1\\-2&-3&-1\end{bmatrix}\boldsymbol{V}_1=\boldsymbol{0}$$

we can only find $n-\text{rank}(\lambda_1 \boldsymbol{I}-\boldsymbol{A})=1$ linearly independent eigenvector of $\boldsymbol{A}$ corresponding to $\boldsymbol{\lambda}_{1,2}$. Obviously

$$\boldsymbol{V}_{11}=\begin{bmatrix}1\\-1\\1\end{bmatrix}$$

can be selected.

We can find 1 linearly independent generalized eigenvector of $\boldsymbol{A}$ corresponding to $\lambda_{1,2}$ from

$$(\lambda_1\boldsymbol{I}-\boldsymbol{A})\boldsymbol{V}_{12}=\begin{bmatrix}-1 & -1 & 0\\0 & -1 & -1\\-2 & -3 & -1\end{bmatrix}\boldsymbol{V}_{12}=-\boldsymbol{V}_{11}=\begin{bmatrix}-1\\1\\-1\end{bmatrix}$$

Consequently

$$\boldsymbol{V}_{12}=\begin{bmatrix}1\\0\\-1\end{bmatrix}$$

can be selected.

When $\lambda_3=2$ from the equation

$$(\lambda_3\boldsymbol{I}-\boldsymbol{A})\boldsymbol{V}_3=\begin{bmatrix}2 & -1 & 0\\0 & 2 & -1\\-2 & -3 & 2\end{bmatrix}\boldsymbol{V}_3=\boldsymbol{0}$$

the solution vector

$$\boldsymbol{V}_3=\begin{bmatrix}1\\2\\4\end{bmatrix}$$

can be selected as the eigenvector of $\boldsymbol{A}$ corresponding to λ_3.

Then the nonsingular matrix is constructed as

$$\boldsymbol{P}=[\boldsymbol{V}_{11}\quad \boldsymbol{V}_{12}\quad \boldsymbol{V}_3]=\begin{bmatrix}1 & 1 & 1\\-1 & 0 & 2\\1 & -1 & 4\end{bmatrix}$$

and

$$\boldsymbol{P}^{-1}=\frac{1}{9}\begin{bmatrix}2 & -5 & 2\\6 & 3 & -3\\1 & 2 & 1\end{bmatrix}$$

So the Jordan canonical form can be obtained by the state transformation $\boldsymbol{X}(t)=\boldsymbol{P}\overline{\boldsymbol{X}}(t)$.

$$\dot{\overline{\boldsymbol{X}}}=\boldsymbol{P}^{-1}\boldsymbol{AP}\overline{\boldsymbol{X}}+\boldsymbol{P}^{-1}\boldsymbol{b}u=\begin{bmatrix}-1 & 1 & 0\\0 & -1 & 0\\0 & 0 & 2\end{bmatrix}\overline{\boldsymbol{X}}+\begin{bmatrix}2/9\\-1/3\\1/9\end{bmatrix}u$$

Problems

1.1 Describe the dynamic for the system, which is shown in Figure P1.1, with the state space description.

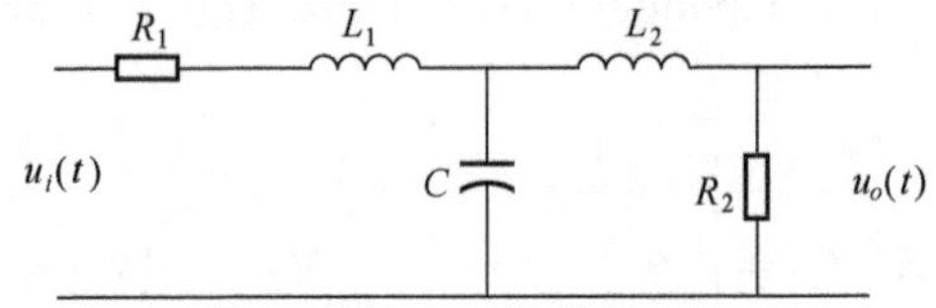

Figure P1.1 RLC Circuit

1.2 Describe the dynamic for the system, which is shown in Figure P1.2, with the state space description.

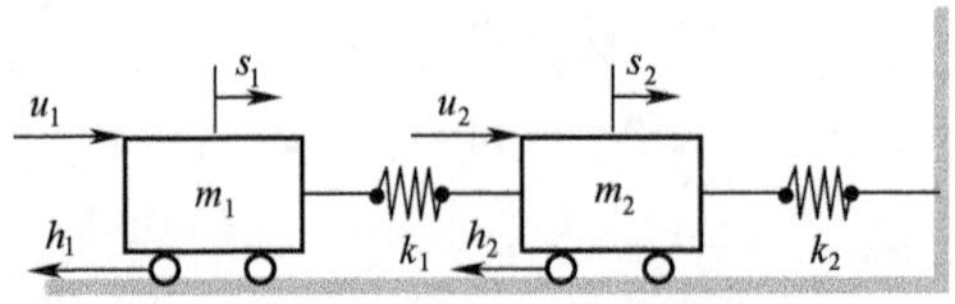

Figure P1.2 Moving Body System

1.3 Obtain the state space description and draw the simulation diagram for the systems, which are described by the following differential equations.

(1) $\frac{d^3y}{dt^3}+4\frac{d^2y}{dt^2}+6\frac{dy}{dt}+8y=20u(t)$

(2) $\frac{d^3y}{dt^3}-11\frac{d^2y}{dt^2}+38\frac{dy}{dt}-40y=2\frac{d^2u}{dt^2}+6\frac{du}{dt}+u$

(3) $\frac{d^3y}{dt^3}+5\frac{d^2y}{dt^2}+12\frac{dy}{dt}+8y=2\frac{d^3u}{dt^3}+6\frac{d^2u}{dt^2}+12\frac{du}{dt}+6u$

1.4 Obtain the state space description and draw the simulation diagram for the systems, which are described by the following transfer functions, by two different methods: direct decomposition and parallel decomposition.

(1) $\frac{7}{(s+1)(s+7)}$; (2) $\frac{12}{s(s+2)(s+6)}$; (3) $\frac{4s^2+17s+16}{(s+3)(s+2)^2}$; (4) $\frac{4s^2+17s+16}{(s+3)(s+2)}$

1.5 Obtain the state space description for the system shown in Figure P1.5.

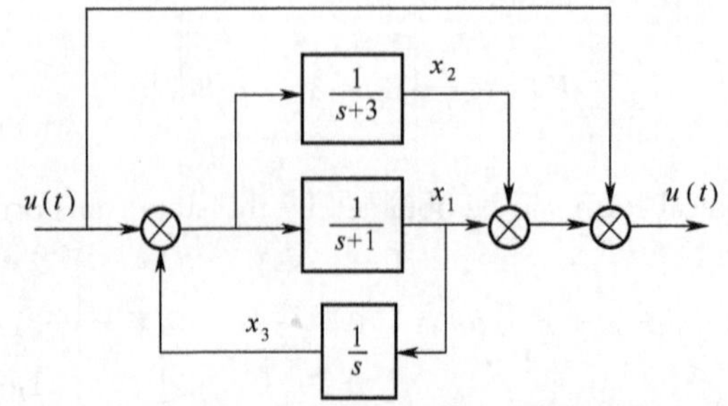

Figure P1.5 The Block Diagram of a Control System

1.6 Determine the transfer function matrix $\boldsymbol{G}(s)$ for the systems, which are described by the following state space descriptions.

(1) $\dot{\boldsymbol{X}}=\begin{bmatrix}-6 & -1\\ 5 & 0\end{bmatrix}\boldsymbol{X}+\begin{bmatrix}1\\ 0\end{bmatrix}u, y=\begin{bmatrix}1 & 0\end{bmatrix}\boldsymbol{X}$

(2) $\dot{\boldsymbol{X}}=\begin{bmatrix}0 & 1 & 0\\ 0 & 0 & 1\\ -1 & -1 & -2\end{bmatrix}\boldsymbol{X}+\begin{bmatrix}0 & 0\\ 1 & 0\\ 0 & 1\end{bmatrix}\boldsymbol{u}$

$\boldsymbol{y}=\begin{bmatrix}1 & 0 & 0\\ 0 & 0 & 1\end{bmatrix}\boldsymbol{X}+\begin{bmatrix}0 & 1\\ 1 & 1\end{bmatrix}\boldsymbol{u}$

1.7 A composite system is built with the following two subsystems by two different forms: parallel connection and series connection shown in Figure 1.15 and Figure 1.16.

$$\Sigma_1: \begin{array}{l}\dot{\boldsymbol{X}}=\begin{bmatrix}-6 & -1\\ 5 & 0\end{bmatrix}\boldsymbol{X}+\begin{bmatrix}1\\ 0\end{bmatrix}u\\ y=\begin{bmatrix}1 & 0\end{bmatrix}\boldsymbol{X}\end{array}, \quad \Sigma_2: \begin{array}{l}\dot{\boldsymbol{X}}=\begin{bmatrix}-2 & -1 & -1\\ 0 & -1 & 0\\ 0 & 2 & 1\end{bmatrix}\boldsymbol{X}+\begin{bmatrix}7\\ 2\\ 3\end{bmatrix}u\\ y=\begin{bmatrix}1 & 0 & 1\end{bmatrix}\boldsymbol{X}\end{array}$$

Determine the state space description for the composite systems.

1.8 Determine the transfer function matrix $\boldsymbol{G}(s)$ for the feedback connection system shown in Figure 1.17 with the following two subsystems.

$$\boldsymbol{G}_1(s)=\begin{bmatrix}\frac{1}{2s+1} & 0\\ 1 & \frac{1}{s+1}\end{bmatrix}, \boldsymbol{G}_2(s)=\begin{bmatrix}1 & 0\\ 0 & 1\end{bmatrix}$$

1.9 Transform the following state space description into the diagonal canonical form by a linear nonsingular state transformation $\boldsymbol{X}(t)=\boldsymbol{P}\overline{\boldsymbol{X}}(t)$.

$$\dot{\boldsymbol{X}}=\begin{bmatrix}0 & 1 & -1\\ -6 & -11 & 6\\ -6 & -11 & 5\end{bmatrix}\boldsymbol{X}+\begin{bmatrix}0\\ 0\\ 1\end{bmatrix}u, \quad y=\begin{bmatrix}1 & 0 & 0\end{bmatrix}\boldsymbol{X}$$

1.10 Transform the following state space description into the Jordan canonical form by a linear nonsingular state transformation $\boldsymbol{X}(t)=\boldsymbol{P}\overline{\boldsymbol{X}}(t)$.

$$\dot{\boldsymbol{X}}=\begin{bmatrix}1 & 1 & 2\\ 0 & 1 & 3\\ 0 & 0 & 2\end{bmatrix}\boldsymbol{X}+\begin{bmatrix}0\\ 0\\ 1\end{bmatrix}u, \quad y=\begin{bmatrix}1 & 0 & 0\end{bmatrix}\boldsymbol{X}$$

Richard Ernest Bellman (1920—1984) was an American applied mathematician, celebrated for his invention of *dynamic programming* in 1953, and important contributions in other fields of mathematics.

A *Bellman equation*, also known as a dynamic programming equation, is a necessary condition for optimality associated with the mathematical optimization method known as dynamic programming. Almost any problem which can be solved using optimal control theory can also be solved by analyzing the appropriate Bellman equation. The Bellman equation was first applied to engineering control theory and to other topics in applied mathematics, and subsequently became an important tool in economic theory.

The *Hamilton-Jacobi-Bellman* (HJB) *equation* is a partial differential equation which is central to optimal control theory. The solution of the HJB equation is the "value function", which gives the optimal cost-to-go for a given dynamical system with an associated cost function. Classical variational problems, for example, the brachistochrone problem can be solved using this method as well.

The "Curse of dimensionality" is a term coined by Bellman to describe the problem caused by the exponential increase in volume associated with adding extra dimensions to a (mathematical) space. One implication of the curse of dimensionality is that some methods for numerical solution of the Bellman equation require vastly more computer time when there are more state variables in the value function.

The Bellman-Ford algorithm is sometimes referred to as the Label Correcting Algorithm, computes single-source shortest paths in a weighted digraph (where some of the edge weights may be negative). Dijkstra's algorithm accomplishes the same problem with a lower running time, but requires edge weights to be non-negative. Thus, Bellman-Ford algorithm is usually used only when there are negative edge weights.

Chapter2 Time Response of the LTI System

2.1 Time Response of the LTI Homogeneous System

It is often desirable to obtain the time response of the state variables of a control system and thus examine the performance of the system. The state equation of the LTI system can be described by

$$\dot{\boldsymbol{X}}(t)=\boldsymbol{A}\boldsymbol{X}(t)+\boldsymbol{B}\boldsymbol{u}(t) \tag{2.1}$$

Where $\boldsymbol{u}(t)$ is the input vector of the system. The first term on the right hand of the state equation (2.1) is known as the homogenous part of the state equation. If $\boldsymbol{u}(t)=\boldsymbol{0}$, the system is called a LTI **homogeneous system**, or a free motion system.

In this section, we focus on the simple case. Consider the response of the LTI homogeneous system

$$\dot{\boldsymbol{X}}(t)=\boldsymbol{A}\boldsymbol{X}(t) \tag{2.2}$$

with the initial condition $\boldsymbol{X}(t_0)=\boldsymbol{X}(0)$. The solution of Eq. (2.2) can be obtained in a manner similar to the approach we utilize for solving a first-order differential equation.

Consider the first-order differential equation

$$\dot{x}(t)=ax(t) \tag{2.3}$$

Where $x(t)$ is a scalar function of time. Taking Laplace transform of Eq. (2.3), we have

$$sx(s)-x(0)=ax(s) \tag{2.4}$$

Therefore

$$x(s)=\frac{1}{s-a}x(0) \tag{2.5}$$

The inverse Laplace transform of Eq. (2.5) results in the solution of Eq. (2.3)

$$x(t)=\mathrm{e}^{at}x(0) \tag{2.6}$$

The **power series** of the exponential function e^{at} can be written as

$$\mathrm{e}^{at}=1+at+\frac{a^2t^2}{2!}+\frac{a^3t^3}{3!}+\cdots \tag{2.7}$$

By the comparison with Eq. (2.7), the **matrix exponential function** is defined as

$$\mathrm{e}^{\boldsymbol{A}t}=\boldsymbol{I}+\boldsymbol{A}t+\frac{\boldsymbol{A}^2t^2}{2!}+\frac{\boldsymbol{A}^3t^3}{3!}+\cdots \tag{2.8}$$

Obviously, the state vector

$$\boldsymbol{X}(t)=\mathrm{e}^{\boldsymbol{A}t}\boldsymbol{X}(0) \tag{2.9}$$

is the solution of the LTI homogenous system (2.2).

If the initial condition of the LTI homogenous system is the more general case $\boldsymbol{X}(t_0)$, the solution of Eq. (2.2) will be

$$\boldsymbol{X}(t)=\mathrm{e}^{\boldsymbol{A}(t-t_0)}\boldsymbol{X}(t_0) \tag{2.10}$$

2.2 State Transition Matrix

2.2.1 Definition

Definition 2.1 The **state transition matrix** $\boldsymbol{\Phi}(t-t_0)$ is defined as a matrix that satisfies the conditions shown as

$$\dot{\boldsymbol{\Phi}}(t-t_0)=\boldsymbol{A\Phi}(t-t_0),\boldsymbol{\Phi}(t_0)=\boldsymbol{I},t\geqslant t_0\geqslant 0 \tag{2.11}$$

When the initial time $t_0=0$, The state transition matrix can be written as $\boldsymbol{\Phi}(t)$.

Based on the definition of the matrix exponential function $\mathrm{e}^{\boldsymbol{A}t}$, the derivation of it can be obtained as

$$\begin{aligned}\frac{\mathrm{d}}{\mathrm{d}t}[\mathrm{e}^{\boldsymbol{A}t}]&=\frac{\mathrm{d}}{\mathrm{d}t}[\boldsymbol{I}+\boldsymbol{A}t+\frac{(\boldsymbol{A}t)^2}{2!}+\cdots+\frac{(\boldsymbol{A}t)^k}{k!}+\cdots]\\&=\boldsymbol{A}+\boldsymbol{A}^2t+\frac{\boldsymbol{A}^3t^2}{2!}+\cdots+\frac{\boldsymbol{A}^kt^{k-1}}{(k-1)!}+\cdots\\&=\boldsymbol{A}[\boldsymbol{I}+\frac{\boldsymbol{A}t}{1!}+\frac{(\boldsymbol{A}t)^2}{2!}+\cdots+\frac{(\boldsymbol{A}t)^k}{k!}+\cdots]\\&=\boldsymbol{A}\mathrm{e}^{\boldsymbol{A}t}\end{aligned}$$

Similarly, we also have

$$\frac{\mathrm{d}}{\mathrm{d}t}[\mathrm{e}^{\boldsymbol{A}(t-t_0)}]=\boldsymbol{A}\mathrm{e}^{\boldsymbol{A}(t-t_0)} \tag{2.12}$$

Visibly

$$\mathrm{e}^{\boldsymbol{A}(t_0-t_0)}=\boldsymbol{I} \tag{2.13}$$

Comparing (2.12), (2.13) with (2.11), we obtain another expression of the state transition matrix for the LTI system as

$$\boldsymbol{\Phi}(t-t_0)=\mathrm{e}^{\boldsymbol{A}(t-t_0)} \tag{2.14}$$

or

$$\boldsymbol{\Phi}(t)=\mathrm{e}^{\boldsymbol{A}t} \tag{2.15}$$

Consequently, the solution of the LTI homogenous system (2.2) can be represented as

$$\boldsymbol{X}(t)=\boldsymbol{\Phi}(t-t_0)\boldsymbol{X}(t_0) \tag{2.16}$$

or

$$\boldsymbol{X}(t)=\boldsymbol{\Phi}(t)\boldsymbol{X}(0) \tag{2.17}$$

Since the state transition matrix satisfies the homogenous state equation, it represents the free response of the system. In other words, it governs the response that is excited by the initial conditions only. In view of Eq. (2.14), the transition matrix is dependent only upon the system matrix $\boldsymbol{A}$, and, therefore, is sometimes referred to as the transition matrix of $\boldsymbol{A}$. As the name implies, the transition matrix $\boldsymbol{\Phi}(t-t_0)$ completely defines the transition of the state from the initial time t_0 to any time t when the inputs are zero.

2.2.2 Properties of the State Transition Matrix

It is observed from the last section that $\boldsymbol{\Phi}(t)$ plays a key role in finding the solution for a given LTI system. This section will present some properties of the state transition matrix.

1. $\boldsymbol{\Phi}^{-1}(t)=\boldsymbol{\Phi}(-t)$ (2.18)

Proof. Since $\boldsymbol{\Phi}(t)\boldsymbol{\Phi}(-t)=\mathrm{e}^{\boldsymbol{A}t}\cdot\mathrm{e}^{-\boldsymbol{A}t}=\boldsymbol{I}$

Thus $$\boldsymbol{\Phi}(-t)=\boldsymbol{\Phi}^{-1}(t)$$

2. $\boldsymbol{\Phi}(t_1+t_2)=\boldsymbol{\Phi}(t_1)\boldsymbol{\Phi}(t_2)$ (2.19)

Proof. $$\boldsymbol{\Phi}(t_1+t_2)=\mathrm{e}^{\boldsymbol{A}(t_1+t_2)}=\mathrm{e}^{\boldsymbol{A}t_1}\cdot\mathrm{e}^{\boldsymbol{A}t_2}=\boldsymbol{\Phi}(t_1)\boldsymbol{\Phi}(t_2)$$

3. $\boldsymbol{\Phi}(t_2-t_1)\boldsymbol{\Phi}(t_1-t_0)=\boldsymbol{\Phi}(t_2-t_0)$ for any t_0, t_1, t_2 (2.20)

Proof. $\boldsymbol{\Phi}(t_2-t_1)\boldsymbol{\Phi}(t_1-t_0)=\mathrm{e}^{\boldsymbol{A}(t_2-t_1)}\cdot\mathrm{e}^{\boldsymbol{A}(t_1-t_0)}=\mathrm{e}^{\boldsymbol{A}(t_1-t_0)}=\boldsymbol{\Phi}(t_2-t_0)$

4. $[\boldsymbol{\Phi}(t)]^k=\boldsymbol{\Phi}(kt)$ (2.21)

Proof. $$[\boldsymbol{\Phi}(t)]^k=\mathrm{e}^{\boldsymbol{A}t}\cdot\mathrm{e}^{\boldsymbol{A}t}\cdots\mathrm{e}^{\boldsymbol{A}t}=\mathrm{e}^{k\boldsymbol{A}t}=\boldsymbol{\Phi}(kt)$$

2.3 Calculation of the Matrix Exponential Function

From the previous discussion, we have known that the matrix exponential function plays an important role in the study of the response for the LTI system. In this section, we will introduce four methods to calculate the matrix exponential function.

2.3.1 Direct Method

Utilizing the infinite power series

$$\mathrm{e}^{\boldsymbol{A}t}=\boldsymbol{I}+\boldsymbol{A}t+\frac{1}{2!}\boldsymbol{A}^2t^2+\cdots$$

we can obtain the matrix exponential function theoretically. It should be noted that this power series will not, generally, give a closed-form solution and it is mainly used on computer simulation.

2.3.2 Laplace Transform Method

Consider the homogeneous state equation shown as Eq. (2.2). Taking the Laplace transform on both side of Eq. (2.2), we have

$$s\boldsymbol{X}(s)-\boldsymbol{X}(0)=\boldsymbol{A}\boldsymbol{X}(s)$$

Where $\boldsymbol{X}(0)$ denotes the initial state vector evaluated at $t=0$. Grouping the terms $\boldsymbol{X}$

(s) yields

$$(s\boldsymbol{I}-\boldsymbol{A})\boldsymbol{X}(s)=\boldsymbol{X}(0)$$

Moreover, solving for $\boldsymbol{X}(s)$ yields

$$\boldsymbol{X}(s)=(s\boldsymbol{I}-\boldsymbol{A})^{-1}\boldsymbol{X}(0) \tag{2.22}$$

The free response of the LTI homogeneous system (2.2) can be obtained by taking the inverse Laplace transform on both side of Eq. (2.22).

$$\boldsymbol{X}(t)=L^{-1}[(s\boldsymbol{I}-\boldsymbol{A})^{-1}]\boldsymbol{X}(0) \tag{2.23}$$

Comparing (2.23) with the solution (2.9), we can obtain anther expression of the matrix exponential function as

$$\mathrm{e}^{\boldsymbol{A}t}=L^{-1}[(s\boldsymbol{I}-\boldsymbol{A})^{-1}] \tag{2.24}$$

Example 2.1 Calculate $\mathrm{e}^{\boldsymbol{A}t}$ by using the Laplace transform method.

$$\boldsymbol{A}=\begin{bmatrix}0 & 1\\ -2 & -3\end{bmatrix}$$

Solution The characteristic matrix and its inverse matrix are calculated as

$$(s\boldsymbol{I}-\boldsymbol{A})=\begin{bmatrix}s & -1\\ 2 & s+3\end{bmatrix}$$

and $$(s\boldsymbol{I}-\boldsymbol{A})^{-1}=\frac{1}{s(s+3)+2}\begin{bmatrix}s+3 & 1\\ -2 & s\end{bmatrix}=\begin{bmatrix}\dfrac{s+3}{(s+1)(s+2)} & \dfrac{1}{(s+1)(s+2)}\\ \dfrac{-2}{(s+1)(s+2)} & \dfrac{s}{(s+1)(s+2)}\end{bmatrix}$$

Taking the inverse Laplace transform, the matrix exponential function can be obtained as

$$\mathrm{e}^{\boldsymbol{A}t}=L^{-1}[(s\boldsymbol{I}-\boldsymbol{A})^{-1}]$$

$$=L^{-1}\begin{bmatrix}\dfrac{(s+3)}{(s+1)(s+2)} & \dfrac{1}{(s+1)(s+2)}\\ \dfrac{-2}{(s+1)(s+2)} & \dfrac{s}{(s+1)(s+2)}\end{bmatrix}=L^{-1}\begin{bmatrix}\dfrac{2}{s+1}+\dfrac{-1}{s+2} & \dfrac{1}{s+1}+\dfrac{-1}{s+2}\\ \dfrac{-2}{s+1}+\dfrac{2}{s+2} & \dfrac{-1}{s+1}+\dfrac{2}{s+2}\end{bmatrix}$$

$$=\begin{bmatrix}2\mathrm{e}^{-t}-\mathrm{e}^{-2t} & \mathrm{e}^{-t}-\mathrm{e}^{-2t}\\ -2\mathrm{e}^{-t}+2\mathrm{e}^{-2t} & -\mathrm{e}^{-t}+2\mathrm{e}^{-2t}\end{bmatrix}$$

2.3.3 Similarity Transformation Method

Case 1 Consider a n dimension system, governed by state space description

$$\begin{aligned}\dot{\boldsymbol{X}}&=\boldsymbol{A}\boldsymbol{X}+\boldsymbol{B}u\\ y&=\boldsymbol{C}\boldsymbol{X}+\boldsymbol{D}u\end{aligned} \tag{2.25}$$

If $\boldsymbol{A}$ has distinct eigenvalues $\lambda_1, \lambda_2, \cdots, \lambda_n$, as we have known that there is a nonsingular transformation

$$\boldsymbol{X}(t)=\boldsymbol{P}\overline{\boldsymbol{X}}(t) \tag{2.26}$$

which transform the general state description, such as (2.25), into the diagonal canonical form, such as

$$\dot{\overline{X}}=\overline{A}\,\overline{X}+\overline{B}u$$
$$y=\overline{C}\,\overline{X}+\overline{D}u \tag{2.27}$$

Where $\overline{A}=P^{-1}AP=\begin{bmatrix}\lambda_1 & & 0\\ & \ddots & \\ 0 & & \lambda_n\end{bmatrix}$ is a diagonal matrix and

$$e^{\overline{A}t}=\begin{bmatrix}e^{\lambda_1 t} & & & 0\\ & e^{\lambda_2 t} & & \\ & & \ddots & \\ 0 & & & e^{\lambda_n t}\end{bmatrix} \tag{2.28}$$

In this case, the matrix exponential function e^{At} can be calculated as

$$e^{At}=P\cdot e^{\overline{A}t}\cdot P^{-1}=P\cdot\begin{bmatrix}e^{\lambda_1 t} & & & 0\\ & e^{\lambda_2 t} & & \\ & & \ddots & \\ 0 & & & e^{\lambda_n t}\end{bmatrix}\cdot P^{-1} \tag{2.29}$$

Example 2.2 Calculate e^{At} by using the similarity transform method.

$$A=\begin{bmatrix}0 & 1\\ -2 & -3\end{bmatrix}$$

Solution The characteristic equation $|\lambda I-A|=0$ yields the roots

$$\lambda_1=-1,\lambda_2=-2$$

Since the eigenvalues are distinct and the matrix A is a companion matrix, the Vandermonde matrix

$$P=\begin{bmatrix}1 & 1\\ \lambda_1 & \lambda_2\end{bmatrix}=\begin{bmatrix}1 & 1\\ -1 & -2\end{bmatrix}$$

can be constructed as the transformation matrix to transform A into the diagonal matrix by a similar transformation, that is

$$\overline{A}=P^{-1}AP=\begin{bmatrix}-1 & 0\\ 0 & -2\end{bmatrix}$$

Where

$$P^{-1}=\begin{bmatrix}2 & 1\\ -1 & -1\end{bmatrix}$$

Therefore, the matrix exponential function is obtained as

$$e^{At}=P\cdot e^{\overline{A}t}\cdot P^{-1}=P\cdot\begin{bmatrix}e^{-t} & 0\\ 0 & e^{-2t}\end{bmatrix}\cdot P^{-1}$$

$$=\begin{bmatrix}2e^{-t}-e^{-2t} & e^{-t}-e^{-2t}\\ -2e^{-t}+2e^{-2t} & -e^{-t}+2e^{-2t}\end{bmatrix}$$

Case 2 Consider a n dimension system such as (2.25). As we have known that there is a nonsingular transformation

$$X(t)=P\bar{X}(t)$$

which transforms the state description, such as (2.25), into the Jordan canonical form, such as (2.27).

For $J_i=\begin{bmatrix}\lambda_i & 1 & & 0\\ & \lambda_i & \ddots & \\ & & \ddots & 1\\ 0 & & & \lambda_i\end{bmatrix}$ is a Jordan block and

$$e^{J_i t}=\begin{bmatrix} e^{\lambda_i t} & te^{\lambda_i t} & \frac{t^2}{2!}e^{\lambda_i t} & \cdots & \frac{1}{(n-1)!}t^{n-1}e^{\lambda_i t}\\ & e^{\lambda_i t} & te^{\lambda_i t} & \ddots & \vdots\\ & & e^{\lambda_i t} & \ddots & \frac{t^2}{2!}e^{\lambda_i t}\\ & & & \ddots & te^{\lambda_i t}\\ 0 & & & & e^{\lambda_i t}\end{bmatrix} \tag{2.30}$$

In this case, the matrix exponential function e^{At} can be calculated as

$$e^{At}=P\cdot e^{Jt}\cdot P^{-1}$$

$$=P\cdot\begin{bmatrix} e^{\lambda_1 t} & te^{\lambda_1 t} & \frac{t^2}{2!}e^{\lambda_1 t} & \cdots & \frac{1}{(n-1)!}t^{n-1}e^{\lambda_1 t}\\ & e^{\lambda_1 t} & te^{\lambda_1 t} & \ddots & \vdots\\ & & e^{\lambda_1 t} & \ddots & \frac{t^2}{2!}e^{\lambda_1 t}\\ & & & \ddots & te^{\lambda_1 t}\\ 0 & & & & e^{\lambda_1 t}\end{bmatrix}\cdot P^{-1} \tag{2.31}$$

For a more general case

$$J=P^{-1}AP=\begin{bmatrix} J_1 & & & \mathbf{0}\\ & J_2 & & \\ & & \ddots & \\ \mathbf{0} & & & J_m\end{bmatrix} \tag{2.32}$$

Where J is a Jordan matrix and $J_i\,(i=1,\cdots,m)$ are those Jordan blocks within the Jordan matrix, the matrix exponential function e^{At} can be calculated as

$$e^{At}=P\cdot e^{Jt}\cdot P^{-1}=P\cdot\begin{bmatrix} e^{J_1 t} & & & \mathbf{0}\\ & e^{J_2 t} & & \\ & & \ddots & \\ \mathbf{0} & & & e^{J_m t}\end{bmatrix}\cdot P^{-1} \tag{2.33}$$

Example 2.3 Calculate e^{At} by the similarity transformation method.

$$A=\begin{bmatrix}0 & 6 & -5\\ 1 & 0 & 2\\ 3 & 2 & 4\end{bmatrix}$$

Solution Solving the characteristic equation

$$
|\lambda \boldsymbol{I}-\boldsymbol{A}|=\begin{vmatrix} \lambda & -6 & 5 \\ -1 & \lambda & -2 \\ -3 & -2 & \lambda-4 \end{vmatrix}=(\lambda-1)^2(\lambda-2)=0
$$

we obtain that $\lambda_1=\lambda_2=1$ and $\lambda_3=2$.

Calculating their eigenvectors and generalized eigenvector (for λ_1) produces

$$
\boldsymbol{V}_{11}=\begin{bmatrix} 1 \\ -3/7 \\ -5/7 \end{bmatrix},\boldsymbol{V}_{12}=\begin{bmatrix} 1 \\ -22/49 \\ -46/49 \end{bmatrix},\boldsymbol{V}_3=\begin{bmatrix} 2 \\ -1 \\ -2 \end{bmatrix}
$$

Then, the nonsingular matrix can be constructed as

$$
\boldsymbol{P}=[\boldsymbol{V}_{11} \quad \boldsymbol{V}_{12} \quad \boldsymbol{V}_3]=\begin{bmatrix} 1 & 1 & 2 \\ -3/7 & -22/49 & -1 \\ -5/7 & -46/49 & -2 \end{bmatrix}
$$

and

$$
\boldsymbol{P}^{-1}=\begin{bmatrix} 2 & -6 & 5 \\ 7 & 28 & -7 \\ -4 & -11 & 1 \end{bmatrix}
$$

So the Jordan matrix is obtained by the similarly transformation as

$$
\boldsymbol{J}=\boldsymbol{P}^{-1}\boldsymbol{A}\boldsymbol{P}=\left[\begin{array}{cc:c} 1 & 1 & 0 \\ 0 & 1 & 0 \\ \hdashline 0 & 0 & 2 \end{array}\right]=\begin{bmatrix} \boldsymbol{J}_1 & \boldsymbol{0} \\ \boldsymbol{0} & \boldsymbol{J}_2 \end{bmatrix}
$$

Hence

$$
\begin{aligned}
\mathrm{e}^{\boldsymbol{A}t}&=\boldsymbol{P}\mathrm{e}^{\boldsymbol{J}t}\boldsymbol{P}^{-1}=\boldsymbol{P}\begin{bmatrix} \mathrm{e}^{\boldsymbol{J}_1 t} & 0 \\ 0 & \mathrm{e}^{\boldsymbol{J}_2 t} \end{bmatrix}\boldsymbol{P}^{-1} \\
&=\begin{bmatrix} 1 & 1 & 2 \\ -3/7 & -22/49 & -1 \\ -5/7 & -46/49 & -2 \end{bmatrix}\left[\begin{array}{cc:c} \mathrm{e}^t & t\mathrm{e}^t & 0 \\ 0 & \mathrm{e}^t & 0 \\ \hdashline 0 & 0 & \mathrm{e}^{2t} \end{array}\right]\begin{bmatrix} 2 & -6 & 5 \\ 7 & 28 & -7 \\ -4 & -11 & 1 \end{bmatrix} \\
&=\begin{bmatrix} 9\mathrm{e}^t+7t\mathrm{e}^t-8\mathrm{e}^{2t} & 22\mathrm{e}^t+28t\mathrm{e}^t-22\mathrm{e}^{2t} & -2\mathrm{e}^t-7t\mathrm{e}^t+2\mathrm{e}^{2t} \\ -4\mathrm{e}^t-3t\mathrm{e}^t+4\mathrm{e}^{2t} & -10\mathrm{e}^t-12t\mathrm{e}^t+11\mathrm{e}^{2t} & \mathrm{e}^t+3t\mathrm{e}^t-\mathrm{e}^{2t} \\ -8\mathrm{e}^t-5t\mathrm{e}^t+8\mathrm{e}^{2t} & -22\mathrm{e}^t-20t\mathrm{e}^t+22\mathrm{e}^{2t} & 3\mathrm{e}^t+5t\mathrm{e}^t-2\mathrm{e}^{2t} \end{bmatrix}
\end{aligned}
$$

2.3.4 Cayley-Hamilton Theorem Method

According to **Cayley-Hamilton Theorem**, if the characteristic polynomial of A

$$
|\lambda \boldsymbol{I}-\boldsymbol{A}|=\lambda^n+a_1\lambda^{n-1}+\cdots+a_{n-1}\lambda+a_n=0 \tag{2.34}
$$

then

$$
\boldsymbol{A}^n+a_1\boldsymbol{A}^{n-1}+\cdots+a_{n-1}\boldsymbol{A}+a_n\boldsymbol{I}=\boldsymbol{0} \tag{2.35}
$$

For any $k\geqslant n$, it can be proved easily that $\boldsymbol{A}^k$ can be linearly expressed with $\boldsymbol{A}^i, i=0,\cdots,n-1$, as the matrix equation (2.36)

$$
\boldsymbol{A}^k=c_0\boldsymbol{I}+c_1\boldsymbol{A}+\cdots+c_{n-1}\boldsymbol{A}^{n-1} \tag{2.36}
$$

Where $c_0, \cdots, c_{n-1}$ are n constants.

From the definition of the matrix exponential function

$$e^{At} = I + At + \frac{1}{2!}A^2t^2 + \cdots$$

we find that e^{At} can also be linearly expressed with $A^i, i=0,\cdots,n-1$, as

$$e^{At} = \alpha_0(t)I + \alpha_1(t)A + \cdots + \alpha_{n-1}(t)A^{n-1} = \sum_{i=0}^{n-1}\alpha_i(t)A^i \tag{2.37}$$

Eq. (2.37) indicates that e^{At} can be expressed as an $(n-1)$-th order polynomial of A and the coefficients $\alpha_0(t), \alpha_1(t), \cdots, \alpha_{n-1}(t)$ are functions of the time and can be determined as follows.

Case 1 Matrix A has n distinct eigenvalues

Suppose $\lambda_1, \lambda_2, \cdots, \lambda_n$ are n distinct eigenvalues of A. As we have known that there is a nonsingular matrix P, by which A can be transformed into the diagonal matrix with the following similarly transformation

$$\bar{A} = P^{-1}AP = \begin{bmatrix} \lambda_1 & & 0 \\ & \ddots & \\ 0 & & \lambda_n \end{bmatrix}$$

and

$$e^{At} = P \cdot e^{\bar{A}t} \cdot P^{-1} = P\begin{bmatrix} e^{\lambda_1 t} & & & 0 \\ & e^{\lambda_2 t} & & \\ & & \ddots & \\ 0 & & & e^{\lambda_n t} \end{bmatrix}P^{-1} \tag{2.38}$$

Substituting Eq. (2.37) into Eq. (2.38) yields

$$\begin{bmatrix} e^{\lambda_1 t} & & & 0 \\ & e^{\lambda_2 t} & & \\ & & \ddots & \\ 0 & & & e^{\lambda_n t} \end{bmatrix} = P^{-1}\left(\sum_{i=0}^{n-1}\alpha_i(t)A^i\right)P = \sum_{i=0}^{n-1}\alpha_i(t) \cdot (P^{-1}A^iP)$$

$$= \sum_{i=0}^{n-1}\alpha_i(t) \cdot \bar{A}^i = \sum_{i=0}^{n-1}\alpha_i(t)\begin{bmatrix} \lambda_1 & & & 0 \\ & \lambda_2 & & \\ & & \ddots & \\ 0 & & & \lambda_n \end{bmatrix}^i$$

$$= \sum_{i=0}^{n-1}\alpha_i(t)\begin{bmatrix} \lambda_1^i & & & 0 \\ & \lambda_2^i & & \\ & & \ddots & \\ 0 & & & \lambda_n^i \end{bmatrix} \tag{2.39}$$

It can be verified from Eq. (2.39) that $e^{\lambda_i t}, i=1,2,\cdots,n$, satisfies the following equations.

$$e^{\lambda_1 t} = \alpha_0(t) + \alpha_1(t)\lambda_1 + \cdots + \alpha_{n-1}(t)\lambda_1^{n-1}$$

$$e^{\lambda_2 t}=\alpha_0(t)+\alpha_1(t)\lambda_2+\cdots+\alpha_{n-1}(t)\lambda_2^{n-1}$$

$$\vdots$$

$$e^{\lambda_n t}=\alpha_0(t)+\alpha_1(t)\lambda_n+\cdots+\alpha_{n-1}(t)\lambda_n^{n-1} \tag{2.40}$$

Solving the set of equations (2.40), the coefficients $\alpha_0(t), \alpha_1(t), \cdots, \alpha_{n-1}(t)$ can be obtained as

$$\begin{bmatrix} \alpha_0(t) \\ \alpha_1(t) \\ \vdots \\ \alpha_{n-1}(t) \end{bmatrix} = \begin{bmatrix} 1 & \lambda_1 & \lambda_1^2 & \cdots & \lambda_1^{n-1} \\ 1 & \lambda_2 & \lambda_2^2 & \cdots & \lambda_2^{n-1} \\ \vdots & \vdots & \vdots & & \vdots \\ 1 & \lambda_n & \lambda_n^2 & \cdots & \lambda_n^{n-1} \end{bmatrix}^{-1} \begin{bmatrix} e^{\lambda_1 t} \\ e^{\lambda_2 t} \\ \vdots \\ e^{\lambda_n t} \end{bmatrix} \tag{2.41}$$

Example 2.4 Calculate $e^{\boldsymbol{A}t}$ by using the Cayley-Hamilton Theorem method.

$$\boldsymbol{A}=\begin{bmatrix} 0 & 1 \\ -2 & -3 \end{bmatrix}$$

Solution The characteristic equation $|\lambda \boldsymbol{I}-\boldsymbol{A}|=0$ yields the roots

$$\lambda_1=-1, \lambda_2=-2$$

Since the eigenvalues are distinct, $e^{\boldsymbol{A}t}$ can be expressed as

$$e^{\boldsymbol{A}t}=\alpha_0(t)\boldsymbol{I}+\alpha_1(t)\boldsymbol{A} \tag{2.42}$$

Where

$$\begin{bmatrix} \alpha_0(t) \\ \alpha_1(t) \end{bmatrix} = \begin{bmatrix} 1 & \lambda_1 \\ 1 & \lambda_2 \end{bmatrix}^{-1} \begin{bmatrix} e^{\lambda_1 t} \\ e^{\lambda_2 t} \end{bmatrix} = \begin{bmatrix} 1 & -1 \\ 1 & -2 \end{bmatrix}^{-1} \begin{bmatrix} e^{-t} \\ e^{-2t} \end{bmatrix} = \begin{bmatrix} 2e^{-t}-e^{-2t} \\ e^{-t}-e^{-2t} \end{bmatrix}$$

Substituting $\alpha_0(t)$ and $\alpha_1(t)$ into (2.42) yields

$$\begin{aligned} e^{\boldsymbol{A}t} &= \alpha_0(t)\boldsymbol{I}+\alpha_1(t)\boldsymbol{A} \\ &= (2e^{-t}-e^{-2t})\begin{bmatrix} 1 & 0 \\ 0 & 1 \end{bmatrix} + (e^{-t}-e^{-2t})\begin{bmatrix} 0 & 1 \\ -2 & -3 \end{bmatrix} \\ &= \begin{bmatrix} 2e^{-t}-e^{-2t} & e^{-t}-e^{-2t} \\ -2e^{-t}+2e^{-2t} & -e^{-t}+2e^{-2t} \end{bmatrix} \end{aligned}$$

Case 2 Matrix *A* has multiplicity eigenvalue

For convenience, we assume $\lambda_1=\lambda_2=\cdots=\lambda_n=\lambda$. Using the similar method to (2.39), we can obtain only one equation as

$$e^{\lambda t}=\alpha_0(t)+\alpha_1(t)\lambda+\cdots+\alpha_{n-1}(t)\lambda^{n-1} \tag{2.43}$$

In order to determine $\alpha_0(t), \alpha_1(t), \cdots, \alpha_{n-1}(t)$, we need another $n-1$ independent equations. These equations can be set up by differentiating Eq. (2.43) until $n-1$ times with respect to λ, which gives

$$t e^{\lambda t}=\alpha_1(t)+2\alpha_2(t)\lambda+\cdots+(n-1)\alpha_{n-1}(t)\lambda^{n-2}$$

$$t^2 e^{\lambda t}=2\alpha_2(t)+6\alpha_3(t)\lambda+\cdots+(n-1)(n-2)\alpha_{n-1}(t)\lambda^{n-3}$$

$$\vdots$$

$$t^{n-1}e^{\lambda t}=(n-1)!\ \alpha_{n-1}(t) \tag{2.44}$$

From (2.43) and (2.44), we may deduce that

$$\begin{bmatrix} \alpha_0(t) \\ \alpha_1(t) \\ \alpha_2(t) \\ \vdots \\ \alpha_{n-1}(t) \end{bmatrix} = \begin{bmatrix} 1 & \lambda & \lambda^2 & \cdots & \lambda^{n-1} \\ 0 & 1 & 2\lambda & \cdots & (n-1)\lambda^{n-2} \\ 0 & 0 & 2 & \cdots & (n-1)(n-2)\lambda^{n-3} \\ \vdots & \vdots & \vdots & & \vdots \\ 0 & 0 & 0 & \cdots & (n-1)! \end{bmatrix}^{-1} \begin{bmatrix} e^{\lambda t} \\ te^{\lambda t} \\ t^2 e^{\lambda t} \\ \vdots \\ t^{n-1} e^{\lambda t} \end{bmatrix} \tag{2.45}$$

Example 2.5 Calculate e^{At} by using the Cayley-Hamilton Theorem method.

$$\boldsymbol{A} = \begin{bmatrix} 0 & 1 & 0 \\ 0 & 0 & 1 \\ 2 & 3 & 0 \end{bmatrix}$$

Solution Solving the characteristic equation

$$|\lambda \boldsymbol{I} - \boldsymbol{A}| = (\lambda+1)^2(\lambda-2) = 0$$

yields $\lambda_1 = \lambda_2 = -1$ and $\lambda_3 = 2$.

Therefore, e^{At} can be expressed as

$$e^{At} = \alpha_0(t)\boldsymbol{I} + \alpha_1(t)\boldsymbol{A} + \alpha_2(t)\boldsymbol{A}^2 \tag{2.46}$$

Where

$$\begin{bmatrix} \alpha_0(t) \\ \alpha_1(t) \\ \alpha_2(t) \end{bmatrix} = \begin{bmatrix} 1 & \lambda_1 & \lambda_1^2 \\ 0 & 1 & 2\lambda_1 \\ 1 & \lambda_3 & \lambda_3^2 \end{bmatrix}^{-1} \begin{bmatrix} e^{\lambda_1 t} \\ te^{\lambda_1 t} \\ e^{\lambda_3 t} \end{bmatrix} = \begin{bmatrix} 1 & -1 & 1 \\ 0 & 1 & -2 \\ 1 & 2 & 4 \end{bmatrix}^{-1} \begin{bmatrix} e^{-t} \\ te^{-t} \\ e^{2t} \end{bmatrix}$$

$$= \begin{bmatrix} \frac{1}{9}(8e^{-t} + 6te^{-t} + e^{2t}) \\ \frac{1}{9}(-2e^{-t} + 3te^{-t} + 2e^{2t}) \\ \frac{1}{9}(-e^{-t} - 3te^{-t} + e^{2t}) \end{bmatrix}$$

Substituting $\alpha_0(t)$、$\alpha_1(t)$ and $\alpha_2(t)$ into (2.46) yields

$$e^{At} = \alpha_0(t)\boldsymbol{I} + \alpha_1(t)\boldsymbol{A} + \alpha_2(t)\boldsymbol{A}^2$$

$$= \frac{1}{9}\begin{bmatrix} (8+6t)e^{-t} + e^{2t} & -(2-3t)e^{-t} + 2e^{2t} & -(1+3t)e^{-t} + e^{2t} \\ -(2+6t)e^{-t} + 2e^{2t} & (5-3t)e^{-t} + 4e^{2t} & -(2-3t)e^{-t} + 2e^{2t} \\ (6t-4)e^{-t} + 4e^{2t} & (3t-8)e^{-t} + 8e^{2t} & (5-3t)e^{-t} + 4e^{2t} \end{bmatrix}$$

2.4 Time Response of the LTI System

Consider the LTI system described by (2.1) with the initial condition $\boldsymbol{X}(t_0) = \boldsymbol{X}(0)$. The solution of Eq. (2.1) can be obtained in a manner similar to the approach we have utilized for solving the homogenous LTI system.

Taking Laplace transform of Eq. (2.1), we have

$$s\boldsymbol{X}(s) - \boldsymbol{X}(0) = \boldsymbol{AX}(s) + \boldsymbol{BU}(s) \tag{2.47}$$

Therefore

$$\boldsymbol{X}(s)=(s\boldsymbol{I}-\boldsymbol{A})^{-1}\boldsymbol{X}(0)+(s\boldsymbol{I}-\boldsymbol{A})^{-1}\boldsymbol{B}\boldsymbol{U}(s) \tag{2.48}$$

Note that $L[\mathrm{e}^{\boldsymbol{A}t}]=(s\boldsymbol{I}-\boldsymbol{A})^{-1}$ and $L[\int_0^t f(\tau)g(t-\tau)\mathrm{d}\tau]=F(s)G(s)$. So, the inverse Laplace transform of Eq. (2.48) results in the solution of Eq. (2.1).

$$\boldsymbol{X}(t)=\mathrm{e}^{\boldsymbol{A}t}\boldsymbol{X}(0)+\int_0^t \mathrm{e}^{\boldsymbol{A}(t-\tau)}\cdot\boldsymbol{B}\cdot\boldsymbol{u}(\tau)\mathrm{d}\tau \tag{2.49}$$

If the initial condition of the LTI system is the more general case $\boldsymbol{X}(t_0)$, the solution of Eq. (2.1) will be

$$\boldsymbol{X}(t)=\mathrm{e}^{\boldsymbol{A}(t-t_0)}\boldsymbol{X}(t_0)+\int_{t_0}^t \mathrm{e}^{\boldsymbol{A}(t-\tau)}\cdot\boldsymbol{B}\cdot\boldsymbol{u}(\tau)\mathrm{d}\tau \tag{2.50}$$

Based on the relationship of the matrix exponential function and the state transition matrix

$$\boldsymbol{\Phi}(t-t_0)=\mathrm{e}^{\boldsymbol{A}(t-t_0)}$$

the solution of Eq. (2.1) can also be written as the more general case

$$\boldsymbol{X}(t)=\boldsymbol{\Phi}(t-t_0)\boldsymbol{X}(t_0)+\int_{t_0}^t \boldsymbol{\Phi}(t-\tau)\cdot\boldsymbol{B}\cdot\boldsymbol{u}(\tau)\mathrm{d}\tau \tag{2.51}$$

It is clear that the solution of the LTI system is composed of two terms. By the comparison with Eq. (2.10), we have that the first term on the right-hand side of Eq. (2.51) is the solution of the LTI homogenous system (2.2) and it can be called the force-free response or zero-input response of the LTI system.

Let the initial state $\boldsymbol{X}(t_0)=\boldsymbol{0}$, in other words, the LTI system is taken on the zero-state situation. Based on the definition of the matrix exponential function, the derivation can be obtained as

$$\begin{aligned}\frac{\mathrm{d}}{\mathrm{d}t}[\mathrm{e}^{-\boldsymbol{A}(t-t_0)}\boldsymbol{X}(t)]&=-\boldsymbol{A}\mathrm{e}^{-\boldsymbol{A}(t-t_0)}\boldsymbol{X}(t)+\mathrm{e}^{-\boldsymbol{A}(t-t_0)}\dot{\boldsymbol{X}}(t)\\&=-\mathrm{e}^{-\boldsymbol{A}(t-t_0)}\boldsymbol{A}\boldsymbol{X}(t)+\mathrm{e}^{-\boldsymbol{A}(t-t_0)}\dot{\boldsymbol{X}}(t)\\&=\mathrm{e}^{-\boldsymbol{A}(t-t_0)}[\dot{\boldsymbol{X}}(t)-\boldsymbol{A}\boldsymbol{X}(t)]=\mathrm{e}^{-\boldsymbol{A}(t-t_0)}\boldsymbol{B}\boldsymbol{u}(t)\end{aligned} \tag{2.52}$$

Furthermore, Eq. (2.52) can be written as

$$\mathrm{d}[\mathrm{e}^{-\boldsymbol{A}(t-t_0)}\boldsymbol{X}(t)]=\mathrm{e}^{-\boldsymbol{A}(t-t_0)}\boldsymbol{B}\boldsymbol{u}(t)\mathrm{d}t \tag{2.53}$$

Taking definite integral on both side of Eq. (2.53), we have

$$\int_{t_0}^t \mathrm{d}[\mathrm{e}^{-\boldsymbol{A}(t-t_0)}\boldsymbol{X}(t)]=\mathrm{e}^{-\boldsymbol{A}(t-t_0)}\boldsymbol{X}(t)=\int_{t_0}^t \mathrm{e}^{-\boldsymbol{A}(\tau-t_0)}\boldsymbol{B}\boldsymbol{u}(\tau)\mathrm{d}\tau \tag{2.54}$$

Therefore

$$\boldsymbol{X}(t)=\int_{t_0}^t \mathrm{e}^{\boldsymbol{A}(t-\tau)}\boldsymbol{B}\boldsymbol{u}(\tau)\mathrm{d}\tau \tag{2.55}$$

Obviously, it is the second term on the right-hand side of Eq. (2.50) and it can be called the forced response or zero-state response of the LTI system.

Example 2.6 Determine the solution of the LTI system described by

$$\dot{\boldsymbol{X}}=\begin{bmatrix}0 & 1\\-2 & -3\end{bmatrix}\boldsymbol{X}+\begin{bmatrix}0\\1\end{bmatrix}u, t\geqslant 0$$

Where $u(t)=1(t)$ is the unit step function and $\boldsymbol{X}(0)=[x_1(0)\quad x_2(0)]^{\mathrm{T}}$.

Solution From Example 2.1 we have obtained that

$$\boldsymbol{\Phi}(t)=e^{\boldsymbol{A}t}=\begin{bmatrix} 2e^{-t}-e^{-2t} & e^{-t}-e^{-2t} \\ -2e^{-t}+2e^{-2t} & -e^{-t}+2e^{-2t} \end{bmatrix}$$

Therefore, the solution of the LTI system can also be calculated as

$$\boldsymbol{X}(t)=\boldsymbol{\Phi}(t)\boldsymbol{X}(0)+\int_0^t \boldsymbol{\Phi}(t-\tau)\boldsymbol{B}u(\tau)\mathrm{d}\tau$$

$$=\begin{bmatrix} 2e^{-t}-e^{-2t} & e^{-t}-e^{-2t} \\ -2e^{-t}+2e^{-2t} & -e^{-t}+2e^{-2t} \end{bmatrix}\begin{bmatrix} x_1(0) \\ x_2(0) \end{bmatrix}$$

$$+\int_0^t \begin{bmatrix} 2e^{-(t-\tau)}-e^{-2(t-\tau)} & e^{-(t-\tau)}-e^{-2(t-\tau)} \\ -2e^{-(t-\tau)}+2e^{-2(t-\tau)} & -e^{-(t-\tau)}+2e^{-2(t-\tau)} \end{bmatrix}\begin{bmatrix} 0 \\ 1 \end{bmatrix}\mathrm{d}\tau$$

$$=\begin{bmatrix} (2e^{-t}-e^{-2t})x_1(0)+(e^{-t}-e^{-2t})x_2(0) \\ (-2e^{-t}+2e^{-2t})x_1(0)+(-e^{-t}+2e^{-2t})x_2(0) \end{bmatrix}+\int_0^t \begin{bmatrix} e^{-(t-\tau)}-e^{-2(t-\tau)} \\ -e^{-(t-\tau)}+2e^{-2(t-\tau)} \end{bmatrix}\mathrm{d}\tau$$

$$=\begin{bmatrix} (2e^{-t}-e^{-2t})x_1(0)+(e^{-t}-e^{-2t})x_2(0) \\ (-2e^{-t}+2e^{-2t})x_1(0)+(-e^{-t}+2e^{-2t})x_2(0) \end{bmatrix}+\begin{bmatrix} \frac{1}{2}-e^{-t}+\frac{1}{2}e^{-2t} \\ e^{-t}-e^{-2t} \end{bmatrix}$$

$$=\begin{bmatrix} \frac{1}{2}+[2x_1(0)+x_2(0)-1]e^{-t}-[x_1(0)+x_2(0)-\frac{1}{2}]e^{-2t} \\ -[2x_1(0)+x_2(0)-1]e^{-t}+[2x_1(0)+2x_2(0)-1]e^{-2t} \end{bmatrix}$$

Problems

2.1 Calculate the matrix exponential function $e^{\boldsymbol{A}t}$ with three different methods.

$$\boldsymbol{A}=\begin{bmatrix} 0 & 6 \\ -1 & -5 \end{bmatrix}$$

2.2 A LTI system has the following state transition matrix

$$\boldsymbol{\Phi}(t)=\begin{bmatrix} 2e^{-t}-e^{-2t} & e^{-t}-e^{-2t} \\ -2e^{-t}+2e^{-2t} & -e^{-t}+2e^{-2t} \end{bmatrix}$$

Determine the system matrix $\boldsymbol{A}$ of the system.

2.3 Consider the LTI homogeneous system $\dot{\boldsymbol{X}}(t)=\boldsymbol{A}\boldsymbol{X}(t)$. The zero-input response of it is different in different initial condition. For example, when the initial state $\boldsymbol{X}(0)=\begin{bmatrix} 1 \\ -4 \end{bmatrix}$, the zero-input response of the system is $\boldsymbol{X}(t)=\begin{bmatrix} e^{-3t} \\ -4e^{-3t} \end{bmatrix}$; however, when $\boldsymbol{X}(0)=\begin{bmatrix} 2 \\ -1 \end{bmatrix}$, the zero-input response $\boldsymbol{X}(t)=\begin{bmatrix} 2e^{-2t} \\ -e^{-2t} \end{bmatrix}$. Determine the system matrix $\boldsymbol{A}$ of the system.

2.4 Find the solution of the given LTI homogeneous system

$$\dot{\boldsymbol{X}}=\begin{bmatrix} 1 & -2 \\ 2 & -3 \end{bmatrix}\boldsymbol{X},\boldsymbol{X}(0)=\begin{bmatrix} 10 \\ 10 \end{bmatrix}$$

2.5 Consider the LTI system described by

$$\dot{\boldsymbol{X}}=\begin{bmatrix}-6 & 4\\ -2 & 0\end{bmatrix}\boldsymbol{X}+\begin{bmatrix}0\\ 1\end{bmatrix}u,\boldsymbol{X}(0)=\begin{bmatrix}2\\ 0\end{bmatrix}$$

$$y=[1 \quad 0]\boldsymbol{X}$$

(1) Calculate the unit step response of the system.

(2) Calculate the unit impulse response of the system.

Aleksandr Mikhailovich Lyapunov (1857—1918) was a Russian mathematician, mechanician and physicist. His surname is sometimes romanized as Ljapunov, Liapunov or Ljapunow.

In 1892, Lyapunov defended his doctoral thesis "*The general problem of the stability of motion*".

Lyapunov contributed to several fields, including differential equations, potential theory, dynamical systems and probability theory. His main preoccupations were the stability of equilibria and the motion of mechanical systems, the model theory for the stability of uniform turbulent liquid, and the study of particles under the influence of gravity. His work in the field of mathematical physics regarded the boundary value problem of the equation of Laplace. In the theory of potential, his work from 1897 "*On some questions connected with Dirichlets problem*" clarified several important aspects of the theory. His work in this field is in close connection with the work of Steklov. Lyapunov developed many important approximation methods. His methods, which he developed in 1899, make it possible to define the stability of sets of ordinary differential equations. He created the modern theory of the stability of a dynamic system. In the theory of probability, he generalized the works of Chebyshev and Markov, and proved the Central Limit Theorem under more general conditions than his predecessors. The method he used for the proof found later widespread use in probability theory.

Chapter3 Stability of the control System

3.1 The Basics of Stability Theory in Mathematics

The concept of stability is extremely important because almost every workable system is designed to be stable. If a system is not stable, it is usually of no use in practice.

The most important approach for studying the stability of control system is the Lyapunov stability theory, which is introduced by the Russian mathematician Alexandr Mikhailovich Lyapunov in the late 19th century. Before discussing the Lyapunov stability theory, we need to review some relevant mathematical knowledge, for example: the norm and the quadratic form function.

Definition 3.1 The **norm** of a vector $\boldsymbol{X}$ is a real-valued function $\|\boldsymbol{X}\|$ with properties:

(1) $\|\boldsymbol{X}\| \geqslant 0$ for all $\boldsymbol{X} \in \mathbf{R}^n$ with $\|\boldsymbol{X}\| = 0$ if and only if $\boldsymbol{X} = \mathbf{0}$.

(2) $\|\alpha \boldsymbol{X}\| = |\alpha| \cdot \|\boldsymbol{X}\|$ for all $\alpha \in \mathbf{R}$ and $\boldsymbol{X} \in \mathbf{R}^n$.

(3) Triangle inequality $\|\boldsymbol{X}+\boldsymbol{Y}\| \leqslant \|\boldsymbol{X}\| + \|\boldsymbol{Y}\|$ holds true for $\forall \boldsymbol{X}, \boldsymbol{Y} \in \mathbf{R}^n$.

There are many norms satisfy the conditions of Definition 3.1. The most commonly used norm is **Euclidean norm**, which is defined as

$$\|\boldsymbol{X}\| = \sqrt{x_1^2 + \cdots + x_n^2} \tag{3.1}$$

The Euclidean norm of a vector is the generalization of the idea of length and is a length measurement of a vector in the state space. For example, $\|\boldsymbol{X}_0 - \boldsymbol{X}_e\|$ is used to represent the length from the point $\boldsymbol{X}_0$ to the point $\boldsymbol{X}_e$ in the state space.

Definition 3.2 Definiteness of a scalar function $V(\boldsymbol{X})$.

A scalar function $V(\boldsymbol{X})$ is called **positive definite** as $V(\boldsymbol{X}) > 0$ for all $\boldsymbol{X} \in \mathbf{R}^n$ with $V(\boldsymbol{X}) = 0$ if and only if $\boldsymbol{X} = \mathbf{0}$.

A scalar function $V(\boldsymbol{X})$ is called **positive semi-definite** as $V(\boldsymbol{X}) \geqslant 0$ for all $\boldsymbol{X} \in \mathbf{R}^n$ with $V(\boldsymbol{X}) = 0$ if and only if $\boldsymbol{X} = \mathbf{0}$.

A scalar function $V(\boldsymbol{X})$ is called **negative definite** as $V(\boldsymbol{X}) < 0$ for all $\boldsymbol{X} \in \mathbf{R}^n$ with $V(\boldsymbol{X}) = 0$ if and only if $\boldsymbol{X} = \mathbf{0}$.

A scalar function $V(\boldsymbol{X})$ is called **negative semi-definite** as $V(\boldsymbol{X}) \leqslant 0$ for all $\boldsymbol{X} \in \mathbf{R}^n$ with $V(\boldsymbol{X}) = 0$ if and only if $\boldsymbol{X} = \mathbf{0}$.

A scalar function $V(\boldsymbol{X})$ is called **indefinite** when it presents both positive and negative values for $\boldsymbol{X} \neq \mathbf{0}$.

Definition 3.3 The **quadratic form function** $V(\boldsymbol{X})$ is a real homogeneous polynomial in the real variables $x_1, x_2, \cdots, x_n$ of the form

$$V(\boldsymbol{X}) = \sum_{i=1}^{n}\sum_{j=1}^{n} p_{ij}x_i x_j \tag{3.2}$$

Where p_{ij} are real.

The quadratic form function can also be written with the vector $\boldsymbol{X}=[x_1 x_2 \cdots x_n]^{\mathrm{T}}$ and the matrix $\boldsymbol{P}$ as

$$V(\boldsymbol{X})=\boldsymbol{X}^{\mathrm{T}}\boldsymbol{P}\boldsymbol{X} \tag{3.3}$$

Where

$$\boldsymbol{P}=\begin{bmatrix} p_{11} & p_{12} & \cdots & p_{1n} \\ p_{12} & p_{22} & \cdots & p_{2n} \\ \vdots & \vdots & \ddots & \vdots \\ p_{1n} & p_{2n} & \cdots & p_{nn} \end{bmatrix} \tag{3.4}$$

is a real symmetric matrix and it is called the matrix of the quadratic form function $V(\boldsymbol{X})$.

Example 3.1 Rewrite the following scalar function in the form such as (3.3).

$$V(X)=x_1^2-3x_2^2+x_3^2-4x_4^2-2x_1x_2+4x_1x_3-8x_1x_4-4x_3x_4$$

Solution Letting $\boldsymbol{X}=[x_1 \quad x_2 \quad x_3 \quad x_4]^{\mathrm{T}}$, the scalar function $V(\boldsymbol{X})$ can be written as

$$V(\boldsymbol{X})=\boldsymbol{X}^{\mathrm{T}}\boldsymbol{P}\boldsymbol{X}=[x_1 \quad x_2 \quad x_3 \quad x_4]\begin{bmatrix} 1 & -1 & 2 & -4 \\ -1 & -3 & 0 & 0 \\ 2 & 0 & 1 & -2 \\ -4 & 0 & -2 & -4 \end{bmatrix}\begin{bmatrix} x_1 \\ x_2 \\ x_3 \\ x_4 \end{bmatrix}$$

Where $\boldsymbol{P}$ is a real symmetric matrix and is called the matrix of the quadratic form function $V(\boldsymbol{X})$.

Definition 3.4 Given a $n\times n$ matrix $\boldsymbol{P}$ such as (3.4), let $\boldsymbol{P}_k$ denote the matrix formed by deleting the last $n-k$ rows and columns of $\boldsymbol{P}$, where $k=1,2,\cdots,n$. The determinant

$$\Delta_k = |\boldsymbol{P}_k| \tag{3.5}$$

is called the **leading principal minor determinant** about $\boldsymbol{P}$ of order k.

Obviously, the $n\times n$ matrix $\boldsymbol{P}$ such as (3.4) has n leading principal minor determinant and every leading principal minor determinant must contain the element p_{11}.

For example, the 1-order leading principal minor determinant of $\boldsymbol{P}$ is obtained by

$$\Delta_1 = |p_{11}|. \tag{3.6}$$

The 2-order leading principal minor determinant of $\boldsymbol{P}$ is obtained by

$$\Delta_2 = \begin{vmatrix} p_{11} & p_{12} \\ p_{12} & p_{22} \end{vmatrix}. \tag{3.7}$$

The k-order leading principal minor determinant of $\boldsymbol{P}$ is obtained by

$$\Delta_k = \begin{vmatrix} p_{11} & p_{12} & \cdots & p_{1k} \\ p_{12} & p_{22} & \cdots & p_{2k} \\ \vdots & \vdots & \ddots & \vdots \\ p_{1k} & p_{2k} & \cdots & p_{kk} \end{vmatrix} \tag{3.8}$$

So

$$
\Delta_n = \begin{vmatrix} p_{11} & p_{12} & \cdots & p_{1n} \\ p_{12} & p_{22} & \cdots & p_{2n} \\ \vdots & \vdots & \ddots & \vdots \\ p_{1n} & p_{2n} & \cdots & p_{nn} \end{vmatrix} = |\boldsymbol{P}| \tag{3.9}
$$

is called the n-order leading principal minor determinant of $\boldsymbol{P}$.

Definition 3.5 Given a $n \times n$ matrix $\boldsymbol{P}$ such as (3.4). let $\widetilde{\boldsymbol{P}}_k$ denote the matrix formed by deleting $n-k$ rows and columns, which have the same sequence, of $\boldsymbol{P}$, where $k=1,2,\cdots,n$. The determinant

$$
\widetilde{\Delta}_k = |\widetilde{\boldsymbol{P}}_k| \tag{3.10}
$$

is called the **principal minor determinant** about $\boldsymbol{P}$ of order k .

Comparing Definition 3.4 with 3.5, it can be obtained that, for a $n \times n$ matrix $\boldsymbol{P}$, its k-order leading principal minor determinant Δ_k is unique but its k-order principal minor determinant $\widetilde{\Delta}_k$ is not unique.

Example 3.2 Obtain the leading principal minor determinant and the principal minor determinant of the matrix given as

$$
\boldsymbol{P} = \begin{bmatrix} 1 & 1 & 0 \\ 1 & 1 & 0 \\ 0 & 0 & -1 \end{bmatrix}
$$

Solution The 1-order leading principal minor determinant of $\boldsymbol{P}$ is

$$
\Delta_1 = 1
$$

The 2-order leading principal minor determinant of $\boldsymbol{P}$ is

$$
\Delta_2 = \begin{vmatrix} 1 & 1 \\ 1 & 1 \end{vmatrix} = 0
$$

The 3-order leading principal minor determinant of $\boldsymbol{P}$ is

$$
\Delta_3 = \begin{vmatrix} 1 & 1 & 0 \\ 1 & 1 & 0 \\ 0 & 0 & -1 \end{vmatrix} = 0
$$

According to Definition 3.5, $\boldsymbol{P}$ has three 1-order principal minor determinants as

$$
\widetilde{\Delta}_{11} = -1, \widetilde{\Delta}_{12} = 1, \widetilde{\Delta}_{13} = 1
$$

and three 2-order principal minor determinants as

$$
\widetilde{\Delta}_{21} = \begin{vmatrix} 1 & 0 \\ 0 & -1 \end{vmatrix} = -1, \widetilde{\Delta}_{22} = \begin{vmatrix} 1 & 0 \\ 0 & -1 \end{vmatrix} = -1, \widetilde{\Delta}_{23} = \begin{vmatrix} 1 & 1 \\ 1 & 1 \end{vmatrix} = 0
$$

and only one 3-order principal minor determinant as

$$
\widetilde{\Delta}_3 = \begin{vmatrix} 1 & 1 & 0 \\ 1 & 1 & 0 \\ 0 & 0 & -1 \end{vmatrix} = 0
$$

Theorem 3.1 The quadratic form function $V(\boldsymbol{X})$ which is described by (3.3) is positive definite and $\boldsymbol{P}$ is called a positive definite matrix if and only if the k-order leading principal minor determinant of $\boldsymbol{P}$

$$\Delta_k > 0, k=1,2,\cdots,n \tag{3.11}$$

Example 3.3 Determine the definiteness of the following quadratic form function.

$$V(X)=2x_1^2+2x_2^2+2x_3^2+2x_1x_2+2x_1x_3+2x_2x_3$$

Solution By letting $\boldsymbol{X}=[x_1 \quad x_2 \quad x_3]^{\mathrm{T}}$, the quadratic form function $V(\boldsymbol{X})$ can be written as $V(\boldsymbol{X})=\boldsymbol{X}^{\mathrm{T}}\boldsymbol{P}\boldsymbol{X}$ and the matrix of it can be obtained as

$$\boldsymbol{P}=\begin{bmatrix} 2 & 1 & 1 \\ 1 & 2 & 1 \\ 1 & 1 & 2 \end{bmatrix}$$

The 1-order leading principal minor determinant of $\boldsymbol{P}$ is

$$\Delta_1 = 2 > 0$$

The 2-order leading principal minor determinant of $\boldsymbol{P}$ is

$$\Delta_2 = \begin{vmatrix} 2 & 1 \\ 1 & 2 \end{vmatrix} = 3 > 0$$

The 3-order leading principal minor determinant of $\boldsymbol{P}$ is

$$\Delta_3 = \begin{vmatrix} 2 & 1 & 1 \\ 1 & 2 & 1 \\ 1 & 1 & 2 \end{vmatrix} = 2 > 0$$

According to Theorem 3.1, $V(\boldsymbol{X})$ is positive definite.

Theorem 3.2 The quadratic form function $V(\boldsymbol{X})$ which is described by (3.3) is negative definite if and only if

$$(-1)^k \Delta_k > 0, k=1,2,\cdots,n \tag{3.12}$$

Theorem 3.3 The quadratic form function $V(\boldsymbol{X})$ which is described by (3.3) is positive semi-definite and $\boldsymbol{P}$ is called a positive semi-definite matrix if and only if the k-order principal minor determinant of $\boldsymbol{P}$

$$\tilde{\Delta}_k \geqslant 0, k=1,2,\cdots,n \tag{3.13}$$

Example 3.4 Determine the definiteness of the following quadratic form function.

$$V(X)=x_1^2+x_2^2+2x_3^2+2x_1x_3+2x_2x_3$$

Solution By letting $\boldsymbol{X}=[x_1 \quad x_2 \quad x_3]^{\mathrm{T}}$, the quadratic form function $V(\boldsymbol{X})$ can be written as $V(\boldsymbol{X})=\boldsymbol{X}^{\mathrm{T}}\boldsymbol{P}\boldsymbol{X}$ and the matrix of it can be obtained as.

$$\boldsymbol{P}=\begin{bmatrix} 1 & 0 & 1 \\ 0 & 1 & 1 \\ 1 & 1 & 2 \end{bmatrix}$$

$\boldsymbol{P}$ has three 1-order principal minor determinants as

$$\tilde{\Delta}_{11}=2>0, \tilde{\Delta}_{12}=1>0, \tilde{\Delta}_{13}=1>0$$

and three 2-order principal minor determinants as

$$\tilde{\Delta}_{21}=\begin{vmatrix} 1 & 1 \\ 1 & 2 \end{vmatrix}=1>0, \tilde{\Delta}_{22}=\begin{vmatrix} 1 & 1 \\ 1 & 2 \end{vmatrix}=1>0, \tilde{\Delta}_{23}=\begin{vmatrix} 1 & 0 \\ 0 & 1 \end{vmatrix}=1>0$$

and only one 3-order principal minor determinant as

$$\tilde{\Delta}_3=\begin{vmatrix}1 & 0 & 1\\ 0 & 1 & 1\\ 1 & 1 & 2\end{vmatrix}=0$$

According to Theorem 3.3, $V(\boldsymbol{X})$ is positive semi-definite.

Theorem 3.4 The quadratic form function $V(\boldsymbol{X})$ which is described by (3.3) is negative semi-definite and $\boldsymbol{P}$ is called a negative semi-definite matrix if and only if $-V(\boldsymbol{X})$ is positive semi-definite.

Theorem 3.5 The quadratic form function $V(\boldsymbol{X})$ which is described by (3.3) is indefinite and $\boldsymbol{P}$ is called a indefinite matrix if and only if it is not satisfactory for the conditions of Theorem (3.1)~(3.4).

For example, the matrix $\boldsymbol{P}$ which is given in Example 3.2 is an indefinite matrix.

Theorems 3.1~3.5 are sometimes called **the Sylvester criterion.**

3.2 Lyapunov Stability

3.2.1 Equilibrium Point

The concept of stability plays an important role for the system analysis and synthesis. In the general theory, where time varying and nonlinear systems are considered, the definitions of stability are rather involved and the distinctions are subtle.

In the studying for the stability theory, the equilibrium point of a system is an important concept. In fact, Lyapunov introduced the concepts of the stability in the vicinity of an equilibrium point.

Definition 3.6 Suppose an autonomous (or unforced) system is described by

$$\dot{\boldsymbol{X}}(t)=f[\boldsymbol{X}(t),\boldsymbol{0},t] \tag{3.14}$$

or

$$\dot{\boldsymbol{X}}(t)=f[\boldsymbol{X}(t),t] \tag{3.15}$$

The state $\boldsymbol{X}_e$ is called the **equilibrium point** of the system (3.15) if it satisfies

$$f(\boldsymbol{X}_e,t)=\boldsymbol{0} \quad \text{for} \quad t\geqslant t_0 \tag{3.16}$$

It should be noted that the definition is applicable for both linear system and nonlinear system. Consider the LTI system

$$\dot{\boldsymbol{X}}(t)=\boldsymbol{A}X(t) \tag{3.17}$$

if matrix $\boldsymbol{A}$ is nonsingular, then the system has the only equilibrium point $\boldsymbol{X}_e=\boldsymbol{0}$. Otherwise, the system may have many equilibrium points. Generally, the non-linear system may have many equilibrium points also.

Example 3.5 Determine the equilibrium points of the following system.

$$\begin{cases}\dot{x}_1=-x_2\\ \dot{x}_2=-x_1-x_2\end{cases}$$

Solution The system is a LTI system. By letting $\boldsymbol{X}=[x_1 \quad x_2]^{\mathrm{T}}$, the system can be described by

$$\dot{X}=AX=\begin{bmatrix}0 & -1\\ -1 & -1\end{bmatrix}X$$

Obviously, the system matrix A is nonsingular. By letting $\dot{X}=\mathbf{0}$, it can be deduced that the origin $X_e=\mathbf{0}$ of the state space is the only equilibrium point of the system.

Example 3.6 Determine the equilibrium points of the following system.

$$\begin{cases}\dot{x}_1=-x_1\\ \dot{x}_2=x_1+x_2-x_2^3\end{cases}$$

Solution Obviously, the system is a non-linear system.

By letting $X=[x_1 \quad x_2]^T$ and letting $\dot{X}=\mathbf{0}$, three equilibrium points of the system can be obtained as

$$X_{e1}=\begin{bmatrix}0\\0\end{bmatrix},\quad X_{e2}=\begin{bmatrix}0\\1\end{bmatrix},\quad X_{e3}=\begin{bmatrix}0\\-1\end{bmatrix}$$

In fact, the origin of the state space is always an equilibrium point for systems, although it needs not to be the only one. Only isolated equilibrium points will be considered in this chapter and any nonzero isolated equilibrium point can be transferred to the origin by a change of variable

$$\bar{X}=X-X_e \tag{3.18}$$

For this reason it is assumed in the sequel that $X_e=\mathbf{0}$ and it will only be considered in this chapter.

3.2.2 Concepts of Lyapunov Stability

Just as the discussion above, Lyapunov introduced the concepts of the stability in the vicinity of an equilibrium point. For the continuous-time system with zero input described by (3.15), Simply, the studying for the stability deals with the question that whether the state, which is perturbed from its equilibrium point X_e at time t_0, return to X_e, or remain close to X_e, or diverge from it.

Definition 3.7 The equilibrium point X_e is said to be **stable i. s. L** (i. e. in the sense of Lyapunov), if for any given real $\varepsilon>0$, there exists a real $\delta(\varepsilon, t_0)>0$ so that if the initial state satisfies

$$\| X_0-X_e \| \leqslant \delta(\varepsilon, t_0) \tag{3.19}$$

then

$$\| X(t;X_0,t_0)-X_e \| \leqslant \varepsilon \quad \text{for} \quad \forall t \geqslant t_0 \tag{3.20}$$

Lyapunov stability means that we are able to select a bound on initial condition that will result in the state trajectory remains within a chosen finite limit. The geometrical implication of Lyapunov stability is shown in Figure 3.1.

In most of the engineering applications, we expect that the state gradually go back to its original value rather than maintain in a range by the disturbance. So, the concept

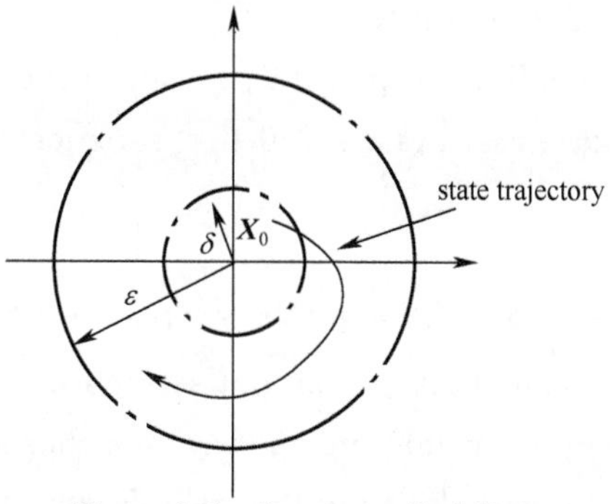

Figure 3. 1 Lyapunov Stability

of the asymptotically stability is more suitable for the engineering requirement.

Definition 3. 8 The equilibrium point $\boldsymbol{X}_e$ is said to be **asymptotically stable i. s. L** if it is stable i. s. L and

$$\lim_{t \to \infty} \| \boldsymbol{X}(t;\boldsymbol{X}_0,t_0) - \boldsymbol{X}_e \| = 0 \tag{3.21}$$

Lyapunov asymptotically stability means that we are able to select a bound on initial condition, that will result in the state trajectory which remains within a chosen finite limit and will return to $\boldsymbol{X}_e$. The geometrical implication of Lyapunov asymptotically stability is shown in Figure 3. 2.

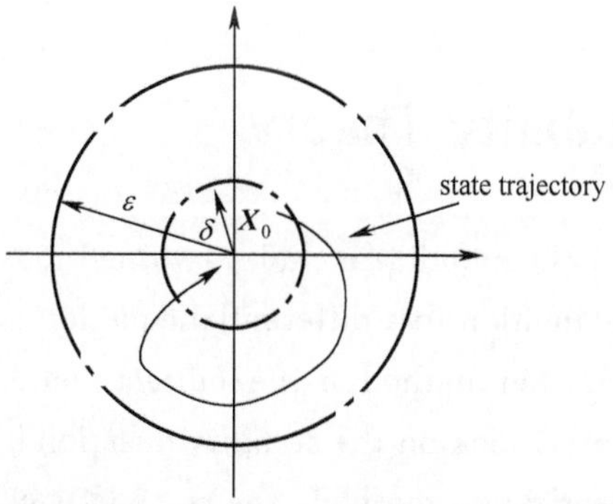

Figure 3. 2 Lyapunov Asymptotically Stability

Definition 3. 9 If δ, which is appear in (3. 19) and indicates the bound on initial condition, is not the function of t_0 and the equilibrium point $\boldsymbol{X}_e$ is stable i. s. L, then $\boldsymbol{X}_e$ is said to be **uniformly stable.**

Definition 3. 10 If δ, which is appear in (3. 19) and indicates the bound on initial condition, is not the function of t_0 and the equilibrium point $\boldsymbol{X}_e$ is asymptotically stable i. s. L, then $\boldsymbol{X}_e$ is said to be **uniformly asymptotically stable.**

Definition 3. 11 If the equilibrium point $\boldsymbol{X}_e$ is asymptotically stable i. s. L for any initial state, then the equilibrium point $\boldsymbol{X}_e$ is said to be **globally asymptotically stable** or asymptotically stable in the large.

Lyapunov globally asymptotically stability means that we are able to select a infinite bound on initial condition that will result in the state trajectory which remains within a chosen finite limit and will return to $\boldsymbol{X}_e$.

Definition 3.12 The equilibrium point $\boldsymbol{X}_e$ is said to be unstable i. s. L, if for any given real $\varepsilon > 0$, there is not a real $\delta(\varepsilon, t_0) > 0$ that satisfies

$$\| \boldsymbol{X}_0 - \boldsymbol{X}_e \| \leqslant \delta(\varepsilon, t_0) \tag{3.22}$$

and

$$\| \boldsymbol{X}(t; \boldsymbol{X}_0, t_0) - \boldsymbol{X}_e \| \leqslant \varepsilon \quad \text{for} \quad \forall t \geqslant t_0 \tag{3.23}$$

Lyapunov unstability means that we can not select a bound on initial condition that will result in the state trajectory which remains within a chosen finite limit. The geometrical implication of Lyapunov unstability is shown in Figure 3.3.

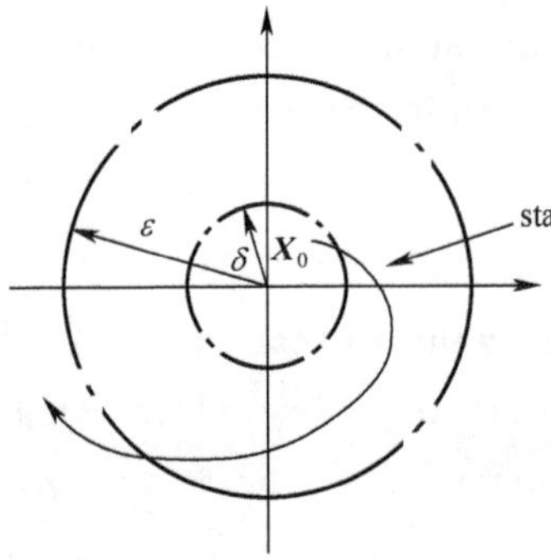

Figure 3.3 Lyapunov Unstability

3.3 Lyapunov Stability Theory

Lyapunov's work about the stability includes two methods. Testing for stability by considering the linear approximation to a differential equation is referred to as Lyapunov first method (i. e. the linearization method or the indirect method). Using the idea of the Lyapunov function for a direct attack on the stability question is Lyapunov second method. Correspond with the linearization method, the method is called the direct method.

3.3.1 Lyapunov First Method

As the discussion above, a nonlinear system may have more than one equilibrium point. The nonlinear system can be expanded in a Taylor series about a equilibrium point (the origin is always selected in this chapter) in a small neighborhood of it.

Assume that the nonlinear system described by (3.15) can be expanded about the equilibrium point $\boldsymbol{X}_e$ in the following **Taylor series**

$$\dot{\boldsymbol{X}} = f(\boldsymbol{X}_e) + \left. \frac{\partial f}{\partial \boldsymbol{X}^{\mathrm{T}}} \right|_{\boldsymbol{X}=\boldsymbol{X}_e} \cdot (\boldsymbol{X} - \boldsymbol{X}_e) + g(\boldsymbol{X}) \tag{3.24}$$

Where $g(\boldsymbol{X})$ is the summation of the higher-order terms in the Taylor series.

Letting

$$\Delta \boldsymbol{X} = \boldsymbol{X} - \boldsymbol{X}_e \tag{3.25}$$

and neglecting the summation of the higher-order terms in the Taylor series yield the linearized differential equation

$$\Delta\dot{\boldsymbol{X}}=\left.\frac{\partial f}{\partial \boldsymbol{X}^{\mathrm{T}}}\right|_{\boldsymbol{X}=\boldsymbol{X}_e}\cdot\Delta\boldsymbol{X}=\boldsymbol{J}\cdot\Delta\boldsymbol{X} \tag{3.26}$$

Where the matrix

$$\boldsymbol{J}=\begin{bmatrix}\dfrac{\partial f_1}{\partial x_1} & \dfrac{\partial f_1}{\partial x_2} & \cdots & \dfrac{\partial f_1}{\partial x_n}\\ \dfrac{\partial f_2}{\partial x_1} & \dfrac{\partial f_2}{\partial x_2} & \cdots & \dfrac{\partial f_2}{\partial x_n}\\ \vdots & \vdots & \ddots & \vdots\\ \dfrac{\partial f_n}{\partial x_1} & \dfrac{\partial f_n}{\partial x_2} & \cdots & \dfrac{\partial f_n}{\partial x_n}\end{bmatrix}_{\boldsymbol{X}=\boldsymbol{X}_e} \tag{3.27}$$

is called a **Jacobi matrix**, and f_i is the ith row of $f(\boldsymbol{X})$.

Theorem 3.6 For a nonlinear system described by (3.15),

(1) If all eigenvalues of the linearized differential equation (3.26) have negative real part, then the equilibrium point $\boldsymbol{X}_e$ is asymptotically stable in a small neighborhood of it.

(2) If there is at least one eigenvalue of the linearized differential equation (3.26) has positive real part, then the equilibrium point $\boldsymbol{X}_e$ is unstable in a small neighborhood of it.

(3) If some eigenvalues of the linearized differential equation (3.26) have zero real part and others have negative real part, the stability of the equilibrium point $\boldsymbol{X}_e$ is related to the summation of the higher-order terms in the Taylor series $g(\boldsymbol{X})$. In the case, the equilibrium point $\boldsymbol{X}_e$ may be stable or unstable in a small neighborhood of it.

Example 3.7 Consider the system described by

$$\begin{cases}\dot{x}_1=x_1-x_1x_2\\ \dot{x}_2=-x_2+x_1x_2\end{cases}$$

Determine the equilibrium points and the stability of the system on them.

Solution Obviously, the system is a non-linear system. By letting $\boldsymbol{X}=[x_1\quad x_2]^{\mathrm{T}}$, the system can be described by.

$$\dot{\boldsymbol{X}}=f(\boldsymbol{X})=\begin{bmatrix}f_1(\boldsymbol{X})\\ f_2(\boldsymbol{X})\end{bmatrix}=\begin{bmatrix}x_1-x_1x_2\\ -x_2+x_1x_2\end{bmatrix}$$

Letting $\dot{\boldsymbol{X}}=\boldsymbol{0}$, two equilibrium points of the system can be obtained as

$$\boldsymbol{X}_{e1}=\begin{bmatrix}0\\0\end{bmatrix},\quad \boldsymbol{X}_{e2}=\begin{bmatrix}1\\1\end{bmatrix}$$

The Jacobi matrix can be calculated as

$$\boldsymbol{J}=\begin{bmatrix}\dfrac{\partial f_1}{\partial x_1} & \dfrac{\partial f_1}{\partial x_2}\\ \dfrac{\partial f_2}{\partial x_1} & \dfrac{\partial f_2}{\partial x_2}\end{bmatrix}=\begin{bmatrix}1-x_2 & -x_1\\ x_2 & -1+x_1\end{bmatrix}$$

The linearized differential equation of the system on $\boldsymbol{X}_{e1}$ can be obtained as

$$\Delta\dot{\boldsymbol{X}}=\begin{bmatrix}1 & 0\\ 0 & -1\end{bmatrix}\cdot\Delta\boldsymbol{X}$$

and the eigenvalues of it are

$$\lambda_1=-1,\quad \lambda_2=1$$

Because $\lambda_2>0$, then the equilibrium point $\boldsymbol{X}_{e1}$ is unstable in a small neighborhood of it.

The linearized differential equation of the system on $\boldsymbol{X}_{e2}$ can be obtained as

$$\Delta\dot{\boldsymbol{X}}=\begin{bmatrix}0 & -1\\ 1 & 0\end{bmatrix}\cdot\Delta\boldsymbol{X}$$

and the eigenvalues of it are

$$\lambda_{1,2}=\pm \mathrm{j}$$

Because the eigenvalues have zero real part, then we can not conclude the stability of the nonlinear system on the equilibrium point $\boldsymbol{X}_{e2}$ by the liearization method.

3.3.2 Lyapunov Second Method

In Lyapunov direct method, the concept of energy is introduced with the property that the system will approach the equilibrium point gradually if the energy is decreased as time is elapsed. Since it is difficult to find the real expression of energy function for a system, Lyapunov proposed to construct a virtual energy function $V(\boldsymbol{X},t)$, so called Lyapunov function, relating to the state variables. Consequently, the decay of the energy can be characterized by $\dot{V}(\boldsymbol{X},t)$. The Lyapunov direct method determines the stability of a system using the definiteness of $V(\boldsymbol{X},t)$ and $\dot{V}(\boldsymbol{X},t)$ without linearization and solving the eigenvalues. The Lyapunov direct method is the most important tool for nonlinear system analysis and design, although it needs much skills when applied.

For a nonlinear system described by (3.15) with the initial condition

$$\boldsymbol{X}(t_0)=\boldsymbol{X}_0,\quad t\geqslant t_0\geqslant 0 \tag{3.28}$$

it is assumed that the origin $\boldsymbol{X}_e=\boldsymbol{0}$ is a equilibrium point of the system. So, several Lyapunov stability theorems will be presented.

Theorem 3.7 If there exists a scalar function $V(\boldsymbol{X},t)$ in a small neighborhood of the equilibrium point $\boldsymbol{X}_e$ such that

(1) $V(\boldsymbol{X},t)$ and $\dot{V}(\boldsymbol{X},t)$ are continuous.

(2) $V(\boldsymbol{X},t)$ is positive definite (i. e. $V(\boldsymbol{X},t)>0$) for $\|\boldsymbol{X}\|\neq 0$, and $V(\boldsymbol{0},t)=0$.

(3) $\dot{V}(\boldsymbol{X},t)$ is negative semi-definite (i. e. $\dot{V}(\boldsymbol{X},t)\leqslant 0$) for $\|\boldsymbol{X}\|\neq 0$.

then the equilibrium point $\boldsymbol{X}_e$ is locally stable i. s. L. The scalar function $V(\boldsymbol{X},t)$ is called the Lyapunov function.

Example 3.8 Consider a continuous time system

$$\begin{cases}\dot{x}_1=kx_2\\ \dot{x}_2=-x_1\end{cases},\qquad k>0$$

Determine the stability of the system at its equilibrium point.

Solution Obviously, the origin $\boldsymbol{X}_e = \mathbf{0}$ is the only equilibrium point of the system. The scalar Lyapunov function is chosen as

$$V(\boldsymbol{X},t)=x_1^2+kx_2^2$$

It is positive definite for $\|\boldsymbol{X}\| \neq 0$ and $V(\mathbf{0})=0$. The derivative

$$\dot{V}(\boldsymbol{X},t)=2x_1\dot{x}_1+2kx_2\dot{x}_2=2kx_1x_2-2kx_1x_2\equiv 0$$

Hence, the equilibrium point $\boldsymbol{X}_e=\mathbf{0}$ is locally stable.

It can be deduced that, condition 1 ensures that $V(\boldsymbol{X},t)$ is a smooth function and generally has the shape of a bowl near the equilibrium point; condition 2 means that $V(\boldsymbol{X},t)>0$, like energy, if any state is different from zero; condition 3 guarantees that any state trajectory moves so as never to climb higher on the bowl than where it started out. If condition 3 is made stronger so that $\dot{V}(\boldsymbol{X},t)<0$ for $\|\boldsymbol{X}\| \neq 0$, then the state trajectory must be drawn to the origin.

Theorem 3.8 If there exists a scalar function $V(\boldsymbol{X},t)$ in a small neighborhood of the equilibrium point $\boldsymbol{X}_e$ such that

(1) $V(\boldsymbol{X},t)$ and $\dot{V}(\boldsymbol{X},t)$ are continuous.

(2) $V(\boldsymbol{X},t)$ is positive definite (i. e. $V(\boldsymbol{X},t)>0$) for $\|\boldsymbol{X}\| \neq 0$, and $V(\mathbf{0},t)=0$.

(3) $\dot{V}(\boldsymbol{X},t)$ is negative definite (i. e. $\dot{V}(\boldsymbol{X},t)<0$) for $\|\boldsymbol{X}\| \neq 0$.

the equilibrium point $\boldsymbol{X}_e$ is locally asymptotically stable i. s. L.

In fact, the condition 3 of Theorem 3.8 can be relaxed as $\dot{V}(\boldsymbol{X},t)\leqslant 0$ if $\dot{V}(\boldsymbol{X},t)$ does not keep zero for any $\|\boldsymbol{X}\| \neq 0$, because the state trajectory will not stay at the points, which satisfies $\dot{V}(\boldsymbol{X},t)=0$, but moves continuously towards the origin of the system.

Theorem 3.9 If there exists a scalar function $V(\boldsymbol{X},t)$ in a small neighborhood of the equilibrium point $\boldsymbol{X}_e$ such that

(1) $V(\boldsymbol{X},t)$ and $\dot{V}(\boldsymbol{X},t)$ are continuous.

(2) $V(\boldsymbol{X},t)$ is positive definite (i. e. $V(\boldsymbol{X},t)>0$) for $\|\boldsymbol{X}\| \neq 0$, and $V(\mathbf{0},t)=0$.

(3) $\dot{V}(\boldsymbol{X},t)$ is negative semi-definite (i. e. $\dot{V}(\boldsymbol{X},t)\leqslant 0$) and $\dot{V}(\boldsymbol{X},t)$ does not keep zero for $\|\boldsymbol{X}\| \neq 0$.

the equilibrium point $\boldsymbol{X}_e$ is locally asymptotically stable i. s. L.

Furthermore, if $V(\boldsymbol{X},t)\to\infty$ as $\|\boldsymbol{X}\| \to\infty$, the equilibrium point $\boldsymbol{X}_e=\mathbf{0}$ is globally asymptotically stable.

Example 3.9 Consider a continuous time system

$$\begin{cases}\dot{x}_1=-x_1(x_1^2+x_2^2)+x_2\\ \dot{x}_2=-x_1-x_2(x_1^2+x_2^2)\end{cases}$$

Determine the stability of the system at its equilibrium point.

Solution Obviously, the origin $\boldsymbol{X}_e=\mathbf{0}$ is the only equilibrium point of the system.

The scalar Lyapunov function is chosen as

$$V(\boldsymbol{X},t)=x_1^2+x_2^2$$

It is positive definite for $\| \boldsymbol{X} \| \neq 0$ and $V(\mathbf{0})=0$. The derivative

$$\dot{V}(\boldsymbol{X},t)=-2(x_1^2+x_2^2)^2$$

is negative definite for $\| \boldsymbol{X} \| \neq 0$ and $\dot{V}(\mathbf{0})=0$. So, the equilibrium point $X_e=\mathbf{0}$ is locally asymptotically stable.

Furthermore, since $V(\boldsymbol{X},t)\to\infty$ as $\| \boldsymbol{X} \| \to\infty$, then the equilibrium point $\boldsymbol{X}_e=\mathbf{0}$ is globally asymptotically stable.

Example 3. 10 Considering a continuous time system

$$\begin{cases}\dot{x}_1=x_2\\ \dot{x}_2=-x_1-x_2(1+x_2)^2\end{cases}$$

Determine the stability of the system at its equilibrium point.

Solution Obviously, the system has the unique equilibrium point at the origin $\boldsymbol{X}_e=\mathbf{0}$.

The scalar Lyapunov function is chosen as

$$V(\boldsymbol{X},t)=x_1^2+x_2^2$$

It is positive definite for $\| \boldsymbol{X} \| \neq 0$ and $V(\mathbf{0})=0$.

Moreover, the derivative

$$\dot{V}(\boldsymbol{X},t)=2x_1\dot{x}_1+2x_2\dot{x}_2=2x_1x_2+2x_2[-x_1-x_2(1+x_2)^2]=-2x_2^2(1+x_2)^2$$

is negative semi-definite for $\| \boldsymbol{X} \| \neq 0$.

It is deduced that only the points, $x_2=0$ and $x_2=-1$ for $\forall x_1$, satisfy $\dot{V}(\boldsymbol{X},t)=0$, and $\dot{V}(\boldsymbol{X},t)$ is always negative definite at the points else on the state trajectory. However, the state trajectory will not stay at the point $x_2=-1$, at which $\dot{V}(\boldsymbol{X},t)=0$, because the point is not the equilibrium point of the system. The state trajectory will moves continuously until it gradually go to the equilibrium point $\boldsymbol{X}_e=\mathbf{0}$. It means that $\dot{V}(\boldsymbol{X},t)$ does not keep zero for $\| \boldsymbol{X} \| \neq 0$. Consequently, the equilibrium point $\boldsymbol{X}_e=\mathbf{0}$ is locally asymptotically stable.

Furthermore, since $V(\boldsymbol{X},t)\to\infty$ as $\| \boldsymbol{X} \| \to\infty$, then the equilibrium point $\boldsymbol{X}_e=\mathbf{0}$ is globally asymptotically stable.

Theorem 3. 10 If there exists a scalar function $V(\boldsymbol{X},t)$ in a small neighborhood of the equilibrium point $\boldsymbol{X}_e$ such that

(1) $V(\boldsymbol{X},t)$ and $\dot{V}(\boldsymbol{X},t)$ are continuous.

(2) $V(\boldsymbol{X},t)$ is positive definite (i. e. $V(\boldsymbol{X},t)>0$) for $\| \boldsymbol{X} \| \neq 0$, and $V(0,t)=0$.

(3) $\dot{V}(\boldsymbol{X},t)$ is positive definite (i. e. $\dot{V}(\boldsymbol{X},t)>0$) for $\| \boldsymbol{X} \| \neq 0$.

then the equilibrium point $\boldsymbol{X}_e$ is globally unstable i. s. L.

Example 3. 11 Consider a continuous time system

$$\begin{cases}\dot{x}_1=x_1+x_2\\ \dot{x}_2=-x_1+x_2\end{cases}$$

Determine the stability of the system at its equilibrium point.

Solution Obviously, the origin $\boldsymbol{X}_e=\mathbf{0}$ is the only equilibrium point of the system.

The scalar Lyapunov function is chosen as.

$$V(\boldsymbol{X},t)=x_1^2+x_2^2$$

It is positive for $\| \boldsymbol{X} \| \neq 0$ and $V(\boldsymbol{0})=0$. The derivative

$$\dot{V}(\boldsymbol{X},t)=2x_1\dot{x}_1+2x_2\dot{x}_2=2x_1^2+2x_2^2$$

is positive definite for $\| \boldsymbol{X} \| \neq 0$ and $\dot{V}(\boldsymbol{0})=0$. So, the equilibrium point $\boldsymbol{X}_e=\boldsymbol{0}$ is globally unstable i. s. L.

Note that Lyapunov stability theorems provide only sufficient conditions, which are not necessary, for the stability of systems. Failure of a Lyapounov function candidate to satisfy the conditions for stability or asymptotically stable does not mean that equilibrium point is not stable or asymptotically stable. It only means that such stability property cannot be established by using this Lyapunov function candidate. Whether the equilibrium point is stable (asymptotically stable) or not can be determined only by further investigation.

3.4 Application of Lyapunov 2nd Method to the LTI System

As mentioned above, the Lyapunov stability theorems provide only the sufficient conditions, which are not necessary, for the stability of the non-linear systems. Yet, for the LTI systems, the Lyapounov stability theorems can be relaxed to the form which is sufficient and necessary.

Theorem 3.11 Consider the LTI homogeneous system described by

$$\dot{\boldsymbol{X}}=\boldsymbol{A}\boldsymbol{X} \tag{3.29}$$

with the initial condition $\boldsymbol{X}(0)=\boldsymbol{X}_0$. The equilibrium point $\boldsymbol{X}_e=\boldsymbol{0}$ is asymptotically stable i. s. L, iff for any symmetric positive definite matrix $\boldsymbol{Q}$, the following **Lyapunov equation**

$$\boldsymbol{A}^{\mathrm{T}}\boldsymbol{P}+\boldsymbol{P}\boldsymbol{A}=-\boldsymbol{Q} \tag{3.30}$$

has the unique symmetric positive definite solution matrix $\boldsymbol{P}$.

Proof. (1) Necessity. Let $\boldsymbol{X}(t;\boldsymbol{X}_0,0)$ denote the solution of system $\dot{\boldsymbol{X}}=\boldsymbol{A}\boldsymbol{X}$ starting from $\boldsymbol{X}_0$ at $t=0$. Since $\dot{\boldsymbol{X}}=\boldsymbol{A}\boldsymbol{X}$ is asymptotically stable i. s. L, we have

$$\lim_{t\to\infty}\boldsymbol{X}(t;\boldsymbol{X}_0,0)=\lim_{t\to\infty}\mathrm{e}^{\boldsymbol{A}t}\boldsymbol{X}_0=\boldsymbol{0} \tag{3.31}$$

It can be deduced from (3.31) that

$$\lim_{t\to\infty}\mathrm{e}^{\boldsymbol{A}t}=\boldsymbol{0} \tag{3.32}$$

For any symmetric positive definite matrix $\boldsymbol{Q}$, the matrix $\boldsymbol{P}$ can be constructed as

$$\boldsymbol{P}=\int_0^{\infty}\mathrm{e}^{\boldsymbol{A}^{\mathrm{T}}t}\boldsymbol{Q}\mathrm{e}^{\boldsymbol{A}t}\,\mathrm{d}t \tag{3.33}$$

Then

$$\begin{aligned}\boldsymbol{A}^{\mathrm{T}}\boldsymbol{P}+\boldsymbol{P}\boldsymbol{A}&=\int_0^{\infty}\boldsymbol{A}^{\mathrm{T}}\mathrm{e}^{\boldsymbol{A}^{\mathrm{T}}t}\boldsymbol{Q}\mathrm{e}^{\boldsymbol{A}t}\,\mathrm{d}t+\int_0^{\infty}\mathrm{e}^{\boldsymbol{A}^{\mathrm{T}}t}\boldsymbol{Q}\mathrm{e}^{\boldsymbol{A}t}\boldsymbol{A}\,\mathrm{d}t=\int_0^{\infty}(\boldsymbol{A}^{\mathrm{T}}\mathrm{e}^{\boldsymbol{A}^{\mathrm{T}}t}\boldsymbol{Q}\mathrm{e}^{\boldsymbol{A}t}+\mathrm{e}^{\boldsymbol{A}^{\mathrm{T}}t}\boldsymbol{Q}\mathrm{e}^{\boldsymbol{A}t}\boldsymbol{A})\,\mathrm{d}t\\&=\int_0^{\infty}\frac{\mathrm{d}}{\mathrm{d}t}(\mathrm{e}^{\boldsymbol{A}^{\mathrm{T}}t}\boldsymbol{Q}\mathrm{e}^{\boldsymbol{A}t})\,\mathrm{d}t=\mathrm{e}^{\boldsymbol{A}^{\mathrm{T}}t}\boldsymbol{Q}\mathrm{e}^{\boldsymbol{A}t}\Big|_0^{\infty}=-\boldsymbol{Q}\end{aligned}$$

It means that the Lyapunov equation (3.30) has the solution matrix $\boldsymbol{P}$.

Based on the symmetric positive definite matrix $\boldsymbol{Q}$, the scalar function constructed as

$$F(\boldsymbol{X})=\boldsymbol{X}^{\mathrm{T}}\boldsymbol{Q}\boldsymbol{X}$$

is positive definite and

$$\begin{aligned}\int_0^{\infty} F(\boldsymbol{X})\mathrm{d}t &= \int_0^{\infty} \boldsymbol{X}^{\mathrm{T}}\boldsymbol{Q}\boldsymbol{X}\mathrm{d}t = \int_0^{\infty} -\boldsymbol{X}^{\mathrm{T}}(\boldsymbol{A}^{\mathrm{T}}\boldsymbol{P}+\boldsymbol{P}\boldsymbol{A})\boldsymbol{X}\mathrm{d}t \\ &= \int_0^{\infty}(-\boldsymbol{X}^{\mathrm{T}}\boldsymbol{A}^{\mathrm{T}}\boldsymbol{P}\boldsymbol{X}-\boldsymbol{X}^{\mathrm{T}}\boldsymbol{P}\boldsymbol{A}\boldsymbol{X})\mathrm{d}t = \int_0^{\infty}(-\dot{\boldsymbol{X}}^{\mathrm{T}}\boldsymbol{P}\boldsymbol{X}-\boldsymbol{X}^{\mathrm{T}}\boldsymbol{P}\dot{\boldsymbol{X}})\mathrm{d}t \\ &= \int_0^{\infty}\frac{\mathrm{d}}{\mathrm{d}t}(-\boldsymbol{X}^{\mathrm{T}}\boldsymbol{P}\boldsymbol{X})\mathrm{d}t = -\boldsymbol{X}^{\mathrm{T}}\boldsymbol{P}\boldsymbol{X}\Big|_0^{\infty} = \boldsymbol{X}_0^{\mathrm{T}}\boldsymbol{P}\boldsymbol{X}_0 > 0\end{aligned}$$

Obviously, $\boldsymbol{P}$ is positive definite.

It is supposed that there is another solution $\overline{\boldsymbol{P}}$ to (3. 30), then

$$\begin{aligned}\boldsymbol{P} &= \int_0^{\infty} \mathrm{e}^{\boldsymbol{A}^{\mathrm{T}}t}\boldsymbol{Q}\mathrm{e}^{\boldsymbol{A}t}\mathrm{d}t = -\int_0^{\infty}\mathrm{e}^{\boldsymbol{A}^{\mathrm{T}}t}(\boldsymbol{A}^{\mathrm{T}}\overline{\boldsymbol{P}}+\overline{\boldsymbol{P}}\boldsymbol{A})\mathrm{e}^{\boldsymbol{A}t}\mathrm{d}t \\ &= -\int_0^{\infty}(\mathrm{e}^{\boldsymbol{A}^{\mathrm{T}}t}\boldsymbol{A}^{\mathrm{T}}\overline{\boldsymbol{P}}\mathrm{e}^{\boldsymbol{A}t}+\mathrm{e}^{\boldsymbol{A}^{\mathrm{T}}t}\overline{\boldsymbol{P}}\boldsymbol{A}\mathrm{e}^{\boldsymbol{A}t})\mathrm{d}t = -\int_0^{\infty}(\boldsymbol{A}^{\mathrm{T}}\mathrm{e}^{\boldsymbol{A}^{\mathrm{T}}t}\overline{\boldsymbol{P}}\mathrm{e}^{\boldsymbol{A}t}+\mathrm{e}^{\boldsymbol{A}^{\mathrm{T}}t}\overline{\boldsymbol{P}}\boldsymbol{A}\mathrm{e}^{\boldsymbol{A}t})\mathrm{d}t \\ &= -\int_0^{\infty}\frac{\mathrm{d}}{\mathrm{d}t}(\mathrm{e}^{\boldsymbol{A}^{\mathrm{T}}t}\overline{\boldsymbol{P}}\mathrm{e}^{\boldsymbol{A}t})\mathrm{d}t = -\mathrm{e}^{\boldsymbol{A}^{\mathrm{T}}t}\overline{\boldsymbol{P}}\mathrm{e}^{\boldsymbol{A}t}\Big|_0^{\infty} = \overline{\boldsymbol{P}}\end{aligned}$$

Hence, the solution of (3. 30) is unique.

Furthermore

$$\boldsymbol{P}^{\mathrm{T}} = \int_0^{\infty}[\mathrm{e}^{\boldsymbol{A}^{\mathrm{T}}t}\boldsymbol{Q}\mathrm{e}^{\boldsymbol{A}t}]^{\mathrm{T}}\mathrm{d}t = \int_0^{\infty}\mathrm{e}^{\boldsymbol{A}^{\mathrm{T}}t}\boldsymbol{Q}^{\mathrm{T}}\mathrm{e}^{\boldsymbol{A}t}\mathrm{d}t = \int_0^{\infty}\mathrm{e}^{\boldsymbol{A}^{\mathrm{T}}t}\boldsymbol{Q}\mathrm{e}^{\boldsymbol{A}t}\mathrm{d}t = \boldsymbol{P}$$

So, matrix $\boldsymbol{P}$ is symmetric.

(2) Sufficiency. Based on the symmetric positive definite matrix $\boldsymbol{P}$, a quadratic form function can be constructed as

$$V(\boldsymbol{X})=\boldsymbol{X}^{\mathrm{T}}\boldsymbol{P}\boldsymbol{X}$$

It is clear that $V(\boldsymbol{X})$ is positive definite for $\|\boldsymbol{X}\|\neq 0$, and $V(\boldsymbol{0})=0$. The derivative

$$\begin{aligned}\dot{V}(\boldsymbol{X}) &= \dot{\boldsymbol{X}}^{\mathrm{T}}\boldsymbol{P}\boldsymbol{X}+\boldsymbol{X}^{\mathrm{T}}\boldsymbol{P}\dot{\boldsymbol{X}} = \boldsymbol{X}^{\mathrm{T}}\boldsymbol{A}^{\mathrm{T}}\boldsymbol{P}\boldsymbol{X}+\boldsymbol{X}^{\mathrm{T}}\boldsymbol{P}\boldsymbol{A}\boldsymbol{X} \\ &= \boldsymbol{X}^{\mathrm{T}}(\boldsymbol{A}^{\mathrm{T}}\boldsymbol{P}+\boldsymbol{P}\boldsymbol{A})\boldsymbol{X} = -\boldsymbol{X}^{\mathrm{T}}\boldsymbol{Q}\boldsymbol{X}\end{aligned}$$

Because the matrix $\boldsymbol{Q}$ is any symmetric positive definite, $\dot{V}(\boldsymbol{X})$ is negative definite for $\|\boldsymbol{X}\|\neq 0$ and $\dot{V}(\boldsymbol{0})=0$. So, the equilibrium point $\boldsymbol{X}_e=\boldsymbol{0}$ is asymptotically stable i. s. L and the quadratic form function $V(\boldsymbol{X})=\boldsymbol{X}^{\mathrm{T}}\boldsymbol{P}\boldsymbol{X}$ is a Lyapunov function.

Example 3. 12 Consider the LTI system

$$\dot{\boldsymbol{X}}=\begin{bmatrix}-1 & 1\\ 2 & 3\end{bmatrix}\boldsymbol{X}$$

Determine the stability of the system at its equilibrium point.

Solution Obviously, the system matrix

$$\boldsymbol{A}=\begin{bmatrix}-1 & 1\\ 2 & -3\end{bmatrix}$$

is nonsingular. So, the origin $\boldsymbol{X}_e=\boldsymbol{0}$ is the only equilibrium point of the system.

Take $\boldsymbol{Q}=\boldsymbol{I}$ and denote $\boldsymbol{P}$ by

$$P=\begin{bmatrix} p_{11} & p_{12} \\ p_{12} & p_{22} \end{bmatrix}$$

From the Lyapunov equation $A^{T}P+PA=-Q$, we have

$$\begin{bmatrix} -1 & 2 \\ 1 & 3 \end{bmatrix}\begin{bmatrix} p_{11} & p_{12} \\ p_{12} & p_{22} \end{bmatrix}+\begin{bmatrix} p_{11} & p_{12} \\ p_{12} & p_{22} \end{bmatrix}\begin{bmatrix} -1 & 1 \\ 2 & 3 \end{bmatrix}=\begin{bmatrix} -1 & 0 \\ 0 & -1 \end{bmatrix}$$

The solution matrix is

$$P=\begin{bmatrix} 7/4 & 5/8 \\ 5/8 & 3/8 \end{bmatrix}$$

and its leading principal minor determinants are

$$\Delta_1=|7/4|>0,\ \Delta_2=\begin{vmatrix} 7/4 & 5/8 \\ 5/8 & 3/8 \end{vmatrix}=\frac{17}{64}>0$$

According to the Sylvester criterion, P is positive definite. Then the equilibrium point $X_e=0$ is asymptotically stable.

Furthermore, $X_e=0$ is the unique equilibrium point of the LTI system, so $X_e=0$ is globally asymptotically stable i. s. L.

The Lyapunov function can be constructed as

$$V(X)=X^{T}PX=\frac{7}{4}x_1^2+\frac{3}{8}x_2^2+\frac{5}{4}x_1x_2$$

3.5 Construction of Lyapunov Function to the Nonlinear System

The Lyapunov direct method is an efficient tool to attack practical nonlinear system. However, an explicit proper Lyapunov function must be constructed before we apply the Lyapunov direct method to a nonlinear system. There is not a general way to construct the Lyapunov function for a practical nonlinear system. The **variable gradient method** is a helpful method to construct the Lyapunov function for nonlinear system.

Consider a nonlinear system described by

$$\dot{X}=f(X,t) \tag{3.34}$$

Where $X=[x_1 \quad x_2 \quad \cdots \quad x_n]^{T}$ is the n dimension state vector. Suppose $V(X)$ is a nonlinear function of X and its gradient is represented as

$$\nabla V(X)=\begin{bmatrix} \frac{\partial V}{\partial x_1} & \frac{\partial V}{\partial x_2} & \cdots & \frac{\partial V}{\partial x_n} \end{bmatrix}^{T}=[\nabla V_1 \quad \nabla V_2 \quad \cdots \quad \nabla V_n]^{T} \tag{3.35}$$

Then the derivative of $V(X)$ can be calculated as

$$\dot{V}(X)=\begin{bmatrix} \frac{\partial V}{\partial x_1} & \frac{\partial V}{\partial x_2} & \cdots & \frac{\partial V}{\partial x_n} \end{bmatrix}\begin{bmatrix} \dot{x}_1 \\ \dot{x}_2 \\ \vdots \\ \dot{x}_n \end{bmatrix}=[\nabla V(X)]^{T}\cdot\dot{X} \tag{3.36}$$

The variable gradient method involves assuming a certain form for the gradient of $V(X)$. Usually, $\nabla V(X)$ is assumed as

$$\nabla V(\boldsymbol{X})=\begin{bmatrix} a_{11}x_1+a_{12}x_2+\cdots+a_{1n}x_n \\ a_{21}x_1+a_{22}x_2+\cdots+a_{2n}x_n \\ \vdots \\ a_{n1}x_1+a_{n2}x_2+\cdots+a_{nn}x_n \end{bmatrix} \tag{3.37}$$

Where the coefficients a_{ij} is real and should be chosen to satisfy the following properties.

(1) By considering the form of $\dot{V}(\boldsymbol{X})$ just as (3.36), a_{ij} should be chosen to make $\dot{V}(\boldsymbol{X})<0$ or $\dot{V}(\boldsymbol{X})\leqslant 0$ in the neighborhood of the origin as large as possible.

(2) Furthermore, based on (3.36), the nonlinear function $V(\boldsymbol{X})$ can be calculated as

$$V(\boldsymbol{X})=\int_0^t \dot{V}(\boldsymbol{X})\mathrm{d}t=\int_0^t [\nabla V(\boldsymbol{X})]^{\mathrm{T}}\cdot\dot{\boldsymbol{X}}\mathrm{d}t=\int_{\boldsymbol{0}}^{\boldsymbol{X}}[\nabla V(\boldsymbol{X})]^{\mathrm{T}}\mathrm{d}\boldsymbol{X} \tag{3.38}$$

In order to simplify the integral operation, a_{ij} should be chosen to satisfy the **curl conditions**, such as

$$\frac{\partial\nabla V_i}{\partial x_j}=\frac{\partial\nabla V_j}{\partial x_i},\qquad i,j=1,2,\cdots,n \tag{3.39}$$

Then, $V(\boldsymbol{X})$ can be calculated simply as

$$V(\boldsymbol{X})=\int_0^{x_1(x_2=x_3=\cdots=x_n=0)}\nabla V_1\mathrm{d}x_1+\int_0^{x_2(x_1=x_1,x_3=x_4=\cdots=x_n=0)}\nabla V_2\mathrm{d}x_2$$
$$+\cdots+\int_0^{x_n(x_1=x_1,x_2=x_2,\cdots,x_{n-1}=x_{n-1})}\nabla V_n\mathrm{d}x_n \tag{3.40}$$

If $V(\boldsymbol{X})>0$, it can be chosen as a Lyapunov function for the nonlinear system, and the system is stable or asymptotically stable i. s. L in the neighborhood of the origin.

Example 3.13 Consider a nonlinear system

$$\begin{cases}\dot{x}_1=-x_1\\ \dot{x}_2=x_1x_2^2-x_2\end{cases}$$

Determine the stability of the system at its equilibrium point.

Solution The system can be described by

$$\dot{\boldsymbol{X}}=f(\boldsymbol{X})=\begin{bmatrix}f_1(\boldsymbol{X})\\ f_2(\boldsymbol{X})\end{bmatrix}=\begin{bmatrix}-x_1\\ x_1x_2^2-x_2\end{bmatrix}$$

By letting $\dot{\boldsymbol{X}}=\boldsymbol{0}$, the origin $\boldsymbol{X}_e=\boldsymbol{0}$ is the only equilibrium point of the system obviously.

Assume that the gradient of a nonlinear function $V(\boldsymbol{X})$ is

$$\nabla V(\boldsymbol{X})=\begin{bmatrix}\nabla V_1\\ \nabla V_2\end{bmatrix}=\begin{bmatrix}a_{11}x_1+a_{12}x_2\\ a_{21}x_1+a_{22}x_2\end{bmatrix}$$

If the coefficients a_{ij} are chosen to be

$$a_{11}=a_{22}=1,\ a_{12}=a_{22}=0$$

$\dot{V}(\boldsymbol{X})$ can be calculated as

$$\dot{V}(\boldsymbol{X})=[\nabla V(\boldsymbol{X})]^{\mathrm{T}}\begin{bmatrix}\dot{x}_1\\ \dot{x}_2\end{bmatrix}=x_1\dot{x}_1+x_2\dot{x}_2=-x_1^2-x_2^2(1-x_1x_2)$$

Obviously, $\dot{V}(\boldsymbol{X})<0$ in the region $(1-x_1x_2)>0$.

Moreover, the curl condition

$$\frac{\partial \nabla V_1}{\partial x_2}=\frac{\partial \nabla V_2}{\partial x_1}=0$$

is satisfied. Then, $V(\boldsymbol{X})$ can be constructed as

$$V(\boldsymbol{X})=\int_0^{x_1(x_2=0)} \nabla V_1 \mathrm{d}x_1+\int_0^{x_2(x_1=x_1)} \nabla V_2 \mathrm{d}x_2=\int_0^{x_1} x_1 \mathrm{d}x_1+\int_0^{x_2} x_2 \mathrm{d}x_2$$

$$=\frac{1}{2}(x_1^2+x_2^2)$$

Because $V(\boldsymbol{X})>0$ in the region $(1-x_1x_2)>0$, so it can be chosen as a Lyapunov function for the nonlinear system, and the system is asymptotically stable i. s. L in the region $(1-x_1x_2)>0$.

Note that the Lyapunov function constructed by the variable gradient method may have many forms.

Problems

3. 1 Determine the definiteness of the following quadratic form functions.

(1) $V(\boldsymbol{X})=-x_1^2-4x_2^2+x_3^2-2x_1x_2-6x_2x_3-2x_1x_3$

(2) $V(\boldsymbol{X})=10x_1^2-3x_2^2-11x_3^2+2x_1x_2-2x_2x_3-4x_1x_3$

3. 2 Consider a nonlinear time invariant system described by

$$\dot{x}_1=x_1+x_1^2-x_1x_2$$

$$\dot{x}_2=x_1-x_2$$

(1) Determine the equilibrium point of the system.

(2) Determine the stability of the system in the neighborhood of its equilibrium point by the Lyapunov 1st Method.

3. 3 Determine the stability of the following LTI systems by the Lyapunov 2nd Method.

(1) $\dot{\boldsymbol{X}}=\begin{bmatrix} 0 & 1 \\ -1 & -1 \end{bmatrix}\boldsymbol{X}+\begin{bmatrix} 1 \\ 1 \end{bmatrix}u$

(2) $\dot{\boldsymbol{X}}=\begin{bmatrix} 1 & 0 & -1 \\ 0 & -2 & 0 \\ -1 & 0 & 2 \end{bmatrix}\boldsymbol{X}+\begin{bmatrix} 0 \\ 0 \\ 1 \end{bmatrix}u$

3. 4 Consider the LTI system described by

$$\dot{\boldsymbol{X}}=\begin{bmatrix} 0 & 1 & 0 \\ 0 & -2 & 1 \\ -K & 0 & -1 \end{bmatrix}\boldsymbol{X}+\begin{bmatrix} 0 \\ 0 \\ K \end{bmatrix}u, \quad K>0$$

Determine the range of K, in which the system is asymptotically stable.

3. 5 Determine the stability of the following nonlinear time invariant system in the neighborhood of its equilibrium point by the variable gradient method.

$$\dot{x}_1=x_2$$

$$\dot{x}_2=x_1+x_1x_2$$

Jorgen Pedersen Gramian (1850—1916) was a Danish actuary and mathematician.

Important papers of his include On series expansions determined by the methods of least squares, and Investigations of the number of primes less than a given number. The mathematical method that bears his name, the *Gram-Schmidt process*, was first published in the former paper, in 1883. The *Gramian matrix* is also named after him.

For number theorists his main fame is the series for the Riemann zeta function (the leading function in Riemann's exact prime-counting function). Instead of using a series of logarithmic integrals, Gram's function uses logarithm powers and the zeta function of positive integers. It has recently been supplanted by a formula of Ramanujan that uses the Bernoulli numbers directly instead of the zeta function.

Gram was the first mathematician to provide a systematic theory of the development of skew frequency curves, showing that the normal symmetric Gaussian error curve was but one special case of a more general class of frequency curves.

Chapter4 Controllability and Observability

The concepts of controllability and observability introduced first by Rudolf Emil Kalman play an important role in both theoretical and practical aspects of modern control theory. The conditions on controllability and observability essentially govern the existence of a solution to an optimal control problem. We shall show that the condition on controllability of a system is closely related to the existence of a solution of state feedback for assigning the eigenvalues of the system arbitrarily. The concept of observability relates to the condition of observing the state variables from the output variables, which are generally measurable. Only the LTI system will be considered in this chapter.

4.1 Controllability of The LTI System

4.1.1 Controllability

Definition 4.1 Consider the LTI system described by

$$\begin{cases}\dot{\boldsymbol{X}}(t)=\boldsymbol{AX}(t)+\boldsymbol{Bu}(t)\\ \boldsymbol{y}(t)=\boldsymbol{CX}(t)+\boldsymbol{Du}(t)\end{cases} \tag{4.1}$$

The non-zero initial state $\boldsymbol{X}(t_0)$ is said to be **controllable**, if there exists a piecewise continuous input $\boldsymbol{u}(t)$ that drives the state to any final state $X(t_1)=X_1$ at a final time $t_1>t_0$. For convenience, the initial time is usually assumed as $t_0=0$.

If any state is controllable, the system is said to be **completely controllable** or, in simply, controllable.

Example 4.1 Consider a bridge circuit shown in Figure 4.1.

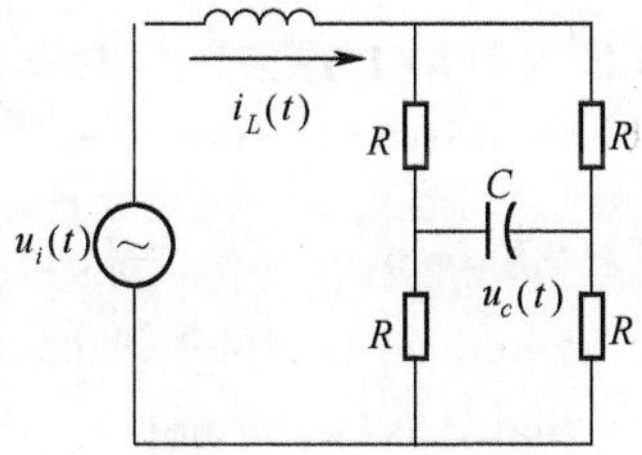

Figure 4.1 Bridge Circuit

The state variables of the system are chosen as the voltage across the capacitor and the current through the inductor, i.e. $x_1(t)=i_L(t)$ and $x_2(t)=u_c(t)$. Furthermore, the input of the system is $u(t)=u_i(t)$ and output is chosen as $y(t)=u_c(t)$.

If the initial value $x_2(t_0)=0$, it can be seen from the bridge circuit that $x_2(t)\equiv 0$ for all $t>t_0$ no matter what input is applied, i. e. the input has no effect on $x_2(t)$, and $x_2(t)$ is said to be uncontrollable. So the system is not completely controllable.

4.1.2 Criteria of Controllability

Theorem 4.1 Gramian Criteria.

The LTI system described by (4.1) is controllable iff the Gramian matrix

$$W_c[0,t_1]=\int_0^{t_1} e^{At}\boldsymbol{B}\boldsymbol{B}^{\mathrm{T}}e^{A^{\mathrm{T}}t}\,dt \tag{4.2}$$

is non-singular.

Proof. (1)Sufficiency. Suppose $\boldsymbol{W}_c[0,t_1]$ is non-singular. The input can be constructed as

$$\boldsymbol{u}(t)=-\boldsymbol{B}^{\mathrm{T}}e^{A^{\mathrm{T}}(t_1-t)}\boldsymbol{W}_c^{-1}[0,t_1](e^{At_1}X_0-X_1),\qquad t\in[0,t_1] \tag{4.3}$$

It is derived that

$$\begin{aligned}X(t_1)&=e^{At_1}\boldsymbol{X}_0+\int_0^{t_1} e^{A(t_1-t)}\boldsymbol{Bu}(t)\,dt\\&=e^{At_1}\boldsymbol{X}_0-\int_0^{t_1} e^{A(t_1-t)}\boldsymbol{B}\boldsymbol{B}^{\mathrm{T}}e^{A^{\mathrm{T}}(t_1-t)}\boldsymbol{W}_c^{-1}[0,t_1](e^{At_1}X_0-X_1)\,dt\\&=e^{At_1}\boldsymbol{X}_0-\int_0^{t_1} e^{A(t_1-t)}\boldsymbol{B}\boldsymbol{B}^{\mathrm{T}}e^{A^{\mathrm{T}}(t_1-t)}\,dt\cdot\boldsymbol{W}_c^{-1}[0,t_1](e^{At_1}X_0-X_1)\\&=e^{At_1}\boldsymbol{X}_0-\int_0^{t_1} e^{At}\boldsymbol{B}\boldsymbol{B}^{\mathrm{T}}e^{A^{\mathrm{T}}t}\,dt\cdot\boldsymbol{W}_c^{-1}[0,t_1](e^{At_1}X_0-X_1)\\&=e^{At_1}\boldsymbol{X}_0-\boldsymbol{W}_c[0,t_1]\boldsymbol{W}_c^{-1}[0,t_1](e^{At_1}X_0-X_1)=X_1\end{aligned}$$

It means that there exists a piecewise continuous input $\boldsymbol{u}(t)$ that drives the state to any final state $X(t_1)=X_1$ at a final time $t_1>t_0$.

(2)Necessity. If the LTI system is controllable, we suppose $\boldsymbol{W}_c[0,t_1]$ is singular preliminarily. Then there must exist a non-zero state $\overline{\boldsymbol{X}}_0$ such that

$$\overline{\boldsymbol{X}}_0{}^{\mathrm{T}}\boldsymbol{W}_c[0,t_1]\overline{\boldsymbol{X}}_0=0 \tag{4.4}$$

Substituting (4.2) into (4.4) yields

$$\overline{\boldsymbol{X}}_0{}^{\mathrm{T}}\int_0^{t_1} e^{At}\boldsymbol{B}\boldsymbol{B}^{\mathrm{T}}e^{A^{\mathrm{T}}t}\,dt\overline{\boldsymbol{X}}_0=\int_0^{t_1}\overline{\boldsymbol{X}}_0{}^{\mathrm{T}}e^{At}\boldsymbol{B}\boldsymbol{B}^{\mathrm{T}}e^{A^{\mathrm{T}}t}\overline{\boldsymbol{X}}_0\,dt=\int_0^{t_1}\|\boldsymbol{B}^{\mathrm{T}}e^{A^{\mathrm{T}}t}\overline{\boldsymbol{X}}_0\|^2\,dt=\boldsymbol{0}$$

Thus

$$\boldsymbol{B}^{\mathrm{T}}e^{A^{\mathrm{T}}t}\overline{\boldsymbol{X}}_0=\boldsymbol{0},\qquad\forall t\in[0,t_1]$$

On the other hand, if the system is completely controllable, there must exists a piecewise continuous input $\boldsymbol{u}(t)$ that drives the non-zero initial state $\boldsymbol{X}(0)=\overline{\boldsymbol{X}}_0$ to the origin of the state space such that

$$\boldsymbol{0}=e^{At_1}\overline{\boldsymbol{X}}_0+\int_0^{t_1} e^{A(t_1-\tau)}\boldsymbol{Bu}(\tau)\,d\tau \tag{4.5}$$

From Eq (4.5), it is derived that

$$\overline{\boldsymbol{X}}_0 = -\int_0^{t_1} \mathrm{e}^{-\boldsymbol{A}\tau}\boldsymbol{B}\boldsymbol{u}(\tau)\mathrm{d}\tau \tag{4.6}$$

It is obtained that

$$\| \overline{\boldsymbol{X}}_0 \|^2 = -\left[\int_0^{t_1} \mathrm{e}^{-\boldsymbol{A}\tau}\boldsymbol{B}\boldsymbol{u}(\tau)\mathrm{d}\tau\right]^{\mathrm{T}}\overline{\boldsymbol{X}}_0 = -\int_0^{t_1} \boldsymbol{u}^{\mathrm{T}}(\tau)\boldsymbol{B}^{\mathrm{T}}\mathrm{e}^{-\boldsymbol{A}^{\mathrm{T}}\tau}\overline{\boldsymbol{X}}_0\mathrm{d}\tau = 0$$

This means if we suppose $\boldsymbol{W}_c[0,t_1]$ being singular, then $\overline{\boldsymbol{X}}_0=\boldsymbol{0}$. It is conflict with our assumption. So, it is concluded that $\boldsymbol{W}_c[0,t_1]$ is non-singular.

Since the matrix $\boldsymbol{A}$ and $\boldsymbol{B}$ are involved, sometimes we say the pair $[\boldsymbol{A},\boldsymbol{B}]$ is controllable. As the need to calculate the matrix exponential function and the integration, the Gramian criteria is difficult to use in practice. But the criteria is valuable in theoretical studies because the following practical criteria can be derived from it.

Theorem 4.2 Rank Criteria.

The LTI system described by (4.1) is controllable iff **the controllability matrix**

$$\boldsymbol{Q}_c = [\boldsymbol{B} \quad \boldsymbol{A}\boldsymbol{B} \quad \cdots \quad \boldsymbol{A}^{n-1}\boldsymbol{B}] \tag{4.7}$$

has full-row rank, i.e. $\mathrm{rank}\boldsymbol{Q}_c = n$.

Proof. (1) Sufficiency. If $\mathrm{rank}\boldsymbol{Q}_c = n$, we assume the system is not completely controllable preliminarily, then the Gramian matrix $\boldsymbol{W}_c[0,t_1]$ must be singular. It means that there exists a non-zero vector $\boldsymbol{\alpha}$ satisfy

$$\boldsymbol{\alpha}^{\mathrm{T}}\boldsymbol{W}_c[0,t_1]\boldsymbol{\alpha} = \boldsymbol{0} \tag{4.8}$$

It is derived that

$$\boldsymbol{\alpha}^{\mathrm{T}}\int_0^{t_1} \mathrm{e}^{\boldsymbol{A}t}\boldsymbol{B}\boldsymbol{B}^{\mathrm{T}}\mathrm{e}^{\boldsymbol{A}^{\mathrm{T}}t}\mathrm{d}t\boldsymbol{\alpha} = \int_0^{t_1} \boldsymbol{\alpha}^{\mathrm{T}}\mathrm{e}^{\boldsymbol{A}t}\boldsymbol{B}\boldsymbol{B}^{\mathrm{T}}\mathrm{e}^{\boldsymbol{A}^{\mathrm{T}}t}\boldsymbol{\alpha}\,\mathrm{d}t = \int_0^{t_1} \| \boldsymbol{\alpha}^{\mathrm{T}}\mathrm{e}^{\boldsymbol{A}t}\boldsymbol{B} \|^2\mathrm{d}t = 0$$

Thus

$$\boldsymbol{\alpha}^{\mathrm{T}}\mathrm{e}^{\boldsymbol{A}t}\boldsymbol{B} = \boldsymbol{0}, \qquad \forall t \in [0,t_1] \tag{4.9}$$

The derivatives can be obtained from (4.9) when $t=0$

$$\boldsymbol{\alpha}^{\mathrm{T}}\mathrm{e}^{\boldsymbol{A}t}\boldsymbol{B}\,|_{t=0} = \boldsymbol{\alpha}^{\mathrm{T}}\boldsymbol{B} = \boldsymbol{0}$$

$$\left.\frac{\mathrm{d}}{\mathrm{d}t}(\boldsymbol{\alpha}^{\mathrm{T}}\mathrm{e}^{\boldsymbol{A}t}\boldsymbol{B})\right|_{t=0} = \boldsymbol{\alpha}^{\mathrm{T}}\boldsymbol{A}\mathrm{e}^{\boldsymbol{A}t}\boldsymbol{B}\,|_{t=0} = \boldsymbol{\alpha}^{\mathrm{T}}\boldsymbol{A}\boldsymbol{B} = \boldsymbol{0}$$

$$\left.\frac{\mathrm{d}^2}{\mathrm{d}t^2}(\boldsymbol{\alpha}^{\mathrm{T}}\mathrm{e}^{\boldsymbol{A}t}\boldsymbol{B})\right|_{t=0} = \boldsymbol{\alpha}^{\mathrm{T}}\boldsymbol{A}^2\mathrm{e}^{\boldsymbol{A}t}\boldsymbol{B}\,|_{t=0} = \boldsymbol{\alpha}^{\mathrm{T}}\boldsymbol{A}^2\boldsymbol{B} = \boldsymbol{0}$$

$$\vdots$$

$$\left.\frac{\mathrm{d}^{n-1}}{\mathrm{d}t^{n-1}}(\boldsymbol{\alpha}^{\mathrm{T}}\mathrm{e}^{\boldsymbol{A}t}\boldsymbol{B})\right|_{t=0} = \boldsymbol{\alpha}^{\mathrm{T}}\boldsymbol{A}^{n-1}\mathrm{e}^{\boldsymbol{A}t}\boldsymbol{B}\,|_{t=0} = \boldsymbol{\alpha}^{\mathrm{T}}\boldsymbol{A}^{n-1}\boldsymbol{B} = \boldsymbol{0}$$

Therefore

$$\boldsymbol{\alpha}^{\mathrm{T}}[\boldsymbol{B} \quad \boldsymbol{A}\boldsymbol{B} \quad \cdots \quad \boldsymbol{A}^{n-1}\boldsymbol{B}] = \boldsymbol{\alpha}^{\mathrm{T}}\boldsymbol{Q}_c = \boldsymbol{0} \tag{4.10}$$

Because $\boldsymbol{\alpha}$ is non-zero vector, we have $\mathrm{rank}\boldsymbol{Q}_c < n$. It is conflict with our assumption. So, it is concluded that the system is completely controllable.

(2) Necessity. If the system is controllable, we assume $\mathrm{rank}\boldsymbol{Q}_c < n$ preliminarily, then there exists a non-zero vector $\boldsymbol{\alpha}$ satisfy (4.10). It is derived that

$$\boldsymbol{\alpha}^{\mathrm{T}}\boldsymbol{A}^i\boldsymbol{B} = \boldsymbol{0}, \qquad i = 0,1,2,\cdots,n-1 \tag{4.11}$$

According to Cayley-Hamilton Theorem, for any $k \geqslant n$, it can be proved easily that $\boldsymbol{A}^k$ can be linearly expressed with $\boldsymbol{A}^i, i=0, \cdots, n-1$. So, Eq (4.11) can be extended such as

$$\boldsymbol{\alpha}^{\mathrm{T}}\boldsymbol{A}^i\boldsymbol{B}=\boldsymbol{0}, \qquad i=0,1,2,\cdots \tag{4.12}$$

and then

$$\boldsymbol{\alpha}^{\mathrm{T}}\mathrm{e}^{\boldsymbol{A}t}\boldsymbol{B}=\boldsymbol{\alpha}^{\mathrm{T}}[\boldsymbol{I}+\boldsymbol{A}t+\frac{1}{2!}\boldsymbol{A}^2t^2+\cdots]\boldsymbol{B}=\boldsymbol{0} \tag{4.13}$$

It can be derived that

$$\boldsymbol{\alpha}^{\mathrm{T}}\boldsymbol{W}_c[0,t_1]\boldsymbol{\alpha}=\int_0^{t_1}\boldsymbol{\alpha}^{\mathrm{T}}\mathrm{e}^{\boldsymbol{A}t}\boldsymbol{B}\boldsymbol{B}^{\mathrm{T}}\mathrm{e}^{\boldsymbol{A}^{\mathrm{T}}t}\boldsymbol{\alpha}\,\mathrm{d}t=\int_0^{t_1}\|\boldsymbol{\alpha}^{\mathrm{T}}\mathrm{e}^{\boldsymbol{A}t}\boldsymbol{B}\|^2\,\mathrm{d}t=0$$

This means if we suppose rank$\boldsymbol{Q}_c<n$, then the Gramian matrix $\boldsymbol{W}_c[0,t_1]$ must be singular and the system is not controllable. It is conflict with our assumption. It is concluded that rank$\boldsymbol{Q}_c=n$.

Note that the rank of the controllability matrix is equal to the number of the controllable state variable if the system is not completely controllable.

Example 4.2 Consider the LTI system described by

$$\dot{\boldsymbol{X}}=\begin{bmatrix}-1 & -4 & -2\\ 0 & 6 & -1\\ 1 & 7 & -1\end{bmatrix}\boldsymbol{X}+\begin{bmatrix}2 & 0\\ 0 & 1\\ 1 & 1\end{bmatrix}\boldsymbol{u}$$

Determine the controllability of the system.

Solution The controllability matrix of the system is

$$\boldsymbol{Q}_c=[\boldsymbol{B}\quad \boldsymbol{AB}\quad \boldsymbol{A}^2\boldsymbol{B}]=\begin{bmatrix}2 & 0 & -4 & -6 & 6 & -26\\ 0 & 1 & -1 & 5 & -7 & 24\\ 1 & 1 & 1 & 6 & -12 & 23\end{bmatrix}$$

and rank$\boldsymbol{Q}_c=3$.

So, the system is completely controllable.

Example 4.3 There is a nonsingular transformation.

$$\boldsymbol{X}(t)=\boldsymbol{P}\,\overline{\boldsymbol{X}}(t)$$

which transforms the description of the LTI system from (4.1) into the form such as

$$\begin{aligned}\dot{\overline{\boldsymbol{X}}}&=\overline{\boldsymbol{A}}\,\overline{\boldsymbol{X}}+\overline{\boldsymbol{B}}\boldsymbol{u}\\ \boldsymbol{y}&=\overline{\boldsymbol{C}}\,\overline{\boldsymbol{X}}+\overline{\boldsymbol{D}}\boldsymbol{u}\end{aligned} \tag{4.14}$$

Where $\overline{\boldsymbol{A}}=\boldsymbol{P}^{-1}\boldsymbol{AP}$; $\overline{\boldsymbol{B}}=\boldsymbol{P}^{-1}\boldsymbol{B}$; $\overline{\boldsymbol{C}}=\boldsymbol{CP}$; $\overline{\boldsymbol{D}}=\boldsymbol{D}$.

Prove that **a change of state by the nonsingular transformation does not change the controllability.**

Proof. The controllability matrix of the system (4.1) is

$$\boldsymbol{Q}_c=[\boldsymbol{B}\quad \boldsymbol{AB}\quad \cdots\quad \boldsymbol{A}^{n-1}\boldsymbol{B}]$$

Furthermore, the controllability matrix of the system (4.14) can be constructed as

$$\begin{aligned}\overline{\boldsymbol{Q}}_c&=[\overline{\boldsymbol{B}}\quad \overline{\boldsymbol{A}}\,\overline{\boldsymbol{B}}\quad \cdots\quad \overline{\boldsymbol{A}}^{n-1}\overline{\boldsymbol{B}}]=[\boldsymbol{P}^{-1}\boldsymbol{B}\quad \boldsymbol{P}^{-1}\boldsymbol{APP}^{-1}\boldsymbol{B}\quad \cdots\quad (\boldsymbol{P}^{-1}\boldsymbol{AP})^{n-1}\boldsymbol{P}^{-1}\boldsymbol{B}]\\ &=[\boldsymbol{P}^{-1}\boldsymbol{B}\quad \boldsymbol{P}^{-1}\boldsymbol{AB}\quad \cdots\quad \boldsymbol{P}^{-1}\boldsymbol{A}^{n-1}\boldsymbol{B}]=\boldsymbol{P}^{-1}[\boldsymbol{B}\quad \boldsymbol{AB}\quad \cdots\quad \boldsymbol{A}^{n-1}\boldsymbol{B}]\end{aligned}$$

Obviously $\text{rank}\boldsymbol{Q}_c = \text{rank}\overline{\boldsymbol{Q}}_c$

It means that the controllability is preserved by the nonsingular transformation.

Theorem 4.3 PBH (Popov-Belevich-Hautus) Rank Criteria.

The LTI system described by (4.1) is controllable iff

$$\text{rank}[\lambda_i \boldsymbol{I} - \boldsymbol{A}, \boldsymbol{B}] = n, \qquad i = 1, 2, \cdots, n \tag{4.15}$$

for all eigenvalues λ_i of $\boldsymbol{A}$.

Proof. (1) Sufficiency. If $\text{rank}[\lambda_i \boldsymbol{I} - \boldsymbol{A}, \boldsymbol{B}] = n$, $i = 1, 2, \cdots, n$, we assume the system is not completely controllable preliminarily, then the controllability matrix $\boldsymbol{Q}_c$ must be singular. It means that there exists a non-zero vector $\boldsymbol{\alpha}^{\mathrm{T}}$ satisfy (4.10) and (4.11).

Because $\text{rank}[\lambda_i \boldsymbol{I} - \boldsymbol{A}, \boldsymbol{B}] = n, \qquad i = 1, 2, \cdots, n$

It is derived that

$$\boldsymbol{\alpha}^{\mathrm{T}}[\lambda_1 \boldsymbol{I} - \boldsymbol{A}, \boldsymbol{B}] = [\boldsymbol{\alpha}^{\mathrm{T}}(\lambda_1 \boldsymbol{I} - \boldsymbol{A}), \boldsymbol{\alpha}^{\mathrm{T}}\boldsymbol{B}] = [\boldsymbol{\alpha}^{\mathrm{T}}(\lambda_1 \boldsymbol{I} - \boldsymbol{A}), \boldsymbol{0}] \neq \boldsymbol{0}$$

Thus

$$\boldsymbol{\alpha}^{\mathrm{T}}(\lambda_1 \boldsymbol{I} - \boldsymbol{A}) \neq \boldsymbol{0} \tag{4.16}$$

and then

$$\begin{aligned}\boldsymbol{\alpha}^{\mathrm{T}}(\lambda_1 \boldsymbol{I} - \boldsymbol{A})[\lambda_2 \boldsymbol{I} - \boldsymbol{A}, \boldsymbol{B}] &= [\boldsymbol{\alpha}^{\mathrm{T}}(\lambda_1 \boldsymbol{I} - \boldsymbol{A})(\lambda_2 \boldsymbol{I} - \boldsymbol{A}), \boldsymbol{\alpha}^{\mathrm{T}}(\lambda_1 \boldsymbol{I} - \boldsymbol{A})\boldsymbol{B}] \\ &= [\boldsymbol{\alpha}^{\mathrm{T}}(\lambda_1 \boldsymbol{I} - \boldsymbol{A})(\lambda_2 \boldsymbol{I} - \boldsymbol{A}), \boldsymbol{0}] \neq \boldsymbol{0}\end{aligned}$$

Therefore

$$\boldsymbol{\alpha}^{\mathrm{T}}(\lambda_1 \boldsymbol{I} - \boldsymbol{A})(\lambda_2 \boldsymbol{I} - \boldsymbol{A}) \neq \boldsymbol{0} \tag{4.17}$$

By analogy, we can obtain (4.18) as

$$\boldsymbol{\alpha}^{\mathrm{T}}(\lambda_1 \boldsymbol{I} - \boldsymbol{A})(\lambda_2 \boldsymbol{I} - \boldsymbol{A}) \cdots (\lambda_n \boldsymbol{I} - \boldsymbol{A}) = \boldsymbol{\alpha}^{\mathrm{T}} \prod_{i=1}^{n} (\lambda_i \boldsymbol{I} - \boldsymbol{A}) \neq \boldsymbol{0} \tag{4.18}$$

Because $\boldsymbol{\alpha}^{\mathrm{T}}$ is a non-zero vector, we have

$$\prod_{i=1}^{n} (\lambda_i \boldsymbol{I} - \boldsymbol{A}) \neq \boldsymbol{0} \tag{4.19}$$

However, according to Cayley-Hamilton Theorem, if

$$|\lambda \boldsymbol{I} - \boldsymbol{A}| = \prod_{i=1}^{n} (\lambda - \lambda_i) = 0 \tag{4.20}$$

then

$$\prod_{i=1}^{n} (\lambda_i \boldsymbol{I} - \boldsymbol{A}) = \boldsymbol{0} \tag{4.21}$$

Obviously, Eq. (4.21) is conflict with Eq. (4.19). So, it is concluded that the system is completely controllable.

(2) Necessity. If the system is controllable, we assume preliminarily that there exists λ_i such that $\text{rank}[\lambda_i \boldsymbol{I} - \boldsymbol{A}, \boldsymbol{B}] < n$. Then the row vectors of $[\lambda_i \boldsymbol{I} - \boldsymbol{A}, \boldsymbol{B}]$ are linearly dependent. Hence, there exists a non-zero vector $\boldsymbol{\alpha}$ satisfy

$$\boldsymbol{\alpha}^{\mathrm{T}}[\lambda_i \boldsymbol{I} - \boldsymbol{A}, \boldsymbol{B}] = \boldsymbol{0} \tag{4.22}$$

It follows from above that

$$\boldsymbol{\alpha}^{\mathrm{T}}\boldsymbol{A} = \lambda_i \boldsymbol{\alpha}^{\mathrm{T}}, \qquad \boldsymbol{\alpha}^{\mathrm{T}}\boldsymbol{B} = \boldsymbol{0}$$

and then

$$\boldsymbol{\alpha}^{T}\boldsymbol{AB}=\lambda_i\boldsymbol{\alpha}^{T}\boldsymbol{B}=\boldsymbol{0},\boldsymbol{\alpha}^{T}\boldsymbol{A}^2\boldsymbol{B}=\lambda_i\boldsymbol{\alpha}^{T}\boldsymbol{AB}=\lambda_i^2\boldsymbol{\alpha}^{T}\boldsymbol{B}=0,\cdots,\boldsymbol{\alpha}^{T}\boldsymbol{A}^{n-1}\boldsymbol{B}=\lambda_i^{n-1}\boldsymbol{\alpha}^{T}\boldsymbol{B}=\boldsymbol{0}$$

So

$$[\boldsymbol{\alpha}^{T}\boldsymbol{B}\quad \boldsymbol{\alpha}^{T}\boldsymbol{AB}\quad \boldsymbol{\alpha}^{T}\boldsymbol{A}^2\boldsymbol{B}\quad \cdots\quad \boldsymbol{\alpha}^{T}\boldsymbol{A}^{n-1}\boldsymbol{B}]=\boldsymbol{\alpha}^{T}[\boldsymbol{B}\quad \boldsymbol{AB}\quad \boldsymbol{A}^2\boldsymbol{B}\quad \cdots\quad \boldsymbol{A}^{n-1}\boldsymbol{B}]=\boldsymbol{0}$$

This means if we suppose rank$[\lambda_i\boldsymbol{I}-\boldsymbol{A},\boldsymbol{B}]<n$, then the controllability matrix $\boldsymbol{Q}_c$ must be singular and the system is not controllable. It is conflict with our assumption. It is concluded that rank$[\lambda_i\boldsymbol{I}-\boldsymbol{A},\boldsymbol{B}]=n$, $i=1,2,\cdots,n$.

Example 4.4 Consider the LTI system given by

$$\dot{\boldsymbol{X}}=\begin{bmatrix}0 & 1 & 0 & 0\\ 0 & 0 & -1 & 0\\ 0 & 0 & 0 & 1\\ 0 & 0 & 5 & 0\end{bmatrix}\boldsymbol{X}+\begin{bmatrix}0 & 1\\ 1 & 0\\ 0 & 1\\ -2 & 0\end{bmatrix}\boldsymbol{u}$$

Determine the controllability of the system.

Solution The eigenvalues of $\boldsymbol{A}$ are

$$\lambda_{1,2}=0,\qquad \lambda_3=\sqrt{5},\qquad \lambda_4=-\sqrt{5}$$

It is calculated that

$$\text{rank}[\lambda_1\boldsymbol{I}-\boldsymbol{A},\boldsymbol{B}]=\text{rank}\begin{bmatrix}0 & -1 & 0 & 0 & 0 & 1\\ 0 & 0 & 1 & 0 & 1 & 0\\ 0 & 0 & 0 & -1 & 0 & 1\\ 0 & 0 & -5 & 0 & -2 & 0\end{bmatrix}=4$$

$$\text{rank}[\lambda_3\boldsymbol{I}-\boldsymbol{A},\boldsymbol{B}]=\text{rank}\begin{bmatrix}\sqrt{5} & -1 & 0 & 0 & 0 & 1\\ 0 & \sqrt{5} & 1 & 0 & 1 & 0\\ 0 & 0 & \sqrt{5} & -1 & 0 & 1\\ 0 & 0 & -5 & \sqrt{5} & -2 & 0\end{bmatrix}=4$$

$$\text{rank}[\lambda_4\boldsymbol{I}-\boldsymbol{A},\boldsymbol{B}]=\text{rank}\begin{bmatrix}-\sqrt{5} & -1 & 0 & 0 & 0 & 1\\ 0 & -\sqrt{5} & 1 & 0 & 1 & 0\\ 0 & 0 & -\sqrt{5} & -1 & 0 & 1\\ 0 & 0 & -5 & -\sqrt{5} & -2 & 0\end{bmatrix}=4$$

Therefore, the system is completely controllable.

Theorem 4.4 Diagonal Canonical Form Criteria.

If the eigenvalues $\lambda_1,\lambda_2,\cdots,\lambda_n$ of the LTI system described by (4.1) are distinct, the diagonal canonical form (4.14) can be obtained by the nonsingular transformation $\boldsymbol{X}(t)=\boldsymbol{P}\overline{\boldsymbol{X}}(t)$, where

$$\overline{\boldsymbol{A}}=\boldsymbol{P}^{-1}\boldsymbol{AP}=\begin{bmatrix}\lambda_1 & & & \\ & \lambda_2 & & \\ & & \ddots & \\ & & & \lambda_n\end{bmatrix},\quad \overline{\boldsymbol{B}}=\boldsymbol{P}^{-1}\boldsymbol{B}=\begin{bmatrix}\overline{\boldsymbol{b}}_1\\ \overline{\boldsymbol{b}}_2\\ \vdots\\ \overline{\boldsymbol{b}}_n\end{bmatrix}$$

The LTI system is controllable iff $\overline{\boldsymbol{B}}$ has no zero row vector, i. e.

$$\overline{\boldsymbol{b}}_i \neq \boldsymbol{0}, i=1,2,\cdots,n \tag{4.23}$$

Proof. For any eigenvalue λ_i of $\overline{\boldsymbol{A}}$

$$\operatorname{rank}[\lambda_i \boldsymbol{I}-\overline{\boldsymbol{A}}, \overline{\boldsymbol{B}}]=\operatorname{rank}\begin{bmatrix} \lambda_i-\lambda_1 & & & \overline{\boldsymbol{b}}_1 \\ & \ddots & & \vdots \\ & & \lambda_i-\lambda_n & \overline{\boldsymbol{b}}_n \end{bmatrix} \quad i=1,2,\cdots,n$$

Obviously, for any eigenvalue λ_i of $\boldsymbol{A}$, $\operatorname{rank}[\lambda_i \boldsymbol{I}-\overline{\boldsymbol{A}}, \overline{\boldsymbol{B}}]=n$ iff $\overline{b}_i \neq 0$.

Example 4.5 Consider the LTI system described by

$$\dot{\boldsymbol{X}}=\begin{bmatrix} -7 & 0 & 0 \\ 0 & -2 & 0 \\ 0 & 0 & 1 \end{bmatrix}\boldsymbol{X}+\begin{bmatrix} 0 & 2 \\ 4 & 0 \\ 0 & 1 \end{bmatrix}\boldsymbol{u}$$

Determine the controllability of the system.

Solution Since the system matrix $\boldsymbol{A}$ is a diagonal matrix and the input matrix $\boldsymbol{B}$ has no zero row vector, the system is completely controllable.

Example 4.6 Consider the LTI system described by.

$$\dot{\boldsymbol{X}}=\begin{bmatrix} -7 & 0 & 0 \\ 0 & -2 & 0 \\ 0 & 0 & 1 \end{bmatrix}\boldsymbol{X}+\begin{bmatrix} 0 & 2 \\ 0 & 0 \\ 0 & 1 \end{bmatrix}\boldsymbol{u}$$

Determine the controllability of the system.

Solution Since the system matrix $\boldsymbol{A}$ is a diagonal matrix but the row vector $\boldsymbol{b}_2$ within the input matrix $\boldsymbol{B}$ is a zero row vector, the system is not completely controllable.

Moreover, because there are not coupling relationships between the state variable x_2 and other variables, and the input do not affect x_2, so the state variable x_2 is an uncontrollable state variable in the system.

Theorem 4.5 Jordan Canonical Form Criteria.

If the eigenvalues $\lambda_1, \lambda_2, \cdots, \lambda_l$ of the LTI system described by (4.1) are not distinct, the Jordan canonical form (4.14) can be obtained by the nonsingular transformation $\boldsymbol{X}(t)=\boldsymbol{P}\overline{\boldsymbol{X}}(t)$, where

$$\overline{\boldsymbol{A}}=\boldsymbol{P}^{-1}\boldsymbol{A}\boldsymbol{P}=\begin{bmatrix} \boldsymbol{J}_1 & & & \boldsymbol{0} \\ & \boldsymbol{J}_2 & & \\ & & \ddots & \\ \boldsymbol{0} & & & \boldsymbol{J}_l \end{bmatrix}, \ \overline{\boldsymbol{B}}=\boldsymbol{P}^{-1}\boldsymbol{B}=\begin{bmatrix} \overline{\boldsymbol{B}}_1 \\ \overline{\boldsymbol{B}}_2 \\ \vdots \\ \overline{\boldsymbol{B}}_l \end{bmatrix}$$

$$\boldsymbol{J}_i=\begin{bmatrix} \boldsymbol{J}_{i1} & & \boldsymbol{0} \\ & \ddots & \\ \boldsymbol{0} & & \boldsymbol{J}_{i\alpha_i} \end{bmatrix}, \quad \overline{\boldsymbol{B}}_i=\begin{bmatrix} \overline{\boldsymbol{B}}_{i1} \\ \vdots \\ \overline{\boldsymbol{B}}_{i\alpha_i} \end{bmatrix}$$

$$\boldsymbol{J}_{ik}=\begin{bmatrix} \lambda_i & 1 & \boldsymbol{0} \\ & \ddots & 1 \\ \boldsymbol{0} & & \lambda_i \end{bmatrix}, \quad \overline{\boldsymbol{B}}_{ik}=\begin{bmatrix} \overline{\boldsymbol{b}}_{ik1} \\ \vdots \\ \overline{\boldsymbol{b}}_{ikr_k} \end{bmatrix}$$

and α_i is the geometric multiplicity of the eigenvalue λ_i, $i=1,2,\cdots,l$. The matrix $\bar{\boldsymbol{A}}$ is a Jordan matrix, the matrix $\boldsymbol{J}_i$ is called a Jordan block within the Jordan matrix according to the eigenvalue λ_i, and the matrix $\boldsymbol{J}_{ik}$, $k=1,2,\cdots,\alpha_i$, is called the k th sub-Jordan block within the Jordan block $\boldsymbol{J}_i$.

The LTI system is controllable iff those row vectors, which are the last row vectors of $\bar{\boldsymbol{B}}_{ik}$ within $\bar{\boldsymbol{B}}_i$ corresponding to the eigenvalue λ_i, are independent, i. e.

$$\text{rank}\begin{bmatrix}\bar{\boldsymbol{b}}_{i1r_1}\\ \bar{\boldsymbol{b}}_{i2r_2}\\ \vdots\\ \bar{\boldsymbol{b}}_{i\alpha_i r_{\alpha i}}\end{bmatrix}=\alpha_i \tag{4.24}$$

Note that α_i, the geometric multiplicity of the eigenvalue λ_i, is equal to the number of sub-Jordan blocks within each Jordan block corresponding to the eigenvalue λ_i.

Proof. In order to simplify the description, we assume preliminarily that

$$\bar{\boldsymbol{A}}=\left[\begin{array}{ccc:cc:cc:c}\lambda_1 & 1 & & & & & & \\ & \lambda_1 & 1 & & & & & \\ & & \lambda_1 & & & & & \\ \hdashline & & & \lambda_1 & 1 & & & \\ & & & & \lambda_1 & & & \\ \hdashline & & & & & \lambda_2 & 1 & \\ & & & & & & \lambda_2 & \\ \hdashline & & & & & & & \lambda_3\end{array}\right],\quad \bar{\boldsymbol{B}}=\begin{bmatrix}\bar{\boldsymbol{b}}_{111}\\ \bar{\boldsymbol{b}}_{112}\\ \bar{\boldsymbol{b}}_{113}\\ \hdashline \bar{\boldsymbol{b}}_{121}\\ \bar{\boldsymbol{b}}_{122}\\ \hdashline \bar{\boldsymbol{b}}_{211}\\ \bar{\boldsymbol{b}}_{212}\\ \hdashline \bar{\boldsymbol{b}}_{311}\end{bmatrix}$$

For any eigenvalue λ_i ($i=1,2,3$) of $\bar{\boldsymbol{A}}$

$$\text{rank}[\lambda_i\boldsymbol{I}-\bar{\boldsymbol{A}},\bar{\boldsymbol{B}}]$$

$$=\text{rank}\left[\begin{array}{ccc:cc:cc:c:c}\lambda_i-\lambda_1 & -1 & & & & & & & \bar{\boldsymbol{b}}_{111}\\ & \lambda_i-\lambda_1 & -1 & & & & & & \bar{\boldsymbol{b}}_{112}\\ & & \lambda_i-\lambda_1 & & & & & & \bar{\boldsymbol{b}}_{113}\\ \hdashline & & & \lambda_i-\lambda_1 & -1 & & & & \bar{\boldsymbol{b}}_{121}\\ & & & & \lambda_i-\lambda_1 & & & & \bar{\boldsymbol{b}}_{122}\\ \hdashline & & & & & \lambda_i-\lambda_2 & -1 & & \bar{\boldsymbol{b}}_{211}\\ & & & & & & \lambda_i-\lambda_2 & & \bar{\boldsymbol{b}}_{212}\\ \hdashline & & & & & & & \lambda_i-\lambda_3 & \bar{\boldsymbol{b}}_{311}\end{array}\right]$$

When $i=1$, the geometric multiplicity of the eigenvalue λ_1 is $\alpha_1=2$. Then rank $[\lambda_1\boldsymbol{I}-\bar{\boldsymbol{A}},\bar{\boldsymbol{B}}]=8$ iff the row vectors $\bar{\boldsymbol{b}}_{113}$ and $\bar{\boldsymbol{b}}_{122}$ are independent, i. e.

$$\text{rank}\begin{bmatrix}\bar{\boldsymbol{b}}_{113}\\ \bar{\boldsymbol{b}}_{122}\end{bmatrix}=2=\alpha_1$$

When $i=2$, the geometric multiplicity of the eigenvalue λ_2 is $\alpha_2=1$. Then rank

$[\lambda_2 \boldsymbol{I}-\overline{\boldsymbol{A}}, \overline{\boldsymbol{B}}]=8$ iff the row vector $\overline{\boldsymbol{b}}_{212} \neq 0$, i. e.

$$\operatorname{rank}[\overline{\boldsymbol{b}}_{212}]=1=\alpha_2$$

When $i=3$, the geometric multiplicity of the eigenvalue λ_3 is $\alpha_3=1$. Then rank $[\lambda_3 \boldsymbol{I}-\overline{\boldsymbol{A}}, \overline{\boldsymbol{B}}]=8$ iff the row vector $\overline{\boldsymbol{b}}_{311} \neq 0$, i. e.

$$\operatorname{rank}[\overline{\boldsymbol{b}}_{311}]=1=\alpha_3$$

From the extension of the analysis above, the Theorem 4. 5 can be proved.

Example 4. 7 Consider the LTI system described by

$$\dot{\boldsymbol{X}}=\begin{bmatrix} -2 & 1 & 0 \\ 0 & -2 & 0 \\ 0 & 0 & -2 \end{bmatrix}\boldsymbol{X}+\begin{bmatrix} 0 & 2 \\ 4 & 0 \\ 0 & 1 \end{bmatrix}\boldsymbol{u}$$

Determine the controllability of the system.

Solution The system matrix $\boldsymbol{A}$ is a Jordan matrix and has the only eigenvalue $\lambda=-2$. The geometric multiplicity of the eigenvalue λ is $\alpha=2$. Denote the input matrix $\boldsymbol{B}$ as

$$\boldsymbol{B}=\begin{bmatrix} \boldsymbol{b}_{111} \\ \boldsymbol{b}_{112} \\ \boldsymbol{b}_{121} \end{bmatrix}=\begin{bmatrix} 0 & 2 \\ 4 & 0 \\ 0 & 1 \end{bmatrix}$$

Since $\operatorname{rank}\begin{bmatrix} \boldsymbol{b}_{112} \\ \boldsymbol{b}_{121} \end{bmatrix}=2=\alpha$, the system is completely controllable.

Example 4. 8 Consider the LTI system described by

$$\dot{\boldsymbol{X}}=\begin{bmatrix} -2 & 0 & 0 \\ 0 & -2 & 0 \\ 0 & 0 & -2 \end{bmatrix}\boldsymbol{X}+\begin{bmatrix} 0 & 2 \\ 4 & 0 \\ 0 & 1 \end{bmatrix}\boldsymbol{u}$$

Determine the controllability of the system.

Solution The system matrix $\boldsymbol{A}$ is a Jordan matrix and has the only eigenvalue $\lambda=-2$. The geometric multiplicity of the eigenvalue λ is $\alpha=3$. Denote the input matrix $\boldsymbol{B}$ as

$$\boldsymbol{B}=\begin{bmatrix} \boldsymbol{b}_{111} \\ \boldsymbol{b}_{121} \\ \boldsymbol{b}_{131} \end{bmatrix}=\begin{bmatrix} 0 & 2 \\ 4 & 0 \\ 0 & 1 \end{bmatrix}$$

Since $\operatorname{rank}\begin{bmatrix} \boldsymbol{b}_{111} \\ \boldsymbol{b}_{121} \\ \boldsymbol{b}_{131} \end{bmatrix}=2 \neq \alpha$, the system is not completely controllable.

Example 4. 9 Consider the LTI system described by

$$\dot{\boldsymbol{X}}=\begin{bmatrix} -2 & 1 & 0 & 0 & 0 & 0 & 0 \\ 0 & -2 & 0 & 0 & 0 & 0 & 0 \\ 0 & 0 & -2 & 0 & 0 & 0 & 0 \\ 0 & 0 & 0 & -2 & 0 & 0 & 0 \\ 0 & 0 & 0 & 0 & 3 & 1 & 0 \\ 0 & 0 & 0 & 0 & 0 & 3 & 0 \\ 0 & 0 & 0 & 0 & 0 & 0 & 3 \end{bmatrix}\boldsymbol{X}+\begin{bmatrix} 0 & 0 & 0 \\ 1 & 0 & 0 \\ 0 & 4 & 0 \\ 0 & 0 & 7 \\ 0 & 0 & 0 \\ 1 & 1 & 0 \\ 0 & 4 & 1 \end{bmatrix}\boldsymbol{u}$$

Determine the controllability of the system.

Solution The system matrix $\boldsymbol{A}$ is a Jordan matrix and has the eigenvalues $\lambda_1=-2$ and $\lambda_2=3$. The geometric multiplicity of the eigenvalue λ_1 is $\alpha_1=3$ and the geometric multiplicity of the eigenvalue λ_2 is $\alpha_2=2$. Denote the input matrix $\boldsymbol{B}$ as

$$\boldsymbol{B}=\begin{bmatrix} \boldsymbol{b}_{111} \\ \boldsymbol{b}_{112} \\ \boldsymbol{b}_{121} \\ \boldsymbol{b}_{131} \\ \boldsymbol{b}_{211} \\ \boldsymbol{b}_{212} \\ \boldsymbol{b}_{221} \end{bmatrix}=\begin{bmatrix} 0 & 0 & 0 \\ 1 & 0 & 0 \\ 0 & 4 & 0 \\ 0 & 0 & 7 \\ 0 & 0 & 0 \\ 1 & 1 & 0 \\ 0 & 4 & 1 \end{bmatrix}$$

Since $\operatorname{rank}\begin{bmatrix} \boldsymbol{b}_{112} \\ \boldsymbol{b}_{121} \\ \boldsymbol{b}_{131} \end{bmatrix}=3=\alpha_1$ and $\operatorname{rank}\begin{bmatrix} \boldsymbol{b}_{212} \\ \boldsymbol{b}_{221} \end{bmatrix}=2=\alpha_2$, the system is completely controllable.

Specially, when each Jordan block of a Jordan matrix has only one sub-Jordan block, the system is controllable iff those row vectors, which are the rows of $\boldsymbol{B}$ corresponding to the last row of each Jordan block, are not the zero row vectors.

Example 4.10 Consider the LTI system described by

$$\dot{\boldsymbol{X}}=\begin{bmatrix} -2 & 1 & 0 \\ 0 & -2 & 0 \\ 0 & 0 & -1 \end{bmatrix}\boldsymbol{X}+\begin{bmatrix} 0 & 2 \\ 4 & 0 \\ 0 & 1 \end{bmatrix}\boldsymbol{u}$$

Determine the controllability of the system.

Solution The system matrix $\boldsymbol{A}$ is a Jordan matrix and has the eigenvalues $\lambda_1=-2$ and $\lambda_2=-1$. The geometric multiplicity of the eigenvalue λ_1 is $\alpha_1=1$ and the geometric multiplicity of the eigenvalue λ_2 is $\alpha_2=1$. Denote the input matrix $\boldsymbol{B}$ as

$$\boldsymbol{B}=\begin{bmatrix} \boldsymbol{b}_{111} \\ \boldsymbol{b}_{112} \\ \boldsymbol{b}_{211} \end{bmatrix}=\begin{bmatrix} 0 & 2 \\ 4 & 0 \\ 0 & 1 \end{bmatrix}$$

Since $\boldsymbol{b}_{112}\neq 0$ and $\boldsymbol{b}_{211}\neq 0$, the system is completely controllable.

4.2 Observability of The LTI System

4.2.1 Observability

Definition 4.2 Consider the LTI system described by

$$\begin{cases} \dot{\boldsymbol{X}}(t)=\boldsymbol{A}\boldsymbol{X}(t)+\boldsymbol{B}\boldsymbol{u}(t) \\ \boldsymbol{y}(t)=\boldsymbol{C}\boldsymbol{X}(t)+\boldsymbol{D}\boldsymbol{u}(t) \end{cases} \tag{4.25}$$

The initial state $\boldsymbol{X}(t_0)$ is said to be **observable**, if there exists a final time $t_1>t_0$

such that $\boldsymbol{X}(t_0)$ can be determined uniquely by the knowledge of the output $\boldsymbol{y}(t)$ over $[t_0, t_1]$. For convenience, the initial time are usually assumed as $t_0=0$.

If any state is observable, the system is said to be **completely observable** or, in simply, observable.

The output response of the system (4.25) excited by initial state $\boldsymbol{X}(t_0)$ and the input $\boldsymbol{u}(t)$ can be derived as

$$\boldsymbol{y}(t)=\boldsymbol{C}e^{\boldsymbol{A}(t-t_0)}X(t_0)+\boldsymbol{C}\int_{t_0}^{t}e^{\boldsymbol{A}(t-\tau)}\boldsymbol{B}\boldsymbol{u}(\tau)d\tau+\boldsymbol{D}\boldsymbol{u}(t) \tag{4.26}$$

Since the concept of observability is defined under the assumption that the matrixes $\boldsymbol{A}$, $\boldsymbol{B}, \boldsymbol{C}, \boldsymbol{D}$ and the input $\boldsymbol{u}(t)$ are known, the last terms on the right side of (4.26) are known and can be separated from the total output in the study of observability. For convenience, the input are usually assumed as $\boldsymbol{u}(t)=\boldsymbol{0}$, i.e. it is sufficient to consider the unforced system for studying a necessary and sufficient condition for completely observable.

Example 4.11 Consider a bridge circuit shown in Figure 4.2.

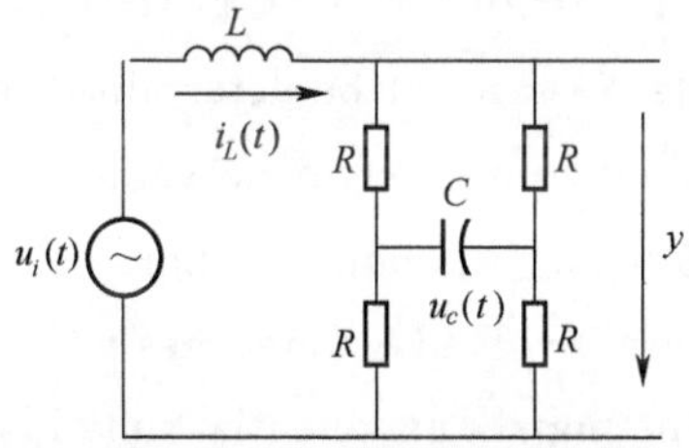

Figure 4.2 Bridge Circuit

The state variables of the system are chosen as the voltage across the capacitor and the current through the inductor, i.e. $x_1(t)=i_L(t)$ and $x_2(t)=u_c(t)$. Furthermore, the input of the system is $u(t)=u_i(t)$ and the output of the two-port network $y(t)$ labeled in Figure 4.2 is chosen as the output of the system.

If the input of the system $u(t)=0$, it can be seen from the bridge circuit that $y(t)\equiv 0$ for all $t>t_0$ no matter what value is the initial value of $x_2(t_0)$, i.e. $x_2(t)$ is said to be unobservable. So the system is not completely observable.

4.2.2 Criteria of Observability

Theorem 4.6 Gramian Criteria.

The LTI system described by (4.25) is completely observable iff the Gramian matrix

$$\boldsymbol{W}_o[0,t_1]=\int_0^{t_1}e^{\boldsymbol{A}^{\mathrm{T}}t}\boldsymbol{C}^{\mathrm{T}}\boldsymbol{C}e^{\boldsymbol{A}t}dt \tag{4.27}$$

is non-singular.

Proof. (1) Sufficiency. Suppose $\boldsymbol{W}_c[0,t_1]$ is non-singular and the initial state is $\boldsymbol{X}(0)=\boldsymbol{X}_0$, then the initial state can be constructed when $\boldsymbol{u}(t)=\boldsymbol{0}$ as

$$\boldsymbol{X}_0 = \boldsymbol{W}_o^{-1}[0,t_1]\boldsymbol{W}_o[0,t_1]\boldsymbol{X}_0 = \boldsymbol{W}_o^{-1}[0,t_1]\int_0^{t_1} \mathrm{e}^{\boldsymbol{A}^{\mathrm{T}}t}\boldsymbol{C}^{\mathrm{T}}\boldsymbol{C}\mathrm{e}^{\boldsymbol{A}t}\,\mathrm{d}t\boldsymbol{X}_0$$
$$= \boldsymbol{W}_o^{-1}[0,t_1]\int_0^{t_1} \mathrm{e}^{\boldsymbol{A}^{\mathrm{T}}t}\boldsymbol{C}^{\mathrm{T}}\boldsymbol{C}\mathrm{e}^{\boldsymbol{A}t}\boldsymbol{X}_0\,\mathrm{d}t = \boldsymbol{W}_o^{-1}[0,t_1]\int_0^{t_1} \mathrm{e}^{\boldsymbol{A}^{\mathrm{T}}t}\boldsymbol{C}^{\mathrm{T}}\boldsymbol{y}(t)\,\mathrm{d}t \tag{4.28}$$

It means that if $\boldsymbol{W}_c[0,t_1]$ is non-singular, the initial state $\boldsymbol{X}_0$ can be determined uniquely by the output $\boldsymbol{y}(t)$ over $[0,t_1]$, i. e. the system is completely observable.

(2) Necessity. If the LTI system is observable, we suppose $\boldsymbol{W}_o[0,t_1]$ is singular preliminarily. Then there must exist a non-zero state $\overline{\boldsymbol{X}}_0$ such that

$$\overline{\boldsymbol{X}}_0^{\mathrm{T}}\boldsymbol{W}_o[0,t_1]\overline{\boldsymbol{X}}_0 = 0 \tag{4.29}$$

Substituting (4. 27) into (4. 29) yields

$$\overline{\boldsymbol{X}}_0^{\mathrm{T}}\int_0^{t_1} \mathrm{e}^{\boldsymbol{A}^{\mathrm{T}}t}\boldsymbol{C}^{\mathrm{T}}\boldsymbol{C}\mathrm{e}^{\boldsymbol{A}t}\,\mathrm{d}t\overline{\boldsymbol{X}}_0 = \int_0^{t_1} \overline{\boldsymbol{X}}_0^{\mathrm{T}}\mathrm{e}^{\boldsymbol{A}^{\mathrm{T}}t}\boldsymbol{C}^{\mathrm{T}}\boldsymbol{C}\mathrm{e}^{\boldsymbol{A}t}\overline{\boldsymbol{X}}_0\,\mathrm{d}t$$
$$= \int_0^{t_1} \|\boldsymbol{C}\mathrm{e}^{\boldsymbol{A}t}\overline{\boldsymbol{X}}_0\|^2\,\mathrm{d}t = \int_0^{t_1} \|\boldsymbol{y}(t)\|^2\,\mathrm{d}t = 0 \tag{4.30}$$

Thus

$$\boldsymbol{y}(t) \equiv \boldsymbol{0}, \qquad \forall t \in [0,t_1] \tag{4.31}$$

This means the initial state $\boldsymbol{X}_0$ can not be determined uniquely by the output $\boldsymbol{y}(t)$ over $[0,t_1]$, i. e. the system is not completely observable. It is conflict with our assumption. So, it is concluded that $\boldsymbol{W}_o[0,t_1]$ is non-singular.

Since the matrix $\boldsymbol{A}$ and $\boldsymbol{C}$ are involved, sometimes we say the pair $[\boldsymbol{C},\boldsymbol{A}]$ is observable. As the need to calculate the matrix exponential function and the integration, so the Gramian criteria is difficult to use in practice. But the criteria is valuable in theoretical studies because the following practical criteria can be derived from it.

In addition, the following practical observability criteria can be proved from the controllability criteria discussed above by using the dual principle which will be discussed in section 4. 4.

Theorem 4. 7 Rank Criteria.

The LTI system described by (4. 25) is observable iff **the observability matrix**

$$\boldsymbol{Q}_o = \begin{bmatrix} \boldsymbol{C} \\ \boldsymbol{CA} \\ \vdots \\ \boldsymbol{CA}^{n-1} \end{bmatrix} \tag{4.32}$$

has full-column rank, i. e. rank$\boldsymbol{Q}_o = n$.

Note that the rank of the observability matrix is equal to the number of the observable state variables if the system is not completely observable.

Example 4. 12 Consider the LTI system described by

$$\dot{\boldsymbol{X}} = \begin{bmatrix} -1 & -4 & -2 \\ 0 & 6 & -1 \\ 1 & 7 & -1 \end{bmatrix}\boldsymbol{X} + \begin{bmatrix} 0 & 1 \\ 2 & 1 \\ 1 & 0 \end{bmatrix}\boldsymbol{u}$$

$$y=\begin{bmatrix}0 & 2 & 1\\ 1 & 1 & 0\end{bmatrix}X$$

Determine the observability of the system.

Solution The observability matrix of the system is

$$Q_o=\begin{bmatrix}C\\ CA\\ CA^2\end{bmatrix}=\begin{bmatrix}0 & 2 & 1\\ 1 & 1 & 0\\ 1 & 19 & -3\\ -1 & 2 & -3\\ -4 & 89 & -18\\ -2 & -5 & 3\end{bmatrix}$$

and rank$Q_o=3$. So, the system is completely observable.

Example 4. 13 There is a nonsingular transformation

$$X(t)=P\overline{X}(t)$$

which transforms the description of the LTI system from (4. 25) to the form such as

$$\begin{aligned}\dot{\overline{X}}&=\overline{A}\,\overline{X}+\overline{B}u\\ y&=\overline{C}\,\overline{X}+\overline{D}u\end{aligned}\qquad(4.33)$$

Where $\overline{A}=P^{-1}AP$; $\overline{B}=P^{-1}B$; $\overline{C}=CP$; $\overline{D}=D$.

Prove that **a change of state by the nonsingular transformation does not change the observability.**

Proof. The observability matrix of the system (4. 25) is

$$Q_o=\begin{bmatrix}C\\ CA\\ \vdots\\ CA^{n-1}\end{bmatrix}$$

Furthermore, the observability matrix of the system (4. 33) can be constructed as

$$\overline{Q}_o=\begin{bmatrix}\overline{C}\\ \overline{C}\,\overline{A}\\ \vdots\\ \overline{C}\,\overline{A}^{n-1}\end{bmatrix}=\begin{bmatrix}CP\\ CPP^{-1}AP\\ \vdots\\ CP(P^{-1}AP)^{n-1}\end{bmatrix}=\begin{bmatrix}CP\\ CAP\\ \vdots\\ CA^{n-1}P\end{bmatrix}=\begin{bmatrix}C\\ CA\\ \vdots\\ CA^{n-1}\end{bmatrix}P$$

Obviously $\qquad$ rank$Q_o=$rank$\overline{Q}_o$

It means that the observability is preserved after the nonsingular transformation.

Theorem 4. 8 PBH Rank Criteria.

The LTI system described by (4. 25) is observable iff

$$\text{rank}\begin{bmatrix}\lambda_i I-A\\ C\end{bmatrix}=n,\qquad i=1,2,\cdots,n\qquad(4.34)$$

for all eigenvalues λ_i of A.

Example 4. 14 Consider the LTI system described by.

$$\dot{X}=\begin{bmatrix}0&1&0&0\\0&0&-1&0\\0&0&0&1\\0&0&5&0\end{bmatrix}X+\begin{bmatrix}0&1\\1&0\\0&1\\-2&0\end{bmatrix}u$$

$$y=\begin{bmatrix}0&1&0&-2\\1&0&1&0\end{bmatrix}X$$

Determine the observability of the system.

Solution The eigenvalues of A are

$$\lambda_{1,2}=0, \qquad \lambda_3=\sqrt{5}, \qquad \lambda_4=-\sqrt{5}$$

It is calculated that

$$\operatorname{rank}\begin{bmatrix}\lambda_1 I-A\\C\end{bmatrix}=\operatorname{rank}\begin{bmatrix}0&-1&0&0\\0&0&1&0\\0&0&0&-1\\0&0&-5&0\\0&1&0&-2\\1&0&1&0\end{bmatrix}=4$$

$$\operatorname{rank}\begin{bmatrix}\lambda_3 I-A\\C\end{bmatrix}=\operatorname{rank}\begin{bmatrix}\sqrt{5}&-1&0&0\\0&\sqrt{5}&1&0\\0&0&\sqrt{5}&-1\\0&0&-5&\sqrt{5}\\0&1&0&-2\\1&0&1&0\end{bmatrix}=4$$

$$\operatorname{rank}\begin{bmatrix}\lambda_4 I-A\\C\end{bmatrix}=\operatorname{rank}\begin{bmatrix}-\sqrt{5}&-1&0&0\\0&-\sqrt{5}&1&0\\0&0&-\sqrt{5}&-1\\0&0&-5&-\sqrt{5}\\0&1&0&-2\\1&0&1&0\end{bmatrix}=4$$

Therefore, the system is completely observable.

Theorem 4.9 Diagonal Canonical Form Criteria.

If the eigenvalues $\lambda_1, \lambda_2, \cdots, \lambda_n$ of the LTI system described by (4.25) are distinct, the diagonal canonical form (4.33) can be obtained by the nonsingular transformation $X(t)=P\overline{X}(t)$, where

$$\overline{A}=P^{-1}AP=\begin{bmatrix}\lambda_1&&&\\&\lambda_2&&\\&&\ddots&\\&&&\lambda_n\end{bmatrix}, \quad \overline{C}=CP=[\overline{c}_1 \quad \overline{c}_2 \quad \cdots \quad \overline{c}_n]$$

The LTI system is observable iff $\overline{C}$ has no zero column vector, i. e.

$$\overline{c}_i \neq \mathbf{0}, i=1,2,\cdots,n \tag{4.35}$$

Example 4.15 Consider the LTI system described by

$$\dot{\boldsymbol{X}} = \begin{bmatrix} -7 & 0 & 0 \\ 0 & -2 & 0 \\ 0 & 0 & 1 \end{bmatrix} \boldsymbol{X} + \begin{bmatrix} 0 & 2 \\ 4 & 0 \\ 0 & 1 \end{bmatrix} \boldsymbol{u}$$

$$\boldsymbol{y} = \begin{bmatrix} 0 & 4 & 0 \\ 2 & 0 & 1 \end{bmatrix} \boldsymbol{X}$$

Determine the observability of the system.

Solution Since the system matrix $\boldsymbol{A}$ is a diagonal matrix and the output matrix $\boldsymbol{C}$ has no zero column vector, the system is completely observable.

Example 4.16 Consider the LTI system described by

$$\dot{\boldsymbol{X}} = \begin{bmatrix} -7 & 0 & 0 \\ 0 & -2 & 0 \\ 0 & 0 & 1 \end{bmatrix} \boldsymbol{X} + \begin{bmatrix} 0 & 2 \\ 0 & 0 \\ 0 & 1 \end{bmatrix} \boldsymbol{u}$$

$$\boldsymbol{y} = \begin{bmatrix} 0 & 0 & 0 \\ 2 & 0 & 1 \end{bmatrix} \boldsymbol{X}$$

Determine the observability of the system.

Solution Since the system matrix $\boldsymbol{A}$ is a diagonal matrix but the output matrix $\boldsymbol{C}$ has a zero column vector $\boldsymbol{c}_2$, the system is not completely observable.

Moreover, because there are not coupling relationships between the state variable x_2 and other variables, and x_2 does not affect the output, so the state variable x_2 is an unobservable variable in the system.

Theorem 4.10 Jordan Canonical Form Criteria.

If the eigenvalues $\lambda_1, \lambda_2, \cdots, \lambda_l$ of the LTI system described by (4.25) are not distinct, the Jordan canonical form (4.33) can be obtained by the nonsingular transformation $\boldsymbol{X}(t) = \boldsymbol{P}\overline{\boldsymbol{X}}(t)$, where

$$\overline{\boldsymbol{A}} = \boldsymbol{P}^{-1}\boldsymbol{A}\boldsymbol{P} = \begin{bmatrix} \boldsymbol{J}_1 & & & 0 \\ & \boldsymbol{J}_2 & & \\ & & \ddots & \\ 0 & & & \boldsymbol{J}_l \end{bmatrix}, \quad \overline{\boldsymbol{C}} = \boldsymbol{C}\boldsymbol{P} = [\overline{\boldsymbol{C}}_1 \quad \overline{\boldsymbol{C}}_2 \quad \cdots \quad \overline{\boldsymbol{C}}_l]$$

$$\boldsymbol{J}_i = \begin{bmatrix} \boldsymbol{J}_{i1} & & 0 \\ & \ddots & \\ 0 & & \boldsymbol{J}_{i\alpha_i} \end{bmatrix}, \quad \overline{\boldsymbol{C}}_i = [\overline{\boldsymbol{C}}_{i1} \quad \overline{\boldsymbol{C}}_{i2} \quad \cdots \quad \overline{\boldsymbol{C}}_{i\alpha_i}]$$

$$\boldsymbol{J}_{ik} = \begin{bmatrix} \lambda_i & 1 & 0 \\ & \ddots & 1 \\ 0 & & \lambda_i \end{bmatrix}, \quad \overline{\boldsymbol{C}}_{ik} = [\overline{\boldsymbol{c}}_{ik1} \quad \overline{\boldsymbol{c}}_{ik2} \quad \cdots \quad \overline{\boldsymbol{c}}_{ikr_k}]$$

and α_i is the geometric multiplicity of the eigenvalue λ_i, $i=1,2,\cdots,l$.

The system is observable iff those column vectors, which are the first column vectors of $\overline{C}_{ik}, k=1,2,\cdots,\alpha_i$, within $\overline{C}_i$ corresponding to the eigenvalue λ_i, are independent, i. e.

$$\text{rank}[\bar{c}_{i11} \quad \bar{c}_{i21} \quad \cdots \quad \bar{c}_{i\alpha_i 1}] = \alpha_i \tag{4.36}$$

Example 4.17 Consider the LTI system described by

$$\dot{X} = \begin{bmatrix} -2 & 1 & 0 \\ 0 & -2 & 0 \\ 0 & 0 & -2 \end{bmatrix} X + \begin{bmatrix} 0 & 2 \\ 4 & 0 \\ 0 & 1 \end{bmatrix} u$$

$$y = \begin{bmatrix} 0 & 4 & 0 \\ 2 & 0 & 1 \end{bmatrix} X$$

Determine the observability of the system.

Solution The system matrix A is a Jordan matrix and has the only eigenvalue $\lambda = -2$. The geometric multiplicity of the eigenvalue λ is $\alpha = 2$. Denote the output matrix C as

$$C = [c_{111} \quad c_{112} \quad c_{121}] = \begin{bmatrix} 0 & 4 & 0 \\ 2 & 0 & 1 \end{bmatrix}$$

Since $\text{rank}[c_{111} \quad c_{121}] = 1 \neq \alpha$, then the system is not completely observable.

Example 4.18 Consider the LTI system described by.

$$\dot{X} = \begin{bmatrix} -2 & 0 & 0 \\ 0 & -2 & 0 \\ 0 & 0 & -2 \end{bmatrix} X + \begin{bmatrix} 0 & 2 \\ 4 & 0 \\ 0 & 1 \end{bmatrix} u$$

$$y = \begin{bmatrix} 0 & 4 & 0 \\ 2 & 0 & 1 \end{bmatrix} X$$

Determine the controllability of the system.

Solution The system matrix A is a Jordan matrix and has the only eigenvalue $\lambda = -2$. The geometric multiplicity of the eigenvalue λ is $\alpha = 3$. Denote the output matrix C as

$$C = [c_{111} \quad c_{121} \quad c_{131}] = \begin{bmatrix} 0 & 4 & 0 \\ 2 & 0 & 1 \end{bmatrix}$$

Since $\text{rank}[c_{111} \quad c_{121} \quad c_{131}] = 2 \neq \alpha$, the system is not completely observable.

Example 4.19 Consider the LTI system described by

$$\dot{X} = \begin{bmatrix} -2 & 1 & 0 & 0 & 0 & 0 & 0 \\ 0 & -2 & 0 & 0 & 0 & 0 & 0 \\ 0 & 0 & -2 & 0 & 0 & 0 & 0 \\ 0 & 0 & 0 & -2 & 0 & 0 & 0 \\ 0 & 0 & 0 & 0 & 3 & 1 & 0 \\ 0 & 0 & 0 & 0 & 0 & 3 & 0 \\ 0 & 0 & 0 & 0 & 0 & 0 & 3 \end{bmatrix} X + \begin{bmatrix} 0 & 0 & 0 \\ 1 & 0 & 0 \\ 0 & 4 & 0 \\ 0 & 0 & 7 \\ 0 & 0 & 0 \\ 1 & 1 & 0 \\ 0 & 4 & 1 \end{bmatrix} u$$

$$y = \begin{bmatrix} 0 & 1 & 0 & 0 & 1 & 0 & 0 \\ 0 & 0 & 4 & 0 & 1 & 0 & 4 \\ 0 & 0 & 0 & 7 & 0 & 0 & 1 \end{bmatrix} X$$

Determine the observability of the system.

Solution The system matrix $\boldsymbol{A}$ is a Jordan matrix and has the eigenvalues $\lambda_1=-2$ and $\lambda_2=3$. The geometric multiplicity of the eigenvalue λ_1 is $\alpha_1=3$ and the geometric multiplicity of the eigenvalue λ_2 is $\alpha_2=2$. Denote the output matrix $\boldsymbol{C}$ as

$$\boldsymbol{C}=[\boldsymbol{c}_{111} \quad \boldsymbol{c}_{112} \quad \boldsymbol{c}_{121} \quad \boldsymbol{c}_{131} \quad \boldsymbol{c}_{211} \quad \boldsymbol{c}_{212} \quad \boldsymbol{c}_{221}] = \begin{bmatrix} 0 & 1 & 0 & 0 & 1 & 0 & 0 \\ 0 & 0 & 4 & 0 & 1 & 0 & 4 \\ 0 & 0 & 0 & 7 & 0 & 0 & 1 \end{bmatrix}$$

Since $\text{rank}[\boldsymbol{c}_{111} \quad \boldsymbol{c}_{121} \quad \boldsymbol{c}_{131}]=2\neq\alpha_1$ and $\text{rank}[\boldsymbol{c}_{211} \quad \boldsymbol{c}_{221}]=2=\alpha_2$, the system is not completely observable.

Specially, when each Jordan block of a Jordan matrix has only one sub-Jordan block, the system is observable iff those column vectors, which are the columns of $\boldsymbol{B}$ corresponding to the first column of each Jordan block, are not the zero column vector.

Example 4.20 Consider the LTI system described by

$$\dot{\boldsymbol{X}}=\begin{bmatrix} -2 & 1 & 0 \\ 0 & -2 & 0 \\ 0 & 0 & -1 \end{bmatrix}\boldsymbol{X}+\begin{bmatrix} 0 & 2 \\ 4 & 0 \\ 0 & 1 \end{bmatrix}\boldsymbol{u}$$

$$\boldsymbol{y}=\begin{bmatrix} 0 & 4 & 0 \\ 2 & 0 & 1 \end{bmatrix}\boldsymbol{X}$$

Determine the observability of the system.

Solution The system matrix $\boldsymbol{A}$ is a Jordan matrix and has the eigenvalues $\lambda_1=-2$ and $\lambda_2=-1$. The geometric multiplicity of the eigenvalue λ_1 is $\alpha_1=1$ and the geometric multiplicity of the eigenvalue λ_2 is $\alpha_2=1$. Denote the output matrix $\boldsymbol{C}$ as

$$\boldsymbol{C}=[\boldsymbol{c}_{111} \quad \boldsymbol{c}_{112} \quad \boldsymbol{c}_{211}]=\begin{bmatrix} 0 & 4 & 0 \\ 2 & 0 & 1 \end{bmatrix}$$

Since $\boldsymbol{c}_{111}\neq 0$ and $\boldsymbol{c}_{211}\neq 0$, the system is completely observable.

4.3 Duality

Consider two LTI systems

$$\begin{aligned} \dot{\boldsymbol{X}}&=\boldsymbol{AX}+\boldsymbol{Bu} \\ \boldsymbol{y}&=\boldsymbol{CX}+\boldsymbol{Du} \end{aligned} \tag{4.37}$$

and

$$\begin{aligned} \dot{\boldsymbol{Z}}&=\boldsymbol{A}^{\mathrm{T}}\boldsymbol{Z}+\boldsymbol{C}^{\mathrm{T}}\boldsymbol{v} \\ \boldsymbol{w}&=\boldsymbol{B}^{\mathrm{T}}\boldsymbol{Z}+\boldsymbol{D}^{\mathrm{T}}\boldsymbol{v} \end{aligned} \tag{4.38}$$

There are some relationships between the systems above. The input matrix of system (4.38) is the **transpose** of the output matrix of system (4.37) and the output matrix of system (4.38) is the transpose of the input matrix of system (4.37). Furthermore, the system matrix and the forward matrix of system (4.38) are the transpose of the system matrix and the forward matrix of system (4.37).

Definition 4.3 If two LTI systems have the relationships above, they are called a pair of **dual systems.**

Theorem 4.11 Duality Principle

The system (4.37) is completely controllable (or observable) iff its dual system, system (4.38) is completely observable (or controllable).

Proof. The controllability Gramian matrix of the system (4.37) is

$$\boldsymbol{W}_c[0,t_1]=\int_0^{t_1} \mathrm{e}^{\boldsymbol{A}t}\boldsymbol{B}\boldsymbol{B}^{\mathrm{T}}\mathrm{e}^{\boldsymbol{A}^{\mathrm{T}}t}\,\mathrm{d}t$$

Furthermore, the observability Gramian matrix of the system (4.38) is

$$\overline{\boldsymbol{W}}_o[0,t_1]=\int_0^{t_1} \mathrm{e}^{\boldsymbol{A}t}\boldsymbol{B}\boldsymbol{B}^{\mathrm{T}}\mathrm{e}^{\boldsymbol{A}^{\mathrm{T}}t}\,\mathrm{d}t=\boldsymbol{W}_c[0,t_1]$$

It means that the controllability of the system (4.37) is equivalent to the observability of the system (4.38).

By the similar method, we can obtain that the observability of the system (4.37) is equivalent to the controllability of the system (4.38).

4.4 Obtaining the Controllable and Observable Canonical Form by State Transformation

4.4.1 Obtaining the Controllable Canonical Form by State Transformation

Consider the SISO LTI system described by

$$\begin{aligned}\dot{\boldsymbol{X}}(t)&=\boldsymbol{A}\boldsymbol{X}(t)+\boldsymbol{b}u(t)\\ y(t)&=\boldsymbol{c}\boldsymbol{X}(t)+\boldsymbol{d}u(t)\end{aligned} \tag{4.39}$$

There is a nonsingular transformation

$$\boldsymbol{X}(t)=\boldsymbol{P}\overline{\boldsymbol{X}}(t)$$

which transforms the description of the LTI system from (4.39) into the form such as

$$\begin{aligned}\dot{\overline{\boldsymbol{X}}}(t)&=\overline{\boldsymbol{A}}\,\overline{\boldsymbol{X}}(t)+\overline{\boldsymbol{b}}u(t)\\ y(t)&=\overline{\boldsymbol{c}}\,\overline{\boldsymbol{X}}(t)+\overline{\boldsymbol{d}}u(t)\end{aligned} \tag{4.40}$$

Where $\overline{\boldsymbol{A}}=\boldsymbol{P}^{-1}\boldsymbol{A}\boldsymbol{P}$; $\overline{\boldsymbol{b}}=\boldsymbol{P}^{-1}\boldsymbol{b}$; $\overline{\boldsymbol{c}}=\boldsymbol{c}\boldsymbol{P}$; $\overline{\boldsymbol{d}}=\boldsymbol{d}$.

Theorem 4.12 The LTI system (4.39) is completely controllable iff it is transformable into the controllable canonical form by a nonsingular transformation.

Proof. (1) Sufficiency. If the LTI system (4.39) is transformable into controllable canonical form such as (4.40), we assume

$$\bar{A}=\begin{bmatrix} 0 & 1 & 0 & \cdots & 0 \\ 0 & 0 & 1 & \cdots & 0 \\ \vdots & \vdots & \vdots & \ddots & \vdots \\ 0 & 0 & 0 & \cdots & 1 \\ -a_n & -a_{n-1} & -a_{n-2} & \cdots & -a_1 \end{bmatrix}, \quad \bar{b}=\begin{bmatrix} 0 \\ 0 \\ \vdots \\ 0 \\ 1 \end{bmatrix}$$

So, the characteristic polynomial of the system is

$$|sI-A|=|sI-\bar{A}|=s^n+a_1s^{n-1}+\cdots+a_{n-1}s+a_n \tag{4.41}$$

The controllability matrix of the controllable canonical form (4.40) can be constructed as

$$\bar{Q}_c=[\bar{b} \quad \bar{A}\bar{b} \quad \cdots \quad \bar{A}^{n-1}\bar{b}]=\begin{bmatrix} 0 & 0 & \cdots & 0 & 1 \\ 0 & 0 & \cdots & 1 & * \\ \vdots & \vdots & ⋰ & * & \vdots \\ 0 & 1 & ⋰ & \vdots & \vdots \\ 1 & * & \cdots & * & * \end{bmatrix} \tag{4.42}$$

It can be seen that $\bar{Q}_c$ is a triangular form matrix, and rank $\bar{Q}_c=n$. Since a change of state by the nonsingular transformation does not change the controllability, it is concluded that the system (4.39) is completely controllable.

(2) Necessity. If the LTI system (4.39) is controllable, a nonsingular transformation matrix can be constructed as

$$P=[p_1 \quad p_2 \quad \cdots \quad p_n]$$

$$=[A^{n-1}b \quad A^{n-2}b \quad \cdots \quad b]\begin{bmatrix} 1 & 0 & \cdots & \cdots & \cdots & 0 \\ a_1 & 1 & \ddots & & & \vdots \\ a_2 & a_1 & \ddots & 0 & & \vdots \\ \vdots & a_2 & \ddots & 1 & 0 & \vdots \\ a_{n-2} & \vdots & \ddots & a_1 & 1 & 0 \\ a_{n-1} & a_{n-2} & \cdots & a_2 & a_1 & 1 \end{bmatrix} \tag{4.43}$$

For $\bar{A}=P^{-1}AP$, it can be obtained that

$$P\bar{A}=AP=[Ap_1 \quad Ap_2 \quad \cdots \quad Ap_n] \tag{4.44}$$

According to Cayley-Hamilton Theorem and (4.43), the following equations can be derived that

$$Ap_1=A(A^{n-1}b+a_1A^{n-2}b+\cdots+a_{n-2}Ab+a_{n-1}b)$$
$$=(A^n+a_1A^{n-1}+\cdots+a_{n-2}A^2+a_{n-1}A+a_nI)b-a_nb=-a_np_n$$
$$Ap_2=A(A^{n-2}b+a_1A^{n-3}b+\cdots+a_{n-3}Ab+a_{n-2}b)$$
$$=(A^{n-1}+a_1A^{n-2}+\cdots+a_{n-2}A+a_{n-1}I)b-a_{n-1}b=p_1-a_{n-1}p_n$$
$$\vdots$$
$$Ap_{n-1}=A(Ab+a_1b)=(A^2+a_1A+a_2I)b-a_2b$$
$$=p_{n-2}-a_2p_n$$
$$Ap_n=Ab=(A+a_1I)b-a_1b=p_{n-1}-a_1p_n$$

So, Eq. (4.44) can be rewritten as

$$\boldsymbol{AP}=[\boldsymbol{Ap}_1 \quad \boldsymbol{Ap}_2 \quad \cdots \quad \boldsymbol{Ap}_n]=[-a_n\boldsymbol{p}_n \quad \boldsymbol{p}_1-a_{n-1}\boldsymbol{p}_n \quad \cdots \quad \boldsymbol{p}_{n-2}-a_2\boldsymbol{p}_n \quad \boldsymbol{p}_{n-1}-a_1\boldsymbol{p}_n]$$

$$=[\boldsymbol{p}_1 \quad \boldsymbol{p}_2 \quad \cdots \quad \boldsymbol{p}_{n-1} \quad \boldsymbol{p}_n]\begin{bmatrix} 0 & 1 & 0 & \cdots & 0 \\ 0 & 0 & 1 & \cdots & 0 \\ \vdots & \vdots & \vdots & \ddots & \vdots \\ 0 & 0 & 0 & \cdots & 1 \\ -a_n & -a_{n-1} & -a_{n-2} & \cdots & -a_1 \end{bmatrix}=\boldsymbol{P}\overline{\boldsymbol{A}}$$

and

$$\overline{\boldsymbol{A}}=\boldsymbol{P}^{-1}\boldsymbol{AP}=\begin{bmatrix} 0 & 1 & 0 & \cdots & 0 \\ 0 & 0 & 1 & \cdots & 0 \\ \vdots & \vdots & \vdots & \ddots & \vdots \\ 0 & 0 & 0 & \cdots & 1 \\ -a_n & -a_{n-1} & -a_{n-2} & \cdots & -a_1 \end{bmatrix}$$

is a companion matrix.

Since $\overline{\boldsymbol{b}}=\boldsymbol{P}^{-1}\boldsymbol{b}$,

$$\boldsymbol{P}\overline{\boldsymbol{b}}=\boldsymbol{b}=\boldsymbol{p}_n=[\boldsymbol{p}_1 \quad \boldsymbol{p}_2 \quad \cdots \quad \boldsymbol{p}_n]\begin{bmatrix} 0 \\ \vdots \\ 0 \\ 1 \end{bmatrix}=\boldsymbol{P}\begin{bmatrix} 0 \\ \vdots \\ 0 \\ 1 \end{bmatrix}$$

It can be obtained that

$$\overline{\boldsymbol{b}}=\begin{bmatrix} 0 \\ \vdots \\ 0 \\ 1 \end{bmatrix}$$

Consequently, the system (4.39) can be transformed into controllable canonical form by a nonsingular transformation if it is completely controllable.

Example 4.21 Consider the SISO LTI system described by

$$\dot{\boldsymbol{X}}=\begin{bmatrix} 1 & 2 & 0 \\ 3 & -1 & 1 \\ 0 & 2 & 0 \end{bmatrix}\boldsymbol{X}+\begin{bmatrix} 2 \\ 1 \\ 1 \end{bmatrix}u, \quad y=[0 \quad 0 \quad 1]\boldsymbol{X}$$

Obtain the controllable canonical form of the system by the nonsingular transformation $\boldsymbol{X}=\boldsymbol{P}\overline{\boldsymbol{X}}$.

Solution The controllability matrix

$$\boldsymbol{Q}_c=[\boldsymbol{b} \quad \boldsymbol{Ab} \quad \boldsymbol{A}^2\boldsymbol{b}]=\begin{bmatrix} 2 & 4 & 16 \\ 1 & 6 & 8 \\ 1 & 2 & 12 \end{bmatrix}$$

and $\text{rank}\boldsymbol{Q}_c=3=n$.

So, the system is completely controllable and it can be transformed into the controllable canonical form by a nosingular state transformation.

The characteristic polynomial of the system is

$$|sI-A|=\begin{vmatrix} s-1 & -2 & 0 \\ -3 & s+1 & -1 \\ 0 & -2 & s \end{vmatrix}=s^3-9s+2=s^3+a_1s^2+a_2s+a_3$$

and the coefficients of it are

$$a_1=0, a_2=-9, a_3=2$$

The nonsingular transformation matrix can be constructed as

$$\boldsymbol{P}=[\boldsymbol{A}^2\boldsymbol{b} \quad \boldsymbol{A}\boldsymbol{b} \quad \boldsymbol{b}]\begin{bmatrix} 1 & 0 & 0 \\ a_1 & 1 & 0 \\ a_2 & a_1 & 1 \end{bmatrix}=\begin{bmatrix} 16 & 4 & 2 \\ 8 & 6 & 1 \\ 12 & 2 & 1 \end{bmatrix}\begin{bmatrix} 1 & 0 & 0 \\ 0 & 1 & 0 \\ -9 & 0 & 1 \end{bmatrix}=\begin{bmatrix} -2 & 4 & 2 \\ -1 & 6 & 1 \\ 3 & 2 & 1 \end{bmatrix}$$

and its inverse matrix is

$$\boldsymbol{P}^{-1}=\begin{bmatrix} -\frac{1}{8} & 0 & \frac{1}{4} \\ -\frac{1}{8} & \frac{1}{4} & 0 \\ \frac{5}{8} & -\frac{1}{2} & \frac{1}{4} \end{bmatrix}$$

Then the system can be transformed into the controllable canonical form as

$$\dot{\overline{\boldsymbol{X}}}=\overline{\boldsymbol{A}}\,\overline{\boldsymbol{X}}+\overline{\boldsymbol{b}}u$$
$$y=\overline{\boldsymbol{c}}\,\overline{\boldsymbol{X}}$$

Where $\overline{\boldsymbol{A}}=\boldsymbol{P}^{-1}\boldsymbol{A}\boldsymbol{P}=\begin{bmatrix} 0 & 1 & 0 \\ 0 & 0 & 1 \\ -2 & 9 & 0 \end{bmatrix}$; $\overline{\boldsymbol{b}}=\boldsymbol{P}^{-1}\boldsymbol{b}=\begin{bmatrix} 0 \\ 0 \\ 1 \end{bmatrix}$; $\overline{\boldsymbol{c}}=\boldsymbol{c}\boldsymbol{P}=[3 \quad 2 \quad 1]$.

4.4.2 Obtaining the Observable Canonical Form by State Transformation

Theorem 4.13 The system (4.39) is completely observable if it is transformable into the observable canonical form by a nonsingular transformation.

Theorem 4.13 can be proved by using the similar approach proved Theorem 4.12. It can also be proved by using the dual principle.

In fact, if a nonsingular transformation matrix is constructed as

$$\boldsymbol{P}^{-1}=\begin{bmatrix} 1 & a_1 & a_2 & \cdots & a_{n-2} & a_{n-1} \\ 0 & 1 & a_1 & a_2 & \ddots & a_{n-2} \\ 0 & 0 & 1 & \ddots & \ddots & \vdots \\ \vdots & & 0 & \ddots & a_1 & a_2 \\ \vdots & & & \ddots & 1 & a_1 \\ 0 & \cdots & \cdots & \cdots & 0 & 1 \end{bmatrix}\begin{bmatrix} \boldsymbol{c}\boldsymbol{A}^{n-1} \\ \boldsymbol{c}\boldsymbol{A}^{n-2} \\ \vdots \\ \boldsymbol{c}\boldsymbol{A} \\ \boldsymbol{c} \end{bmatrix} \tag{4.45}$$

we can obtain the observable canonical form description of the LTI system such as (4.40) from (4.39) by taking the nonsingular transformation $\boldsymbol{X}(t)=\boldsymbol{P}\overline{\boldsymbol{X}}(t)$ (or $\overline{\boldsymbol{X}}(t)=\boldsymbol{P}^{-1}\boldsymbol{X}(t)$), and where

$$\bar{A}=\begin{bmatrix} 0 & 0 & \cdots & 0 & -a_n \\ 1 & 0 & \cdots & 0 & -a_{n-1} \\ 0 & 1 & \ddots & \vdots & \vdots \\ \vdots & \ddots & 1 & 0 & -a_2 \\ 0 & \cdots & 0 & 1 & -a_1 \end{bmatrix},\quad \bar{c}=[0 \ \cdots \ 0 \ 1]$$

Example 4.22 Consider the SISO LTI system described by

$$\dot{X}=\begin{bmatrix} 1 & 0 & 2 \\ 2 & 1 & 1 \\ 1 & 0 & -2 \end{bmatrix}X+\begin{bmatrix} 1 \\ 2 \\ 1 \end{bmatrix}u,\ y=[0 \ 1 \ 1]X$$

Obtain the observable canonical form of the system by the nonsingular transformation $X=P\bar{X}$.

Solution The observability matrix

$$Q_o=\begin{bmatrix} c \\ cA \\ cA^2 \end{bmatrix}=\begin{bmatrix} 0 & 1 & 1 \\ 3 & 1 & -1 \\ 4 & 1 & 9 \end{bmatrix}$$

and $\operatorname{rank}Q_o=3=n$.

So, the system is completely observable and it can be transformed into the observable canonical form by a nonsingular state transformation.

The characteristic polynomial of the system is

$$|sI-A|=\begin{vmatrix} s-1 & 0 & -2 \\ -2 & s-1 & -1 \\ -1 & 0 & s+2 \end{vmatrix}=s^3-5s+4=s^3+a_1s^2+a_2s+a_3$$

and the coefficients of it are

$$a_1=0,a_2=-5,a_3=4$$

Construct the nonsingular transformation matrix as

$$P^{-1}=\begin{bmatrix} 1 & a_1 & a_2 \\ 0 & 1 & a_1 \\ 0 & 0 & 1 \end{bmatrix}\begin{bmatrix} cA^2 \\ cA \\ c \end{bmatrix}$$

$$=\begin{bmatrix} 1 & 0 & -5 \\ 0 & 1 & 0 \\ 0 & 0 & 1 \end{bmatrix}\begin{bmatrix} 4 & 1 & 9 \\ 3 & 1 & -1 \\ 0 & 1 & 1 \end{bmatrix}=\begin{bmatrix} 4 & -4 & 4 \\ 3 & 1 & -1 \\ 0 & 1 & 1 \end{bmatrix}$$

and its inverse matrix is

$$P=\begin{bmatrix} \frac{1}{16} & \frac{1}{4} & 0 \\ -\frac{3}{32} & \frac{1}{8} & \frac{1}{2} \\ \frac{3}{32} & -\frac{1}{8} & \frac{1}{2} \end{bmatrix}$$

Then the system can be transformed into the observable canonical form as

$$\dot{\overline{X}}=\overline{A}\,\overline{X}+\overline{b}u$$

$$y=\overline{c}\,\overline{X}$$

Where $\overline{A}=P^{-1}AP=\begin{bmatrix}0 & 0 & -4\\ 1 & 0 & 5\\ 0 & 1 & 0\end{bmatrix}$; $\overline{b}=P^{-1}b=\begin{bmatrix}0\\ 4\\ 3\end{bmatrix}$; $\overline{c}=cP=[0\quad 0\quad 1]$.

Note that the controllable (or observable) canonical form of a SISO system obtained by the linear nonsingular transformation is unique, but the controllable (or observable) canonical form of a MIMO system obtained by the linear nonsingular transformation is not unique.

4.5 Canonical Decomposition of the LTI System

Consider the n dimension LTI system described by

$$\begin{aligned}\dot{X}(t)&=AX(t)+Bu(t)\\ y(t)&=CX(t)+Du(t)\end{aligned}\qquad(4.46)$$

If the system is not completely controllable, then some of the state variables are controllable and others are uncontrollable. It means that the state variables can be separated into two parts: one is controllable described by X_c and the other is uncontrollable described by $X_{\bar{c}}$. The subscript c is used to stand for controllable and $\bar{c}$ is used to stand for uncontrollable.

Similarly, if the system is not completely observable, then some of the state variables are observable and others are unobservable. It means that the state variables can be separated into two parts: one is observable described by X_o and the other is unobservable described by $X_{\bar{o}}$. The subscript o is used to stand for observable and $\bar{o}$ is used to stand for unobservable.

Furthermore, if the system is not completely controllable and completely observable, then the state variables can be separated into four parts: controllable and observable part X_{co}, controllable but not observable part $X_{c\bar{o}}$, observable but not controllable part $X_{\bar{c}o}$, and neither controllable nor observable part $X_{\bar{c}\bar{o}}$. The procedure above is called **canonical decomposition.**

4.5.1 Controllable Canonical Decomposition

Theorem 4.14 Consider the n dimension LTI system (4.46). If the rank of its controllability matrix

$$\text{rank}Q_c=k<n$$

then there exists a nonsingular transformation $X=P\overline{X}$, which transforms the description of the LTI system from (4.46) into the form such as controllability staircase form

$$\begin{bmatrix}\dot{\overline{X}}_c \\ \dot{\overline{X}}_{\bar{c}}\end{bmatrix}=\begin{bmatrix}\overline{A}_c & \overline{A}_{12} \\ \mathbf{0} & \overline{A}_{\bar{c}}\end{bmatrix}\begin{bmatrix}\overline{X}_c \\ \overline{X}_{\bar{c}}\end{bmatrix}+\begin{bmatrix}\overline{B}_c \\ \mathbf{0}\end{bmatrix}u$$

$$y=[\overline{C}_c \quad \overline{C}_{\bar{c}}]\begin{bmatrix}\overline{X}_c \\ \overline{X}_{\bar{c}}\end{bmatrix}+Du \tag{4.47}$$

Proof. Suppose the system (4.46) is not completely controllable and

$$\text{rank}Q_c=\text{rank}[B \quad AB \quad \cdots \quad A^{n-1}B]=k<n$$

By letting $p_1, p_2, \cdots, p_k$ be any k linearly independent columns of Q_c, a nonsingular matrix can be constructed as

$$P=[p_1 \quad p_2 \quad \cdots \quad p_k \quad p_{k+1} \quad \cdots \quad p_n] \tag{4.48}$$

Where the last $n-k$ columns of P are entirely arbitrary so long as the matrix P is nonsingular.

Then the inverse matrix of P is existent and can be represented as

$$P^{-1}=Q=\begin{bmatrix}q_1^T \\ q_2^T \\ \vdots \\ q_n^T\end{bmatrix} \tag{4.49}$$

Since $QP=I$, i. e.

$$QP=\begin{bmatrix}q_1^T p_1 & q_1^T p_2 & \cdots & q_1^T p_n \\ q_2^T p_1 & q_2^T p_2 & \cdots & q_2^T p_n \\ \vdots & \vdots & \ddots & \vdots \\ q_n^T p_1 & q_n^T p_2 & \cdots & q_n^T p_n\end{bmatrix}=\begin{bmatrix}1 & 0 & \cdots & 0 \\ 0 & 1 & \ddots & \vdots \\ \vdots & \ddots & \ddots & 0 \\ 0 & \cdots & 0 & 1\end{bmatrix}$$

Therefore

$$q_i^T p_j=0, \quad \text{for} \quad \forall i\neq j \tag{4.50}$$

Since $p_1, p_2, \cdots, p_k$ are linearly independent, then p_j, $j=1, 2, \cdots, k$ can be expressed linearly by $p_1, p_2, \cdots, p_k$. Furthermore, the vector Ap_j can also be expressed linearly by $p_1, \cdots, p_k$. So, we have

$$q_i^T A p_j=\mathbf{0}, \quad \text{for} \quad \forall i=k+1,\cdots,n, \qquad \forall j=1,\cdots,k \tag{4.51}$$

Similarly, since $p_1, p_2, \cdots, p_k$ are linear independently and chosen from

$$Q_c=[B \quad AB \quad \cdots \quad A^{n-1}B]$$

then B can be expressed linearly by $p_1, p_2, \cdots, p_k$ and we have

$$q_i^T B=\mathbf{0}, \quad \text{for} \quad \forall i=k+1,\cdots,n \tag{4.52}$$

Consequently, by the nonsingular transformation $X=P\overline{X}$, the system (4.46) can be transformed into

$$\begin{aligned}\dot{\overline{X}}&=\overline{A}\,\overline{X}+\overline{B}u \\ y&=\overline{C}\,\overline{X}+\overline{D}u\end{aligned} \tag{4.53}$$

Where

$$\overline{A}=P^{-1}AP=QAP$$

$$
=\left[\begin{array}{ccc:ccc}
q_1^T A p_1 & \cdots & q_1^T A p_k & q_1^T A p_{k+1} & \cdots & q_1^T A p_n \\
\vdots & \ddots & \vdots & \vdots & \ddots & \vdots \\
q_k^T A p_1 & \cdots & q_k^T A p_k & q_k^T A p_{k+1} & \cdots & q_k^T A p_n \\
\hdashline
q_{k+1}^T A p_1 & \cdots & q_{k+1}^T A p_k & q_{k+1}^T A p_{k+1} & \cdots & q_{k+1}^T A p_n \\
\vdots & \ddots & \vdots & \vdots & \ddots & \vdots \\
q_n^T A p_1 & \cdots & q_n^T A p_k & q_n^T A p_{k+1} & \cdots & q_n^T A p_n
\end{array}\right]
=\begin{bmatrix} \overline{A}_c & \overline{A}_{12} \\ 0 & \overline{A}_{\bar{c}} \end{bmatrix};
$$

$$
\overline{B}=P^{-1}B=\left[\begin{array}{c} q_1^T B \\ \vdots \\ q_k^T B \\ \hdashline q_{k+1}^T B \\ \vdots \\ q_n^T B \end{array}\right]=\begin{bmatrix} \overline{B}_c \\ 0 \end{bmatrix};
$$

$$
\overline{C}=CP=[Cp_1 \quad \cdots \quad Cp_k \ \vdots\ Cp_{k+1} \quad \cdots Cp_n]=[\overline{C}_c \quad \overline{C}_{\bar{c}}];\ \overline{D}=D.
$$

Since a change of state by the nonsingular transformation does not change the controllability, it can be derived that

$$
\begin{aligned}
\operatorname{rank}\overline{Q}_c &= \operatorname{rank}\begin{bmatrix} \overline{B}_c & \overline{A}_c\overline{B}_c & \cdots & \overline{A}_c^{\,n-1}\overline{B}_c \\ 0 & 0 & \cdots & 0 \end{bmatrix} \\
&= \operatorname{rank}[\overline{B}_c \quad \overline{A}_c\overline{B}_c \quad \cdots \quad \overline{A}_c^{\,n-1}\overline{B}_c]=k
\end{aligned}
$$

It means that the k dimension subsystem is controllable. So, the description (4.47) is called a controllable canonical decomposition of the LTI system (4.46).

The result of controllable canonical decomposition is shown in Figure 4.3.

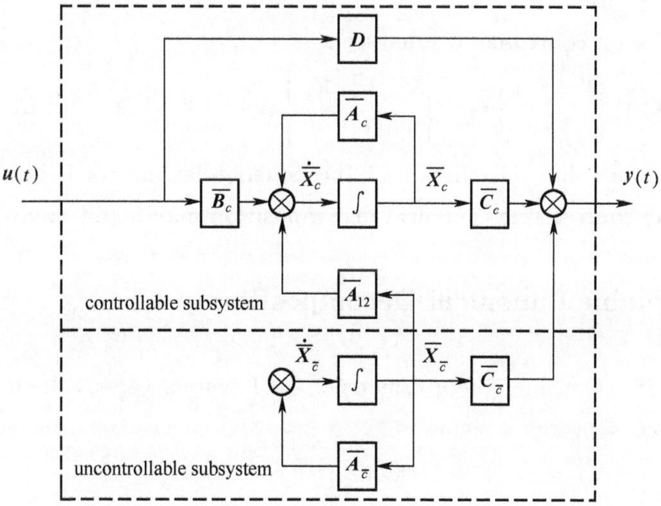

Figure 4.3 Controllable Canonical Decomposition

Example 4.23 Consider the LTI system described by

$$\dot{\boldsymbol{X}}=\begin{bmatrix}1 & 1 & 0\\0 & 1 & 0\\0 & 1 & 1\end{bmatrix}\boldsymbol{X}+\begin{bmatrix}0\\1\\0\end{bmatrix}u,\ y=[1\quad 1\quad 1]\boldsymbol{X}$$

Take the controllable canonical decomposition for the system.

Solution Since

$$\text{rank}\boldsymbol{Q}_c=\text{rank}[\boldsymbol{b}\quad \boldsymbol{Ab}\quad \boldsymbol{A}^2\boldsymbol{b}]=\text{rank}\begin{bmatrix}0 & 1 & 2\\1 & 1 & 1\\0 & 1 & 2\end{bmatrix}=2<3$$

the system is not completely controllable.

A nonsingular matrix can be constructed as

$$\boldsymbol{P}=[\boldsymbol{p}_1\quad \boldsymbol{p}_2\quad \boldsymbol{p}_3]=\begin{bmatrix}0 & 1 & 1\\1 & 1 & 0\\0 & 1 & 0\end{bmatrix}$$

Where $\boldsymbol{p}_1$ and $\boldsymbol{p}_2$ are the front two vectors of $\boldsymbol{Q}_c$ and they are linearly independent.

Furthermore, the inverse matrix of $\boldsymbol{P}$ is

$$\boldsymbol{P}^{-1}=\begin{bmatrix}0 & 1 & -1\\0 & 0 & 1\\1 & 0 & -1\end{bmatrix}$$

By the nonsingular transformation $\boldsymbol{X}=\boldsymbol{P}\overline{\boldsymbol{X}}$, the system can be transformed into

$$\dot{\overline{\boldsymbol{X}}}=\overline{\boldsymbol{A}}\,\overline{\boldsymbol{X}}+\overline{\boldsymbol{b}}u$$
$$y=\overline{\boldsymbol{c}}\,\overline{\boldsymbol{X}}$$

Where $\overline{\boldsymbol{A}}=\boldsymbol{P}^{-1}\boldsymbol{AP}=\left[\begin{array}{cc:c}0 & -1 & 0\\1 & 2 & 0\\ \hdashline 0 & 0 & 1\end{array}\right]$; $\overline{\boldsymbol{b}}=\boldsymbol{P}^{-1}\boldsymbol{b}=\left[\begin{array}{c}1\\0\\ \hdashline 0\end{array}\right]$; $\overline{\boldsymbol{c}}=\boldsymbol{cP}=[1\quad 3\ \vdots\ 1]$.

The 2 dimension controllable subsystem is

$$\dot{\overline{\boldsymbol{X}}}_c=\begin{bmatrix}0 & -1\\1 & 2\end{bmatrix}\overline{\boldsymbol{X}}_c+\begin{bmatrix}0 & 1\\0 & 0\end{bmatrix}\begin{bmatrix}\overline{\boldsymbol{X}}_{\bar{c}}\\ \boldsymbol{u}\end{bmatrix}u,\ y=[1\quad 3]\overline{\boldsymbol{X}}_c+[1]\overline{\boldsymbol{X}}_{\bar{c}}$$

Obviously, since the construction of the nonsingular matrix $\boldsymbol{P}$ is not unique, then the results of the controllable canonical decomposition may be different.

4.5.2 Observable Canonical Decomposition

Theorem 4.15 Consider the n dimension LTI system (4.46). If the rank of its observability matrix is

$$\text{rank}\boldsymbol{Q}_o=k<n$$

then there exists a nonsingular transformation $\boldsymbol{X}=\boldsymbol{P}\overline{\boldsymbol{X}}$, which transforms the description of the LTI system from (4.46) into the form such as

$$\begin{bmatrix}\dot{\overline{X}}_o\\ \dot{\overline{X}}_{\bar{o}}\end{bmatrix}=\begin{bmatrix}\overline{A}_o & \mathbf{0}\\ \overline{A}_{21} & \overline{A}_{\bar{o}}\end{bmatrix}\begin{bmatrix}\overline{X}_o\\ \overline{X}_{\bar{o}}\end{bmatrix}+\begin{bmatrix}\overline{B}_o\\ \overline{B}_{\bar{o}}\end{bmatrix}u$$

$$y=[\overline{C}_o \quad \mathbf{0}]\begin{bmatrix}\overline{X}_o\\ \overline{X}_{\bar{o}}\end{bmatrix}+Du \qquad (4.54)$$

So, the description (4.54) is called a observable canonical decomposition of the LTI system (4.46). The result of observable canonical decomposition is shown in Figure 4.4.

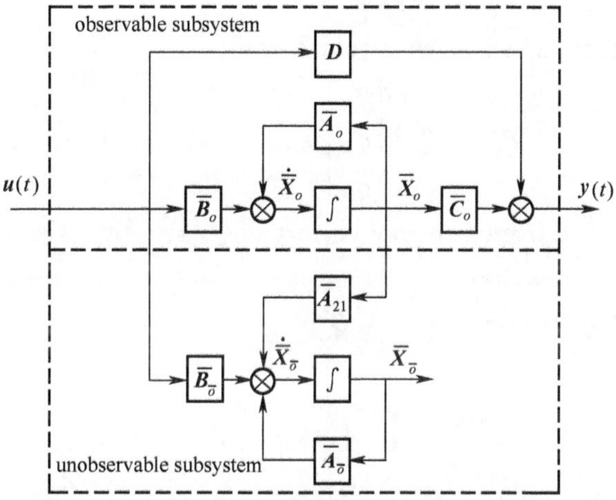

Figure 4.4 Observable Canonical Decomposition

Theorem 4.15 can be proved by using the similar approach which proved Theorem 4.14. In fact, since the system (4.46) is not completely observable and

$$\text{rank}\boldsymbol{Q}_o=\text{rank}\begin{bmatrix}\boldsymbol{C}\\ \boldsymbol{CA}\\ \vdots\\ \boldsymbol{CA}^{n-1}\end{bmatrix}=k<n$$

By letting $\boldsymbol{q}_1, \boldsymbol{q}_2, \cdots, \boldsymbol{q}_k$ be any k linearly independent rows of $\boldsymbol{Q}_o$, the nonsingular matrix can be constructed as

$$\boldsymbol{P}^{-1}=\boldsymbol{Q}=\begin{bmatrix}\boldsymbol{q}_1\\ \vdots\\ \boldsymbol{q}_k\\ \boldsymbol{q}_{k+1}\\ \vdots\\ \boldsymbol{q}_n\end{bmatrix} \qquad (4.55)$$

Where the last $n-k$ rows of $\boldsymbol{P}^{-1}$ are entirely arbitrary so long as the matrix $\boldsymbol{P}^{-1}$ is nonsingular. Then, we can obtain the observable canonical decomposition of the LTI system such as (4.54) from (4.46) by taking the nonsingular transformation $\boldsymbol{X}(t)=\boldsymbol{P}\overline{\boldsymbol{X}}(t)$.

Example 4. 24 Consider the LTI system described by

$$\dot{\boldsymbol{X}}=\begin{bmatrix}0 & 0 & -1\\ 1 & 0 & -3\\ 0 & 1 & -3\end{bmatrix}\boldsymbol{X}+\begin{bmatrix}1\\ 1\\ 0\end{bmatrix}u,\ y=\begin{bmatrix}0 & 1 & -2\end{bmatrix}\boldsymbol{X}$$

Take the observable canonical decomposition for the system.

Solution Since

$$\text{rank}\boldsymbol{Q}_o=\text{rank}\begin{bmatrix}\boldsymbol{C}\\ \boldsymbol{CA}\\ \boldsymbol{CA}^2\end{bmatrix}=\text{rank}\begin{bmatrix}0 & 1 & -2\\ 1 & -2 & 3\\ -2 & 3 & -4\end{bmatrix}=2<3$$

the system is not completely observable.

A nonsingular matrix can be constructed as

$$\boldsymbol{P}^{-1}=\boldsymbol{Q}=\begin{bmatrix}\boldsymbol{q}_1\\ \boldsymbol{q}_2\\ \boldsymbol{q}_3\end{bmatrix}=\begin{bmatrix}0 & 1 & -2\\ 1 & -2 & 3\\ 0 & 0 & 1\end{bmatrix}$$

Where $\boldsymbol{q}_1$ and $\boldsymbol{q}_2$ are the front two row vectors of $\boldsymbol{Q}_o$ and they are linearly independent.

Furthermore, the inverse matrix of $\boldsymbol{Q}$ is

$$\boldsymbol{P}=\begin{bmatrix}2 & 1 & 1\\ 1 & 0 & 2\\ 0 & 0 & 1\end{bmatrix}$$

By the nonsingular transformation $\boldsymbol{X}=\boldsymbol{P}\overline{\boldsymbol{X}}$, the system can be transformed into

$$\dot{\overline{\boldsymbol{X}}}=\overline{\boldsymbol{A}}\,\overline{\boldsymbol{X}}+\overline{\boldsymbol{b}}u$$
$$y=\overline{\boldsymbol{c}}\,\overline{\boldsymbol{X}}$$

Where $\overline{\boldsymbol{A}}=\boldsymbol{P}^{-1}\boldsymbol{AP}=\left[\begin{array}{cc:c}0 & 1 & 0\\ -1 & -2 & 0\\ \hdashline 1 & 0 & -1\end{array}\right]$; $\overline{\boldsymbol{b}}=\boldsymbol{P}^{-1}\boldsymbol{b}=\left[\begin{array}{c}1\\ -1\\ \hdashline 0\end{array}\right]$; $\overline{\boldsymbol{c}}=\boldsymbol{cP}=\left[\begin{array}{cc:c}1 & 0 & 0\end{array}\right]$.

The 2 dimension observable subsystem is

$$\dot{\overline{\boldsymbol{X}}}_o=\begin{bmatrix}0 & 1\\ -1 & -2\end{bmatrix}\overline{\boldsymbol{X}}_o+\begin{bmatrix}1\\ -1\end{bmatrix}u,\ y=\begin{bmatrix}1 & 0\end{bmatrix}\overline{\boldsymbol{X}}_o$$

Similarly, since the construction of the nonsingular matrix $\boldsymbol{Q}$ is not unique, then the results of the observable canonical decomposition may be different also.

4. 5. 3 Canonical Decomposition

Theorem 4. 16 Consider the n dimension LTI system (4. 46). If the system is not completely controllable and not completely observable, then there exists a nonsingular transformation $\boldsymbol{X}=\boldsymbol{P}\,\overline{\boldsymbol{X}}$, which transforms the description of the LTI system from (4. 46) into the form such as

$$\begin{bmatrix} \dot{\bar{X}}_{co} \\ \dot{\bar{X}}_{c\bar{o}} \\ \dot{\bar{X}}_{\bar{c}o} \\ \dot{\bar{X}}_{\bar{c}\bar{o}} \end{bmatrix} = \begin{bmatrix} \bar{A}_{co} & 0 & \bar{A}_{13} & 0 \\ \bar{A}_{21} & \bar{A}_{c\bar{o}} & \bar{A}_{23} & \bar{A}_{24} \\ \multicolumn{2}{c}{\mathbf{0}} & \bar{A}_{\bar{c}o} & 0 \\ & & \bar{A}_{43} & \bar{A}_{\bar{c}\bar{o}} \end{bmatrix} \begin{bmatrix} \bar{X}_{co} \\ \bar{X}_{c\bar{o}} \\ \bar{X}_{\bar{c}o} \\ \bar{X}_{\bar{c}\bar{o}} \end{bmatrix} + \begin{bmatrix} \bar{B}_{co} \\ \bar{B}_{c\bar{o}} \\ \mathbf{0} \\ \mathbf{0} \end{bmatrix} u$$

$$y = [\bar{C}_{co} \quad \mathbf{0} \quad \bar{C}_{\bar{c}o} \quad 0] \begin{bmatrix} \bar{X}_{co} \\ \bar{X}_{c\bar{o}} \\ \bar{X}_{\bar{c}o} \\ \bar{X}_{\bar{c}\bar{o}} \end{bmatrix} + Du \tag{4.56}$$

The description (4.56) is called a canonical decomposition of the LTI system (4.46), and the completely controllable and observable subsystem of (4.56) is

$$\begin{aligned} \dot{\bar{X}}_{co} &= \bar{A}_{co}\bar{X}_{co} + \bar{A}_{13}\bar{X}_{\bar{c}o} + \bar{B}_{co}u \\ y &= \bar{C}_{co}\bar{X}_{co} + \bar{C}_{\bar{c}o}\bar{X}_{\bar{c}o} + Du \end{aligned} \tag{4.57}$$

The result of canonical decomposition is shown in Figure 4.5.

Note that the completely controllable and observable subsystem (4.57) has the same transfer function matrix as (4.46) and (4.56), i.e.

$$G(s) = C(sI - A)^{-1}B + D = \bar{C}_{co}(sI - \bar{A}_{co})^{-1}\bar{B}_{co} + D \tag{4.58}$$

This is the most important relationship between the input-output description and the state space description. Theorem 4.16 tells us why the input-output description is sometimes insufficient to describe a system, for the uncontrollable and/or unobservable parts of the system do not appear in the transfer function matrix description. In other words, the transfer function matrix only describes the characteristic of the completely controllable and completely observable subsystem of the LTI system.

For general systems, the construction of the nonsingular matrix, by which the canonical decomposition can be taken, is difficult. However, if the state space description of system has the diagonal canonical form or the Jordan canonical form, the nonsingular matrix can be constructed easily.

Example 4.25 Consider the LTI system described by

$$\begin{bmatrix} \dot{x}_1 \\ \dot{x}_2 \\ \dot{x}_3 \\ \dot{x}_4 \end{bmatrix} = \begin{bmatrix} -3 & 0 & 0 & 0 \\ 0 & -1 & 0 & 0 \\ 0 & 0 & -2 & 0 \\ 0 & 0 & 0 & -4 \end{bmatrix} \begin{bmatrix} x_1 \\ x_2 \\ x_3 \\ x_4 \end{bmatrix} + \begin{bmatrix} 1 \\ 2 \\ 0 \\ 0 \end{bmatrix} u$$

$$y = [0 \quad 1 \quad 1 \quad 0] \begin{bmatrix} x_1 \\ x_2 \\ x_3 \\ x_4 \end{bmatrix}$$

(1) Take the canonical decomposition for the system.

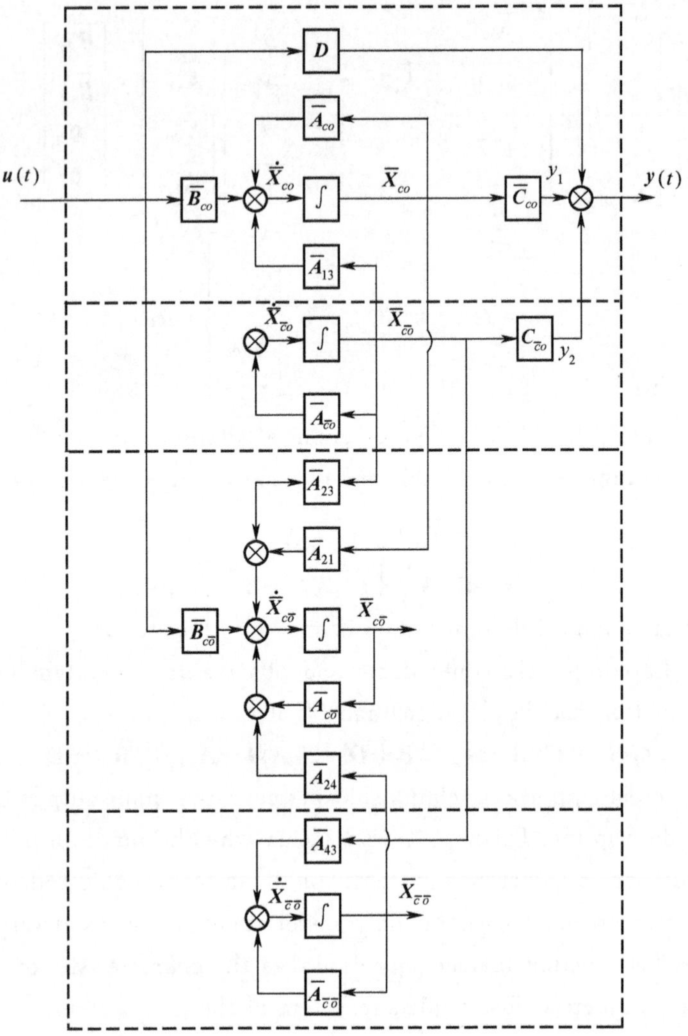

Figure 4.5 Canonical Decomposition

(2) Calculate the transfer function of the system.

Solution According to the diagonal canonical form criteria, the system is not completely controllable and not completely observable. Obviously, the state variable x_2 is controllable and observable, x_1 is controllable but not observable, x_3 is observable but not controllable, and x_4 is not controllable and not observable.

A nonsingular matrix can be constructed as

$$\boldsymbol{P}^{-1}=\begin{bmatrix}0 & 1 & 0 & 0\\1 & 0 & 0 & 0\\0 & 0 & 1 & 0\\0 & 0 & 0 & 1\end{bmatrix}$$

By the nonsingular transformation

$$\begin{bmatrix} \overline{x}_{co} \\ \overline{x}_{c\bar{o}} \\ \overline{x}_{\bar{c}o} \\ \overline{x}_{\bar{c}\bar{o}} \end{bmatrix} = \begin{bmatrix} x_2 \\ x_1 \\ x_3 \\ x_4 \end{bmatrix} = \boldsymbol{P}^{-1} \begin{bmatrix} x_1 \\ x_2 \\ x_3 \\ x_4 \end{bmatrix} = \begin{bmatrix} 0 & 1 & 0 & 0 \\ 1 & 0 & 0 & 0 \\ 0 & 0 & 1 & 0 \\ 0 & 0 & 0 & 1 \end{bmatrix} \begin{bmatrix} x_1 \\ x_2 \\ x_3 \\ x_4 \end{bmatrix}$$

the canonical decomposition of the system can be obtained as

$$\begin{bmatrix} \dot{x}_2 \\ \dot{x}_1 \\ \dot{x}_3 \\ \dot{x}_4 \end{bmatrix} = \begin{bmatrix} -1 & 0 & 0 & 0 \\ 0 & -3 & 0 & 0 \\ 0 & 0 & -2 & 0 \\ 0 & 0 & 0 & -4 \end{bmatrix} \begin{bmatrix} x_2 \\ x_1 \\ x_3 \\ x_4 \end{bmatrix} + \begin{bmatrix} 2 \\ 1 \\ 0 \\ 0 \end{bmatrix} u$$

$$y = [1 \quad 0 \quad 1 \quad 0] \begin{bmatrix} x_2 \\ x_1 \\ x_3 \\ x_4 \end{bmatrix}$$

The completely controllable and completely observable subsystem can be obtained as

$$\dot{x}_2 = -x_2 + 2u, y = x_2 + x_3$$

The transfer function of the original system is

$$G(s) = \boldsymbol{C}(s\boldsymbol{I} - \boldsymbol{A})^{-1}\boldsymbol{B} = \overline{\boldsymbol{C}}(s\boldsymbol{I} - \overline{\boldsymbol{A}})^{-1}\overline{\boldsymbol{B}}$$

$$= [1 \quad 0 \quad 1 \quad 0] \begin{bmatrix} s+1 & 0 & 0 & 0 \\ 0 & s+3 & 0 & 0 \\ 0 & 0 & s+2 & 0 \\ 0 & 0 & 0 & s+4 \end{bmatrix}^{-1} \begin{bmatrix} 2 \\ 1 \\ 0 \\ 0 \end{bmatrix}$$

$$= [1 \quad 0 \quad 1 \quad 0] \begin{bmatrix} \frac{1}{s+1} & 0 & 0 & 0 \\ 0 & \frac{1}{s+3} & 0 & 0 \\ 0 & 0 & \frac{1}{s+2} & 0 \\ 0 & 0 & 0 & \frac{1}{s+4} \end{bmatrix} \begin{bmatrix} 2 \\ 1 \\ 0 \\ 0 \end{bmatrix} = \frac{2}{s+1}$$

Furthermore, the transfer function of the system can be obtained as

$$G(s) = \overline{\boldsymbol{C}}_{co}(s\boldsymbol{I} - \overline{\boldsymbol{A}}_{co})^{-1}\overline{\boldsymbol{B}}_{co} = 1 \cdot (s+1)^{-1} \cdot 2 = \frac{2}{s+1}$$

It means that the transfer function of a system can be obtained by its completely controllable and completely observable subsystem.

4.6 Minimal Realization of the LTI System

4.6.1 Realization Problem

Definition 4.4 Consider the LTI system described by the state space description

$$\begin{cases}\dot{\boldsymbol{X}}(t)=\boldsymbol{A}\boldsymbol{X}(t)+\boldsymbol{B}\boldsymbol{u}(t)\\ \boldsymbol{y}(t)=\boldsymbol{C}\boldsymbol{X}(t)+\boldsymbol{D}\boldsymbol{u}(t)\end{cases} \tag{4.59}$$

If the transfer function matrix of it can be calculated as

$$\boldsymbol{G}(s)=\boldsymbol{C}(s\boldsymbol{I}-\boldsymbol{A})^{-1}\boldsymbol{B}+\boldsymbol{D} \tag{4.60}$$

the state space description (4.59) is call a **realization** of $\boldsymbol{G}(s)$.

In chapter 1, we have discussed the realization problems elementary. In fact, the realization problem is how to get the state space description from an external model.

Theorem 4.17 $\boldsymbol{G}(s)$ can be realized if $\boldsymbol{G}(s)$ is proper.

4.6.2 Realization of SISO System

In chapter 1, we discussed the realization of SISO system. Suppose the transfer function of a SISO system is

$$g(s)=\frac{b_1s^{n-1}+\cdots+b_{n-1}s+b_n}{s^n+a_1s^{n-1}+a_2s^{n-2}+\cdots+a_{n-1}s+a_n} \tag{4.61}$$

and it is strictly proper.

The realization in controllable canonical form of $g(s)$ is

$$\begin{aligned}\begin{bmatrix}\dot{x}_1\\ \dot{x}_2\\ \vdots\\ \dot{x}_n\end{bmatrix}&=\begin{bmatrix}0 & 1 & 0 & 0\\ \vdots & \vdots & \ddots & \vdots\\ 0 & 0 & \cdots & 1\\ -a_n & -a_{n-1} & \cdots & -a_1\end{bmatrix}\begin{bmatrix}x_1\\ x_2\\ \vdots\\ x_n\end{bmatrix}+\begin{bmatrix}0\\ 0\\ \vdots\\ 1\end{bmatrix}u\\ y&=\begin{bmatrix}b_n & b_{n-1} & \cdots & b_1\end{bmatrix}\begin{bmatrix}x_1\\ x_2\\ \vdots\\ x_n\end{bmatrix}\end{aligned} \tag{4.62}$$

and the realization in observable canonical form is

$$\begin{aligned}\begin{bmatrix}\dot{x}_1\\ \dot{x}_2\\ \vdots\\ \dot{x}_n\end{bmatrix}&=\begin{bmatrix}0 & \cdots & 0 & -a_n\\ 1 & \cdots & 0 & -a_{n-1}\\ \vdots & \ddots & \vdots & \vdots\\ 0 & \cdots & 1 & -a_1\end{bmatrix}\begin{bmatrix}x_1\\ x_2\\ \vdots\\ x_n\end{bmatrix}+\begin{bmatrix}b_n\\ b_{n-1}\\ \vdots\\ b_1\end{bmatrix}u\\ y&=\begin{bmatrix}0 & \cdots & 0 & 1\end{bmatrix}\begin{bmatrix}x_1\\ x_2\\ \vdots\\ x_n\end{bmatrix}\end{aligned} \tag{4.63}$$

Example 4.26 Consider a SISO system with the transfer function

$$g(s)=\frac{2s^2+6s+1}{s^3-11s^2+38s-40}$$

Determine the realization in controllable canonical form and in observable canonical form.

Solution From the transfer function, the coefficients are noted as

$$b_1=2,\ b_2=6,\ b_3=1;\ a_1=-11,\ a_2=38, a_3=-40$$

Then the controllable canonical realization can be written as

$$\begin{bmatrix}\dot{x}_1\\ \dot{x}_2\\ \dot{x}_3\end{bmatrix}=\begin{bmatrix}0 & 1 & 0\\ 0 & 0 & 1\\ -a_3 & -a_2 & -a_1\end{bmatrix}\begin{bmatrix}x_1\\ x_2\\ x_3\end{bmatrix}+\begin{bmatrix}0\\ 0\\ 1\end{bmatrix}u=\begin{bmatrix}0 & 1 & 0\\ 0 & 0 & 1\\ 40 & -38 & 11\end{bmatrix}\begin{bmatrix}x_1\\ x_2\\ x_3\end{bmatrix}+\begin{bmatrix}0\\ 0\\ 1\end{bmatrix}u$$

$$y=[b_3\quad b_2\quad b_1]\begin{bmatrix}x_1\\ x_2\\ x_3\end{bmatrix}=[1\quad 6\quad 2]\begin{bmatrix}x_1\\ x_2\\ x_3\end{bmatrix}$$

and the observable canonical realization can be written as

$$\begin{bmatrix}\dot{x}_1\\ \dot{x}_2\\ \dot{x}_3\end{bmatrix}=\begin{bmatrix}0 & 0 & -a_3\\ 1 & 0 & -a_2\\ 0 & 1 & -a_1\end{bmatrix}\begin{bmatrix}x_1\\ x_2\\ x_3\end{bmatrix}+\begin{bmatrix}b_3\\ b_2\\ b_1\end{bmatrix}u=\begin{bmatrix}0 & 0 & 40\\ 1 & 0 & -38\\ 0 & 1 & 11\end{bmatrix}\begin{bmatrix}x_1\\ x_2\\ x_3\end{bmatrix}+\begin{bmatrix}1\\ 6\\ 2\end{bmatrix}u$$

$$y=[0\quad 0\quad 1]\begin{bmatrix}x_1\\ x_2\\ x_3\end{bmatrix}$$

4.6.3 Realization of MIMO System

Similar to the realization of SISO system, we will give the realization of MIMO system. Suppose the transfer function matrix of a MIMO system is

$$\boldsymbol{G}(s)=\begin{bmatrix}g_{11}(s) & \cdots & g_{1m}(s)\\ \vdots & \ddots & \vdots\\ g_{r1}(s) & \cdots & g_{rm}(s)\end{bmatrix}\tag{4.64}$$

and it is strictly proper.

Where

$$g_{ij}(s)=\frac{Y_i(s)}{U_j(s)},\ i=1,2,\cdots,r,\ j=1,2,\cdots,m\tag{4.65}$$

and $Y_i(s)$ and $U_j(s)$ are the Laplace transforms of $y_i(t)$ and $u_j(t)$.

Rewrite $\boldsymbol{G}(s)$ as the following form

$$\boldsymbol{G}(s)=\frac{\boldsymbol{B}_1 s^{n-1}+\cdots+\boldsymbol{B}_{n-1}s+\boldsymbol{B}_n}{s^n+a_1 s^{n-1}+a_2 s^{n-2}+\cdots+a_{n-1}s+a_n}\tag{4.66}$$

The realization in controllable canonical form of $\boldsymbol{G}(s)$ is

$$\begin{gathered}\begin{bmatrix}\dot{x}_1\\ \dot{x}_2\\ \vdots\\ \dot{x}_{n\times m}\end{bmatrix}=\begin{bmatrix}\boldsymbol{0}_m & \boldsymbol{I}_m & \cdots & \boldsymbol{0}_m\\ \vdots & \vdots & \ddots & \vdots\\ \boldsymbol{0}_m & \boldsymbol{0}_m & \cdots & \boldsymbol{I}_m\\ -a_n\boldsymbol{I}_m & -a_{n-1}\boldsymbol{I}_m & \cdots & -a_1\boldsymbol{I}_m\end{bmatrix}\begin{bmatrix}x_1\\ x_2\\ \vdots\\ x_{n\times m}\end{bmatrix}+\begin{bmatrix}\boldsymbol{0}_m\\ \vdots\\ \boldsymbol{0}_m\\ \boldsymbol{I}_m\end{bmatrix}\boldsymbol{u}\\ \boldsymbol{y}=[\boldsymbol{B}_n\quad \boldsymbol{B}_{n-1}\quad \cdots\quad \boldsymbol{B}_1]\begin{bmatrix}x_1\\ x_2\\ \vdots\\ x_{n\times m}\end{bmatrix}\end{gathered}\tag{4.67}$$

and the realization in observable canonical form of $\boldsymbol{G}(s)$ is

$$\begin{bmatrix} \dot{x}_1 \\ \dot{x}_2 \\ \vdots \\ \dot{x}_{n\times r} \end{bmatrix} = \begin{bmatrix} \boldsymbol{0}_r & \cdots & \boldsymbol{0}_r & -a_n\boldsymbol{I}_r \\ \boldsymbol{I}_r & \cdots & \boldsymbol{0}_r & -a_{n-1}\boldsymbol{I}_r \\ \vdots & \ddots & \vdots & \vdots \\ \boldsymbol{0}_r & \cdots & \boldsymbol{I}_r & -a_1\boldsymbol{I}_r \end{bmatrix} \begin{bmatrix} x_1 \\ x_2 \\ \vdots \\ x_{n\times r} \end{bmatrix} + \begin{bmatrix} \boldsymbol{B}_n \\ \boldsymbol{B}_{n-1} \\ \vdots \\ \boldsymbol{B}_1 \end{bmatrix} \boldsymbol{u} \tag{4.68}$$

$$\boldsymbol{y} = [\boldsymbol{0}_r \quad \cdots \quad \boldsymbol{0}_r \quad \boldsymbol{I}_r] \begin{bmatrix} x_1 \\ x_2 \\ \vdots \\ x_{n\times r} \end{bmatrix}$$

Obviously, the dimension of the realization in controllable canonical form is $n\times m$, and the dimension of the realization in observable canonical form is $n\times r$. For the simplification of the realization, the controllable canonical form may be selected as the realization if $m<r$, and the observable canonical form may be selected as the realization if $m>r$.

Example 4.27 Consider the system described by the transfer function matrix.

$$\boldsymbol{G}(s) = \begin{bmatrix} \dfrac{1}{(s+1)(s+2)} & \dfrac{1}{(s+2)(s+3)} \end{bmatrix}$$

Determine the realization in controllable canonical form and in observable canonical form.

Solution The transfer function matrix can be written as

$$\boldsymbol{G}(s) = \frac{1}{(s+1)(s+2)(s+3)}[(s+3) \quad (s+1)]$$

$$= \frac{1}{s^3+6s^2+11s+6}\{[1 \quad 1]s + [3 \quad 1]\}$$

From the transfer function matrix, the coefficients are noted as

$$a_1=6,\ a_2=11,\ a_3=6;$$

$$\boldsymbol{B}_1=[0 \quad 0],\ \boldsymbol{B}_2=[1 \quad 1],\ \boldsymbol{B}_3=[3 \quad 1]$$

The 3 dimension observable canonical form realization is obtained as

$$\begin{bmatrix} \dot{x}_1 \\ \dot{x}_2 \\ \dot{x}_3 \end{bmatrix} = \begin{bmatrix} \boldsymbol{0}_1 & \boldsymbol{0}_1 & -a_3\boldsymbol{I}_1 \\ \boldsymbol{I}_1 & \boldsymbol{0}_1 & -a_2\boldsymbol{I}_1 \\ \boldsymbol{0}_1 & \boldsymbol{I}_1 & -a_1\boldsymbol{I}_1 \end{bmatrix} \begin{bmatrix} x_1 \\ x_2 \\ x_3 \end{bmatrix} + \begin{bmatrix} \boldsymbol{B}_3 \\ \boldsymbol{B}_2 \\ \boldsymbol{B}_1 \end{bmatrix} \boldsymbol{u}$$

$$= \begin{bmatrix} 0 & 0 & -6 \\ 1 & 0 & -11 \\ 0 & 1 & -6 \end{bmatrix} \begin{bmatrix} x_1 \\ x_2 \\ x_3 \end{bmatrix} + \begin{bmatrix} 3 & 1 \\ 1 & 1 \\ 0 & 0 \end{bmatrix} \boldsymbol{u}$$

$$y = [\boldsymbol{0}_1 \quad \boldsymbol{0}_1 \quad \boldsymbol{I}_1] \begin{bmatrix} x_1 \\ x_2 \\ x_3 \end{bmatrix} = [0 \quad 0 \quad 1] \begin{bmatrix} x_1 \\ x_2 \\ x_3 \end{bmatrix}$$

Furthermore, the 6 dimension controllable canonical form realization can be obtained as

$$\begin{bmatrix}\dot{x}_1\\ \dot{x}_2\\ \dot{x}_3\\ \dot{x}_4\\ \dot{x}_5\\ \dot{x}_6\end{bmatrix}=\begin{bmatrix}\mathbf{0}_2 & \boldsymbol{I}_2 & \mathbf{0}_2\\ \mathbf{0}_2 & \mathbf{0}_2 & \boldsymbol{I}_2\\ -a_3\boldsymbol{I}_2 & -a_2\boldsymbol{I}_2 & -a_1\boldsymbol{I}_2\end{bmatrix}\begin{bmatrix}x_1\\ x_2\\ x_3\\ x_4\\ x_5\\ x_6\end{bmatrix}+\begin{bmatrix}\mathbf{0}_2\\ \mathbf{0}_2\\ \boldsymbol{I}_2\end{bmatrix}\boldsymbol{u}$$

$$=\begin{bmatrix}0 & 0 & 1 & 0 & 0 & 0\\ 0 & 0 & 0 & 1 & 0 & 0\\ 0 & 0 & 0 & 0 & 1 & 0\\ 0 & 0 & 0 & 0 & 0 & 1\\ -6 & 0 & -11 & 0 & -6 & 0\\ 0 & -6 & 0 & -11 & 0 & -6\end{bmatrix}\begin{bmatrix}x_1\\ x_2\\ x_3\\ x_4\\ x_5\\ x_6\end{bmatrix}+\begin{bmatrix}0 & 0\\ 0 & 0\\ 0 & 0\\ 0 & 0\\ 1 & 0\\ 0 & 1\end{bmatrix}\boldsymbol{u}$$

$$y=[\boldsymbol{B}_3\quad \boldsymbol{B}_2\quad \boldsymbol{B}_1]\begin{bmatrix}x_1\\ x_2\\ x_3\\ x_4\\ x_5\\ x_6\end{bmatrix}=[3\quad 1\quad 1\quad 1\quad 0\quad 0]\begin{bmatrix}x_1\\ x_2\\ x_3\\ x_4\\ x_5\\ x_6\end{bmatrix}$$

4.6.4 Minimal Realization

From the discussion above, we have that if the LTI system is either uncontrollable or unobservable, there exists a state space description of less dimension that has the same transfer function matrix as the original state space description.

Definition 4.5 Suppose the state space description (4.59) is a realization of $\boldsymbol{G}(s)$. If

$$\dim(\boldsymbol{A})=\deg(\boldsymbol{G}(s)) \tag{4.69}$$

Then the realization (4.59) is called a **minimal realization** of $\boldsymbol{G}(s)$. Where, $\dim(\boldsymbol{A})$ is the dimension of the system matrix $\boldsymbol{A}$, $\deg(\boldsymbol{G}(s))$ is the order of the transfer function matrix $\boldsymbol{G}(s)$. The minimal realization is sometimes called an **irreducible realization.**

Theorem 4.18 The realization (4.59) is a minimal realization of $\boldsymbol{G}(s)$ iff it is completely controllable and completely observable.

Example 4.28 Determine a minimal realization of the following transfer function matrix.

$$\boldsymbol{G}(s)=\begin{bmatrix}\dfrac{s+3}{(s+1)(s+2)}\\ \dfrac{s+4}{s+3}\end{bmatrix}$$

Solution Since

$$\lim_{s\to\infty} \boldsymbol{G}(s)=\begin{bmatrix}0\\1\end{bmatrix}=\boldsymbol{D}$$

so, $\boldsymbol{G}(s)$ is proper and can be decomposed as

$$\boldsymbol{G}(s)=\begin{bmatrix}\dfrac{s+3}{(s+1)(s+2)}\\ \dfrac{1}{s+3}\end{bmatrix}+\begin{bmatrix}0\\1\end{bmatrix}=\boldsymbol{G}'(s)+\boldsymbol{D}$$

Where $\boldsymbol{G}'(s)$ is strictly proper and can be written as

$$\boldsymbol{G}'(s)=\frac{1}{(s+1)(s+2)(s+3)}\begin{bmatrix}(s+3)^2\\(s+1)(s+2)\end{bmatrix}=\frac{1}{s^3+6s^2+11s+6}\begin{bmatrix}s^2+6s+9\\s^2+3s+2\end{bmatrix}$$

$$=\frac{1}{s^3+6s^2+11s+6}\left\{\begin{bmatrix}1\\1\end{bmatrix}s^2+\begin{bmatrix}6\\3\end{bmatrix}s+\begin{bmatrix}9\\2\end{bmatrix}\right\}$$

From the transfer function matrix above, the coefficients are noted as

$$a_1=6, a_2=11, a_3=6;$$

$$\boldsymbol{B}_1=\begin{bmatrix}1\\1\end{bmatrix}, \boldsymbol{B}_2=\begin{bmatrix}6\\3\end{bmatrix}, \boldsymbol{B}_3=\begin{bmatrix}9\\2\end{bmatrix}$$

Since $m<r$, the 3 dimension controllable canonical form is selected as a realization of $\boldsymbol{G}(s)$ such as

$$\begin{bmatrix}\dot{x}_1\\ \dot{x}_2\\ \dot{x}_3\end{bmatrix}=\begin{bmatrix}\boldsymbol{0}_1 & \boldsymbol{I}_1 & \boldsymbol{0}_1\\ \boldsymbol{0}_1 & \boldsymbol{0}_1 & \boldsymbol{I}_1\\ -a_3\boldsymbol{I}_1 & -a_2\boldsymbol{I}_1 & -a_1\boldsymbol{I}_1\end{bmatrix}\begin{bmatrix}x_1\\x_2\\x_3\end{bmatrix}+\begin{bmatrix}\boldsymbol{0}_1\\ \boldsymbol{0}_1\\ \boldsymbol{I}_1\end{bmatrix}u$$

$$=\begin{bmatrix}0 & 1 & 0\\ 0 & 0 & 1\\ -6 & -11 & -6\end{bmatrix}\begin{bmatrix}x_1\\x_2\\x_3\end{bmatrix}+\begin{bmatrix}0\\0\\1\end{bmatrix}u$$

$$\boldsymbol{y}=\begin{bmatrix}\boldsymbol{B}_3 & \boldsymbol{B}_2 & \boldsymbol{B}_1\end{bmatrix}\begin{bmatrix}x_1\\x_2\\x_3\end{bmatrix}+\boldsymbol{D}u=\begin{bmatrix}9 & 6 & 1\\ 2 & 3 & 1\end{bmatrix}\begin{bmatrix}x_1\\x_2\\x_3\end{bmatrix}+\begin{bmatrix}0\\1\end{bmatrix}u$$

Furthermore, the rank of the observability matrix of it is calculated as

$$\text{rank}\boldsymbol{Q}_o=\text{rank}\begin{bmatrix}\boldsymbol{C}\\ \boldsymbol{CA}\\ \boldsymbol{CA}^2\end{bmatrix}=\text{rank}\begin{bmatrix}9 & 6 & 1\\ 2 & 3 & 1\\ -6 & -2 & 0\\ -6 & -9 & -3\\ 0 & -6 & -2\\ 18 & 27 & 9\end{bmatrix}=3$$

It means that the controllable canonical form is completely observable. So, the controllable canonical form above is a minimal realization of $\boldsymbol{G}(s)$.

Example 4.29 Determine a minimal realization of the following transfer function matrix.

$$G(s)=\begin{bmatrix}\dfrac{1}{s+1} & \dfrac{1}{s+3}\\ \dfrac{-1}{s+1} & \dfrac{-1}{s+2}\end{bmatrix}$$

Solution $G(s)$ is strictly proper and can be written as

$$G(s)=\frac{1}{(s+1)(s+2)(s+3)}\begin{bmatrix}(s+2)(s+3) & (s+1)(s+2)\\ -(s+2)(s+3) & -(s+1)(s+3)\end{bmatrix}$$

$$=\frac{1}{s^3+6s^2+11s+6}\begin{bmatrix}s^2+5s+6 & s^2+3s+2\\ -(s^2+5s+6) & -(s^2+4s+3)\end{bmatrix}$$

$$=\frac{1}{s^3+6s^2+11s+6}\left\{\begin{bmatrix}1 & 1\\ -1 & -1\end{bmatrix}s^2+\begin{bmatrix}5 & 3\\ -5 & -4\end{bmatrix}s+\begin{bmatrix}6 & 2\\ -6 & -3\end{bmatrix}\right\}$$

From the transfer function matrix above, the coefficients are noted as

$$a_1=6,\ a_2=11,\ a_3=6;$$

$$\boldsymbol{B}_1=\begin{bmatrix}1 & 1\\ -1 & -1\end{bmatrix},\ \boldsymbol{B}_2=\begin{bmatrix}5 & 3\\ -5 & -4\end{bmatrix},\ \boldsymbol{B}_3=\begin{bmatrix}6 & 2\\ -6 & -3\end{bmatrix}$$

The 6 dimension controllable canonical form can be selected as a realization of $G(s)$.

$$\dot{\boldsymbol{X}}=\begin{bmatrix}\boldsymbol{0}_2 & \boldsymbol{I}_2 & \boldsymbol{0}_2\\ \boldsymbol{0}_2 & \boldsymbol{0}_2 & \boldsymbol{I}_2\\ -a_3\boldsymbol{I}_2 & -a_2\boldsymbol{I}_2 & -a_1\boldsymbol{I}_2\end{bmatrix}\boldsymbol{X}+\begin{bmatrix}\boldsymbol{0}_2\\ \boldsymbol{0}_2\\ \boldsymbol{I}_2\end{bmatrix}\boldsymbol{u}$$

$$=\begin{bmatrix}0 & 0 & 1 & 0 & 0 & 0\\ 0 & 0 & 0 & 1 & 0 & 0\\ 0 & 0 & 0 & 0 & 1 & 0\\ 0 & 0 & 0 & 0 & 0 & 1\\ -6 & 0 & -11 & 0 & -6 & 0\\ 0 & -6 & 0 & -11 & 0 & -6\end{bmatrix}\boldsymbol{X}+\begin{bmatrix}0 & 0\\ 0 & 0\\ 0 & 0\\ 0 & 0\\ 1 & 0\\ 0 & 1\end{bmatrix}\boldsymbol{u}$$

$$\boldsymbol{y}=[\boldsymbol{B}_3\quad \boldsymbol{B}_2\quad \boldsymbol{B}_1]\boldsymbol{X}=\begin{bmatrix}6 & 2 & 5 & 3 & 1 & 1\\ -6 & -3 & -5 & -4 & -1 & -1\end{bmatrix}\boldsymbol{X}$$

Furthermore, the rank of the observability matrix of it is calculated as

$$\text{rank}\boldsymbol{Q}_0=\text{rank}\begin{bmatrix}\boldsymbol{C}\\ \boldsymbol{CA}\\ \boldsymbol{CA}^2\\ \boldsymbol{CA}^3\\ \boldsymbol{CA}^4\\ \boldsymbol{CA}^5\end{bmatrix}=\text{rank}\begin{bmatrix}6 & 2 & 5 & 3 & 1 & 1\\ -6 & -3 & -5 & -4 & -1 & -1\\ -6 & -6 & -5 & -9 & -1 & -3\\ 6 & 6 & 5 & 8 & 1 & 2\\ 6 & 18 & 5 & 27 & 1 & 9\\ -6 & -12 & -5 & -16 & -1 & -4\\ -6 & -54 & -5 & -81 & -1 & -27\\ 6 & 24 & 5 & 32 & 1 & 8\\ 6 & 162 & 5 & 243 & 1 & 81\\ -6 & -48 & -5 & -64 & -1 & -16\\ -6 & -486 & -5 & -729 & -1 & -243\\ 6 & 96 & 5 & 128 & 1 & 32\end{bmatrix}=3<6$$

So, the controllable canonical form is not completely observable.

A nonsingular matrix can be constructed as

$$\boldsymbol{Q}=\boldsymbol{P}^{-1}=\begin{bmatrix}\boldsymbol{q}_1\\ \boldsymbol{q}_2\\ \boldsymbol{q}_3\\ \boldsymbol{q}_4\\ \boldsymbol{q}_5\\ \boldsymbol{q}_6\end{bmatrix}=\begin{bmatrix}6 & 2 & 5 & 3 & 1 & 1\\ -6 & -3 & -5 & -4 & -1 & -1\\ -6 & -6 & -5 & -9 & -1 & -3\\ 1 & 0 & 0 & 0 & 0 & 0\\ 0 & 1 & 0 & 0 & 0 & 0\\ 0 & 0 & 1 & 0 & 0 & 0\end{bmatrix}$$

Where $\boldsymbol{q}_1, \boldsymbol{q}_2, \boldsymbol{q}_3$ are the front three row vectors of $\boldsymbol{Q}_o$ and they are linearly independent.

Furthermore, the inverse matrix of $\boldsymbol{Q}$ is

$$\boldsymbol{P}=\begin{bmatrix}0 & 0 & 0 & 1 & 0 & 0\\ 0 & 0 & 0 & 0 & 1 & 0\\ 0 & 0 & 0 & 0 & 0 & 1\\ -1 & -1 & 0 & 0 & -1 & 0\\ \frac{3}{2} & 0 & \frac{1}{2} & -6 & 0 & -5\\ \frac{5}{2} & 3 & -\frac{1}{2} & 0 & 1 & 0\end{bmatrix}$$

By the nonsingular transformation $\boldsymbol{X}=\boldsymbol{P}\overline{\boldsymbol{X}}$, the system can be transformed into the observable canonical decomposition form such as

$$\dot{\overline{\boldsymbol{X}}}=\overline{\boldsymbol{A}}\,\overline{\boldsymbol{X}}+\overline{\boldsymbol{B}}\boldsymbol{u}$$
$$y=\overline{\boldsymbol{C}}\,\overline{\boldsymbol{X}}$$

Where

$$\overline{\boldsymbol{A}}=\boldsymbol{P}^{-1}\boldsymbol{A}\boldsymbol{P}=\left[\begin{array}{ccc:ccc}0 & 0 & 1 & 0 & 0 & 0\\ -\frac{3}{2} & -2 & -\frac{1}{2} & 0 & 0 & 0\\ -3 & 0 & -4 & 0 & 0 & 0\\ \hdashline 0 & 0 & 0 & 0 & 0 & 1\\ -1 & -1 & 0 & 0 & -1 & 0\\ \frac{3}{2} & 0 & \frac{1}{2} & 0 & 0 & 5\end{array}\right];\overline{\boldsymbol{B}}=\boldsymbol{P}^{-1}\boldsymbol{B}=\left[\begin{array}{cc}1 & 1\\ -1 & -1\\ -1 & -3\\ \hdashline 0 & 0\\ 0 & 0\\ 0 & 0\end{array}\right];$$

$$\overline{\boldsymbol{C}}=\boldsymbol{C}\boldsymbol{P}=\left[\begin{array}{ccc:ccc}1 & 0 & 0 & 0 & 0 & 0\\ 0 & 1 & 0 & 0 & 0 & 0\end{array}\right].$$

Since a change of state by the nonsingular transformation does not change the controllability, then the observable subsystem

$$\dot{\overline{\boldsymbol{X}}}_o=\begin{bmatrix}0 & 0 & 1\\ -\frac{3}{2} & -2 & -\frac{1}{2}\\ -3 & 0 & -4\end{bmatrix}\overline{\boldsymbol{X}}_o+\begin{bmatrix}1 & 1\\ -1 & -1\\ -1 & -3\end{bmatrix}\boldsymbol{u}$$

$$y=\begin{bmatrix}1 & 0 & 0\\0 & 1 & 0\end{bmatrix}\overline{X}_o$$

is completely controllable and completely observable and is a minimal realization of $G(s)$.

Note that the minimal realization of a transfer function matrix $G(s)$ is not unique, and the different minimal realizations are equivalent.

Theorem 4.19 The SISO LTI system is completely controllable and observable iff its transfer function does not have **pole-zero cancellation.**

It means that if the transfer function $G(s)$ has pole-zero cancellation, the realization of $G(s)$ may be uncontrollable and/or unobservable.

Problems

4.1 Determine the controllability and observability of the following systems.

(1) $\dot{X}=\begin{bmatrix}0 & 1\\0 & -3\end{bmatrix}X+\begin{bmatrix}0\\1\end{bmatrix}u, y=[0\quad 2]X$

(2) $\dot{X}=\begin{bmatrix}0 & 1 & 0\\0 & 0 & 1\\0 & -2 & -3\end{bmatrix}X+\begin{bmatrix}0\\1\\1\end{bmatrix}u,\quad y=[1\quad 0\quad 1]X$

4.2 Determine the controllability and observability of the following systems.

(1) $\dot{X}=\begin{bmatrix}-4 & 0\\0 & -1\end{bmatrix}X+\begin{bmatrix}0\\1\end{bmatrix}u, y=[1\quad 0]X$

(2) $\dot{X}=\begin{bmatrix}-3 & 1 & 0\\0 & -3 & 0\\0 & 0 & -1\end{bmatrix}X+\begin{bmatrix}1 & -1\\0 & 0\\3 & 0\end{bmatrix}\boldsymbol{u},\quad y=\begin{bmatrix}1 & 0 & 2\\3 & 0 & 2\end{bmatrix}X$

(3) $\dot{X}=\begin{bmatrix}-4 & 0 & 0\\0 & -4 & 0\\0 & 0 & 1\end{bmatrix}X+\begin{bmatrix}1\\1\\1\end{bmatrix}u,\quad y=[2\quad 1\quad 3]X$

4.3 Consider the LTI system

$$\dot{X}=\begin{bmatrix}a & 1\\0 & b\end{bmatrix}X+\begin{bmatrix}1\\1\end{bmatrix}u, y=[1\quad -1]X$$

Find the region in the a-b plane such that the system is completely controllable and completely observable.

4.4 Consider the LTI system described by

$$\dot{X}=\begin{bmatrix}-1 & 0\\1 & -2\end{bmatrix}X+\begin{bmatrix}1\\-1\end{bmatrix}u, y=[1\quad 1]X$$

(1) Is it possible to transform the state space description into the controllable canonical form ? If possible, try to do it.

(2) Is it possible to transform the state space description into the observable canon-

ical form ? If possible, try to do it.

4.5 Consider the LTI system described by

$$\dot{\boldsymbol{X}}=\begin{bmatrix}-1 & 0 & 0\\ 0 & -1 & 0\\ 0 & -2 & -2\end{bmatrix}\boldsymbol{X}+\begin{bmatrix}0\\1\\1\end{bmatrix}u,\quad y=[1\ \ 1\ \ 0]\boldsymbol{X}$$

(1) Find the controllable subsystem of the system.

(2) Find the observable subsystem of the system.

4.6 Consider the LTI system described by

$$\dot{\boldsymbol{X}}=\begin{bmatrix}-1 & 0 & 0\\ 0 & -2 & 0\\ 0 & 0 & -3\end{bmatrix}\boldsymbol{X}+\begin{bmatrix}0\\1\\1\end{bmatrix}u,y=[1\ \ 1\ \ 0]\boldsymbol{X}$$

Calculate the transfer function of the system.

4.7 Consider the LTI system described by

$$G(s)=\frac{s+1}{s^2+3s+2}$$

Find a realization of the system, which is

(1) controllable but unobservable.

(2) observable but uncontrollable.

(3) neither controllable nor observable.

4.8 Find a minimal realization of the following system.

$$\boldsymbol{G}(s)=\begin{bmatrix}\dfrac{1}{s(s+1)} & \dfrac{2}{s+1}\\ \dfrac{2}{s+1} & \dfrac{1}{s+1}\end{bmatrix}$$

Lev Semenovich Pontryagin (1908—1988) was a Soviet mathematician. He was born in Moscow and lost his eyesight due to a primus stove explosion when he was 14. Despite his blindness he was able to become a mathematician due to the help of his mother Tatyana Andreevna who read mathematical books and papers to him. He made major discoveries in a number of fields of mathematics, including algebraic topology and differential topology.

He worked on duality theory for homology while still a student. He went on to lay foundations for the abstract theory of the Fourier transform, now called *Pontryagin duality*. In topology he posed the basic problem of cobordism theory. This led to the introduction around 1940 of a theory of certain characteristic classes, now called *Pontryagin classes*, designed to vanish on a manifold that is a boundary. In 1942 he introduced the cohomology operations now called *Pontryagin squares*. Moreover, in operator theory there are specific instances of Krein spaces called *Pontryagin spaces*.

Later in his career he worked in *optimal control theory*. His maximum principle is fundamental to the modern theory of optimization. He also introduced the idea of *a bang-bang principle*, to describe situations where either the maximum 'steer' should be applied to a system, or none.

Chapter5 Synthesis of the LTI System

Design with the transfer function based on methods such as root locus and frequency response has been referred to as the classical control design, and design with the state space approach has been referred to as the modern control design, or system synthesis.

The state space design method consists of a sequence of independent steps. First we design the control law as if all of the state variables are measured and available for use in the full-state feedback control law. The control law allows us to assign a set of pole locations for the closed-loop system that will satisfy the dynamic response indexes. Having a satisfactory control law based on state feedback, we introduce the concept of an observer and construct estimates of the state based on the sensed output. We then show these estimates can be used in place of the actual state variables. Finally, we will discuss how the control law and the observer fit together. It will be obtained that the combined control law and observer results in the closed-loop pole locations that are the same as those determined when designing the control and the observer separately. Only the SISO system will be considered in this chapter.

5.1 State Feedback Control of the LTI System

5.1.1 State Feedback

Definition 5.1 Consider the LTI system described by

$$\begin{cases} \dot{\boldsymbol{X}}(t)=\boldsymbol{A}\boldsymbol{X}(t)+\boldsymbol{b}u(t) \\ y(t)=\boldsymbol{c}\boldsymbol{X}(t)+\boldsymbol{d}u(t) \end{cases} \tag{5.1}$$

A linear **state feedback control law** is defined by

$$u(t)=-\boldsymbol{K}\boldsymbol{X}(t)+r(t) \tag{5.2}$$

Where $r(t)$ is the reference input into the system, and

$$\boldsymbol{K}=[k_1 \quad k_2 \quad \cdots \quad k_n] \tag{5.3}$$

is a constant state feedback gain vector.

By introducing the state feedback, a closed-loop system is obtained as

$$\begin{aligned} \dot{\boldsymbol{X}}(t)&=(\boldsymbol{A}-\boldsymbol{b}\boldsymbol{K})\boldsymbol{X}(t)+\boldsymbol{b}r(t) \\ y(t)&=(\boldsymbol{c}-\boldsymbol{d}\boldsymbol{K})\boldsymbol{X}(t)+\boldsymbol{d}r(t) \end{aligned} \tag{5.4}$$

The closed-loop system is shown in Figure 5.1.

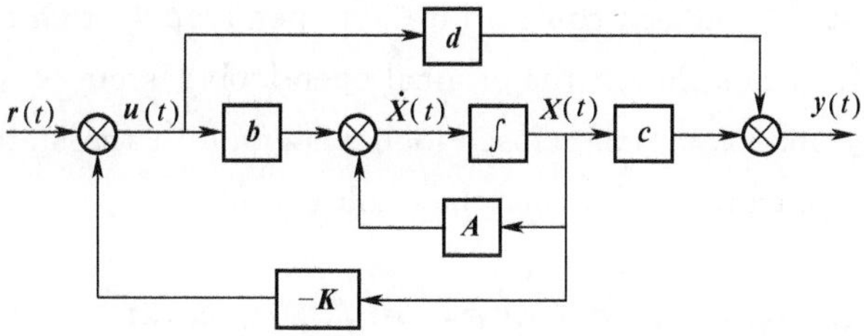

Figure 5.1 Closed-Loop System with State Feedback Control

The transfer function of the closed-loop system is

$$G(s)=(\boldsymbol{c}-\boldsymbol{dK})(s\boldsymbol{I}-\boldsymbol{A}+\boldsymbol{bK})^{-1}\boldsymbol{b}+\boldsymbol{d} \tag{5.5}$$

If $\boldsymbol{G}(s)$ is strictly proper, i. e. $\boldsymbol{d}=\boldsymbol{0}$, we have the closed-loop system as

$$\begin{aligned}\dot{\boldsymbol{X}}(t)&=(\boldsymbol{A}-\boldsymbol{bK})\boldsymbol{X}(t)+\boldsymbol{b}r(t)\\ y(t)&=\boldsymbol{cX}(t)\end{aligned} \tag{5.6}$$

and the transfer function of it is

$$G(s)=\boldsymbol{c}(s\boldsymbol{I}-\boldsymbol{A}+\boldsymbol{bK})^{-1}\boldsymbol{b} \tag{5.7}$$

In the following discussion, we mainly consider the case, $\boldsymbol{d}=\boldsymbol{0}$.

5.1.2 Controllability and Observability of the Closed-Loop System

Theorem 5.1 The closed-loop system (5.4) and (5.6) with the state feedback control is completely controllable iff the original open-loop system (5.1) is completely controllable. In other words, the controllability can be preserved by introducing the state feedback control into the open-loop system. However, the observability may be changed by introducing the state feedback control into the open-loop system.

Proof. The controllability matrix of the open-loop system (5.1) is

$$\boldsymbol{Q}_c=[\boldsymbol{b} \quad \boldsymbol{Ab} \quad \boldsymbol{A}^2\boldsymbol{b} \quad \cdots \quad \boldsymbol{A}^{n-1}\boldsymbol{b}]$$

Furthermore, the controllability matrix of the closed-loop system (5.4) and (5.6) is

$$\begin{aligned}\boldsymbol{Q}_{cf}&=[\boldsymbol{b} \quad (\boldsymbol{A}-\boldsymbol{bK})\boldsymbol{b} \quad (\boldsymbol{A}-\boldsymbol{bK})^2\boldsymbol{b} \quad \cdots \quad (\boldsymbol{A}-\boldsymbol{bK})^{n-1}\boldsymbol{b}]\\ &=[\boldsymbol{b} \quad \boldsymbol{Ab} \quad \boldsymbol{A}^2\boldsymbol{b} \quad \cdots \quad \boldsymbol{A}^{n-1}\boldsymbol{b}]\begin{bmatrix}1 & -\boldsymbol{Kb} & -\boldsymbol{K}(\boldsymbol{A}-\boldsymbol{bK})\boldsymbol{b} & \cdots & * \\ 0 & 1 & -\boldsymbol{Kb} & & \vdots \\ 0 & 0 & 1 & \ddots & \vdots \\ \vdots & \vdots & \ddots & \ddots & \\ 0 & 0 & \cdots & 0 & 1\end{bmatrix}\end{aligned}$$

Obviously, $\quad \operatorname{rank} \boldsymbol{Q}_c=\operatorname{rank}\boldsymbol{Q}_{cf}$

It means that the controllability can be preserved by introducing the state feedback control into the open-loop system.

Suppose the eigenvalues of the closed-loop system are $\lambda_i(\boldsymbol{A}-\boldsymbol{bK}), i=1,2,\cdots,n$. They may be different from the eigenvalues, $\lambda_i(\boldsymbol{A})$, of the original open-loop system.

So, introducing the state feedback control into the open-loop system may yield the pole-zeros cancellation. It means that, if the original open-loop system is completely controllable and completely observable, the closed-loop system with the state feedback control must be completely controllable, but may be unobservable.

5.1.3 Poles Placement by State Feedback Control

An important property of state feedback is that it can be used to assign a set of poles for the closed-loop system which satisfy the dynamic response indexes. The process is called **poles placement** of the control systems.

The pole here refer to that of the transfer function (5.7), which are also the roots of the characteristic equation

$$|s\boldsymbol{I}-\boldsymbol{A}+\boldsymbol{bK}|=0 \tag{5.8}$$

of the closed-loop system (5.6). In other words, they refer to the eigenvalues of the closed-loop system. Knowing the relationship between the closed-loop poles and the system performance, we can effectively carry out the design by specifying the locations of these poles. A natural question would be: Under what condition can the poles be placed arbitrarily?

Theorem 5.2 The poles, or eigenvalues, of the closed-loop system by introducing the state feedback control can be placed arbitrarily iff the original open-loop system is completely controllable.

Proof. (1)Necessity. Consider a SISO LTI system such as (5.9)

$$\begin{aligned}\dot{\boldsymbol{X}}(t)&=\boldsymbol{AX}(t)+\boldsymbol{b}u(t)\\ y(t)&=\boldsymbol{cX}(t)\end{aligned} \tag{5.9}$$

Suppose that the poles of the system can be placed arbitrarily by introducing the state feedback control, but the system is not completely controllable. Then there exists a nonsingular transformation $\boldsymbol{X}=\boldsymbol{P}\bar{\boldsymbol{X}}$, which transforms the description of the LTI system (5.9) into the controllable canonical decomposition form such as

$$\begin{aligned}\begin{bmatrix}\dot{\bar{\boldsymbol{X}}}_c\\ \dot{\bar{\boldsymbol{X}}}_{\bar{c}}\end{bmatrix}&=\begin{bmatrix}\bar{\boldsymbol{A}}_c & \bar{\boldsymbol{A}}_{12}\\ \boldsymbol{0} & \bar{\boldsymbol{A}}_{\bar{c}}\end{bmatrix}\begin{bmatrix}\bar{\boldsymbol{X}}_c\\ \bar{\boldsymbol{X}}_{\bar{c}}\end{bmatrix}+\begin{bmatrix}\bar{\boldsymbol{b}}_c\\ \boldsymbol{0}\end{bmatrix}u\\ y&=\begin{bmatrix}\bar{\boldsymbol{c}}_c & \bar{\boldsymbol{c}}_{\bar{c}}\end{bmatrix}\begin{bmatrix}\bar{\boldsymbol{X}}_c\\ \bar{\boldsymbol{X}}_{\bar{c}}\end{bmatrix}\end{aligned} \tag{5.10}$$

By introducing the state feedback control

$$u(t)=-\bar{\boldsymbol{K}}\bar{\boldsymbol{X}}(t)+r(t) \tag{5.11}$$

into (5.10), the closed-loop system is obtained as

$$\begin{bmatrix}\dot{\bar{\boldsymbol{X}}}_c\\ \dot{\bar{\boldsymbol{X}}}_{\bar{c}}\end{bmatrix}=\begin{bmatrix}\bar{\boldsymbol{A}}_c-\bar{\boldsymbol{b}}_c\bar{\boldsymbol{K}}_c & \bar{\boldsymbol{A}}_{12}-\bar{\boldsymbol{b}}_c\bar{\boldsymbol{K}}_{\bar{c}}\\ \boldsymbol{0} & \bar{\boldsymbol{A}}_{\bar{c}}\end{bmatrix}\begin{bmatrix}\bar{\boldsymbol{X}}_c\\ \bar{\boldsymbol{X}}_{\bar{c}}\end{bmatrix}+\begin{bmatrix}\bar{\boldsymbol{b}}_c\\ \boldsymbol{0}\end{bmatrix}u$$

$$y=[\bar{c}_c \quad \bar{c}_{\bar{c}}]\begin{bmatrix}\bar{X}_c \\ \bar{X}_{\bar{c}}\end{bmatrix} \tag{5.12}$$

Where

$$\bar{K}=[\bar{K}_c \quad \bar{K}_{\bar{c}}] \tag{5.13}$$

The characteristic polynomial of the system can be derived as

$$\begin{vmatrix} sI-(\bar{A}_c-\bar{b}_c\bar{K}_c) & -(\bar{A}_{12}-\bar{b}_c\bar{K}_{\bar{c}}) \\ 0 & sI-\bar{A}_{\bar{c}} \end{vmatrix}=|sI-(\bar{A}_c-\bar{b}_c\bar{K}_c)|\cdot|sI-\bar{A}_{\bar{c}}| \tag{5.14}$$

It means that the eigenvalues of the controllable subsystem of (5.10) can be placed arbitrarily but the eigenvalues of the uncontrollable subsystem can not be placed arbitrarily. It is conflict with our assumption. So, it is concluded that the original open-loop system (5.9) is completely controllable.

(2) Sufficiency.

If the system (5.9) is completely controllable, then it is transformable into the controllable canonical form (5.15) by a nonsingular transformation $X=P\bar{X}$.

$$\begin{aligned}\dot{\bar{X}}(t)&=\bar{A}\bar{X}(t)+\bar{b}u(t)\\ y(t)&=\bar{c}\bar{X}(t)\end{aligned} \tag{5.15}$$

Where

$$\bar{A}=P^{-1}AP=\begin{bmatrix} 0 & 1 & 0 & \cdots & 0 \\ 0 & 0 & 1 & \cdots & 0 \\ \vdots & \vdots & \vdots & \ddots & \vdots \\ 0 & 0 & 0 & \cdots & 1 \\ -a_n & -a_{n-1} & -a_{n-2} & \cdots & -a_1 \end{bmatrix};$$

$$\bar{b}=P^{-1}b=\begin{bmatrix}0\\0\\\vdots\\0\\1\end{bmatrix};\bar{c}=cP=[b_n \quad b_{n-1} \quad \cdots \quad b_2 \quad b_1].$$

Since a change of state by the nonsingular transformation does not change the transfer function, so the transfer function of system (5.9) and (5.15) is

$$g(s)=\frac{b_1s^{n-1}+\cdots+b_{n-1}s+b_n}{s^n+a_1s^{n-1}+a_2s^{n-2}+\cdots+a_{n-1}s+a_n} \tag{5.16}$$

Note that $a_1, a_2, \cdots a_n$ are the coefficients of the characteristic polynomial of the system such as

$$\alpha_0(s)=|sI-A|=|sI-\bar{A}|=s^n+a_1s^{n-1}+\cdots+a_{n-1}s+a_n \tag{5.17}$$

Suppose the expected poles of the system are $s_1^*, s_2^*, \cdots, s_n^*$, the expected characteristic polynomial can be derived as

$$\alpha^*(s)=\prod_{i=1}^{n}(s-s_i^*)=s^n+a_1^*s^{n-1}+\cdots+a_{n-1}^*s+a_n^* \tag{5.18}$$

Obviously, $a_1^*, a_2^*, \cdots, a_n^*$ are the coefficients of the expected characteristic polynomial of the system.

If we introduce a state feedback control such as (5.19) into system (5.15)

$$u(t) = -\overline{\boldsymbol{K}}\,\overline{\boldsymbol{X}}(t) + r(t) \tag{5.19}$$

Where

$$\begin{aligned}\overline{\boldsymbol{K}} &= \boldsymbol{KP} = [\overline{k}_1 \quad \overline{k}_2 \quad \cdots \quad \overline{k}_n] \\ &= [a_n^* - a_n \quad a_{n-1}^* - a_{n-1} \quad \cdots \quad a_1^* - a_1]\end{aligned} \tag{5.20}$$

The closed-loop system can be obtained as

$$\begin{aligned}\dot{\overline{\boldsymbol{X}}}(t) &= (\overline{\boldsymbol{A}} - \overline{\boldsymbol{b}}\,\overline{\boldsymbol{K}})\overline{\boldsymbol{X}}(t) + \overline{\boldsymbol{b}}r(t) \\ y(t) &= \overline{\boldsymbol{c}}\,\overline{\boldsymbol{X}}(t)\end{aligned} \tag{5.21}$$

and the system matrix of it is

$$\overline{\boldsymbol{A}} - \overline{\boldsymbol{b}}\,\overline{\boldsymbol{K}} = \begin{bmatrix} 0 & 1 & \cdots & 0 \\ \vdots & \vdots & \ddots & \vdots \\ 0 & 0 & \cdots & 1 \\ -a_n & -a_{n-1} & \cdots & -a_1 \end{bmatrix} - \begin{bmatrix} 0 \\ \vdots \\ 0 \\ 1 \end{bmatrix} [a_n^* - a_n \quad a_{n-1}^* - a_{n-1} \quad \cdots \quad a_1^* - a_1]$$

$$= \begin{bmatrix} 0 & 1 & \cdots & 0 \\ \vdots & \vdots & \ddots & \vdots \\ 0 & 0 & \cdots & 1 \\ -a_n & -a_{n-1} & \cdots & -a_1 \end{bmatrix} - \begin{bmatrix} 0 & 0 & \cdots & 0 \\ \vdots & \vdots & \cdots & \vdots \\ 0 & 0 & \cdots & 0 \\ a_n^* - a_n & a_{n-1}^* - a_{n-1} & \cdots & a_1^* - a_1 \end{bmatrix}$$

$$= \begin{bmatrix} 0 & 1 & \cdots & 0 \\ \vdots & \vdots & \ddots & \vdots \\ 0 & 0 & \cdots & 1 \\ -a_n^* & -a_{n-1}^* & \cdots & -a_1^* \end{bmatrix}$$

The characteristic polynomial of the system (5.21) can be derived as

$$|s\boldsymbol{I} - \overline{\boldsymbol{A}} + \overline{\boldsymbol{b}}\,\overline{\boldsymbol{K}}| = s^n + a_1^* s^{n-1} + \cdots + a_{n-1}^* s + a_n^* = \alpha^*(s) \tag{5.22}$$

It means that the eigenvalues of the closed-loop system (5.21) can be placed arbitrarily at the expected locations, i. e. $s_1^*, s_2^*, \cdots, s_n^*$, by introducing the state feedback control into the open-loop system (5.15).

Since a change of state by the nonsingular transformation does not change the eigenvalues, we have

$$\begin{aligned}\alpha^*(s) &= |s\boldsymbol{I} - \overline{\boldsymbol{A}} + \overline{\boldsymbol{b}}\,\overline{\boldsymbol{K}}| = |s\boldsymbol{I} - \boldsymbol{P}^{-1}\boldsymbol{AP} + \boldsymbol{P}^{-1}\boldsymbol{bKP}| = |\boldsymbol{P}^{-1}(s\boldsymbol{I} - \boldsymbol{A} + \boldsymbol{bK})\boldsymbol{P}| \\ &= |s\boldsymbol{I} - \boldsymbol{A} + \boldsymbol{bK}| = s^n + a_1^* s^{n-1} + \cdots + a_{n-1}^* s + a_n^*\end{aligned}$$

Consequently, if we introduce the state feedback control

$$u(t) = -\boldsymbol{KX}(t) + r(t) \tag{5.23}$$

into the open-loop system (5.9), the eigenvalues, or poles, of the closed-loop system

$$\begin{aligned}\dot{\boldsymbol{X}}(t) &= (\boldsymbol{A} - \boldsymbol{bK})\boldsymbol{X}(t) + \boldsymbol{b}r(t) \\ y(t) &= \boldsymbol{cX}(t)\end{aligned} \tag{5.24}$$

can also be placed arbitrarily on the expected locations, $s_1^*, s_2^*, \cdots, s_n^*$, and the state feedback gain vector can be constructed as

$$K=\overline{K}P^{-1}=[k_1 \quad k_2 \quad \cdots \quad k_n] \tag{5.25}$$

Note that the proof of sufficiency for Theorem 5.2 tells us incidentally how to construct a state feedback control law based on the controllable canonical form.

Example 5.1 Consider the LTI system described by

$$\dot{X}=\begin{bmatrix}0 & 0 & 0\\ 1 & -6 & 0\\ 0 & 1 & -12\end{bmatrix}X+\begin{bmatrix}1\\0\\0\end{bmatrix}u$$

$$y=[0 \quad 0 \quad 1]X$$

Design the state feedback gain vector K that assigns the set of closed-loop eigenvalues as $\{-2,-1+j,-1-j\}$.

Solution The characteristic polynomial of the original open-loop system is

$$\alpha_0(s)=|sI-A|=\begin{vmatrix}s & 0 & 0\\ -1 & s+6 & 0\\ 0 & -1 & s+12\end{vmatrix}$$

$$=s^3+18s^2+72s=s^3+a_1s^2+a_2s+a_3$$

and the coefficients of the characteristic polynomial are

$$a_1=18, a_2=72, a_3=0$$

Furthermore, the expected characteristic polynomial of the system is

$$\alpha^*(s)=(s+2)(s+1-j)(s+1+j)=s^3+4s^2+6s+4=s^3+a_1^*s^2+a_2^*s+a_3^*$$

and the coefficients of it are

$$a_1^*=4, a_2^*=6, a_3^*=4$$

Let

$$\overline{K}=[a_3^*-a_3 \quad a_2^*-a_2 \quad a_1^*-a_1]=[4 \quad -66 \quad -14]$$

The controllability matrix of the system is

$$Q_c=[b \quad Ab \quad A^2b]=\begin{bmatrix}1 & 0 & 0\\ 0 & 1 & -6\\ 0 & 0 & 1\end{bmatrix}$$

Since rank $Q_c=3$, then the system is completely controllable, and there exists a nonsingular transformation $X=P\overline{X}$, which transforms the system into the controllable canonical form.

The nonsingular transformation matrix can be constructed as

$$P=[A^2b \quad Ab \quad b]\begin{bmatrix}1 & 0 & 0\\ \alpha_1 & 1 & 0\\ \alpha_2 & \alpha_1 & 1\end{bmatrix}$$

$$=\begin{bmatrix}0 & 0 & 1\\ -6 & 1 & 0\\ 1 & 0 & 0\end{bmatrix}\begin{bmatrix}1 & 0 & 0\\ 18 & 1 & 0\\ 72 & 18 & 1\end{bmatrix}=\begin{bmatrix}72 & 18 & 1\\ 12 & 1 & 0\\ 1 & 0 & 0\end{bmatrix}$$

and its inverse matrix is

$$P^{-1}=\begin{bmatrix}0 & 0 & 1\\ 0 & 1 & -12\\ 1 & -18 & 144\end{bmatrix}$$

So, the state feedback gain vector can be constructed as

$$K=\overline{K}P^{-1}=[4 \quad -66 \quad -14]\begin{bmatrix}0 & 0 & 1\\ 0 & 1 & -12\\ 1 & -18 & 144\end{bmatrix}$$

$$=[-14 \quad 186 \quad -1220]$$

By introducing the state feedback control $u(t)=-KX(t)+r(t)$, the poles of the closed-loop system can be placed at the expected locations, $-2, -1+\mathrm{j}, -1-\mathrm{j}$, and the closed-loop system is shown in Figure 5.2.

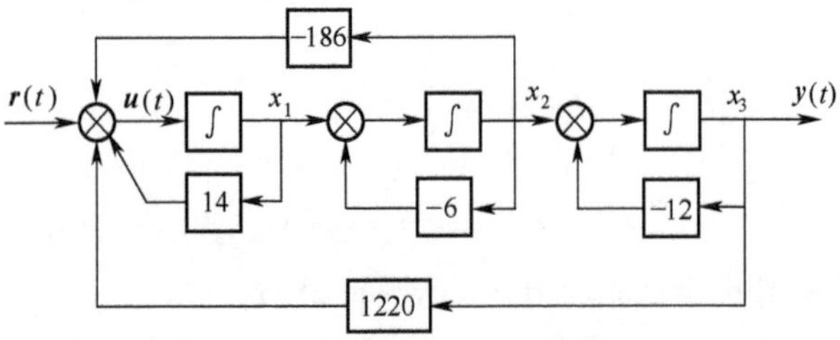

Figure 5.2 Closed-Loop System

In fact, by introducing the feedback control (5.23), the characteristic polynomial

$$\alpha(s)=|sI-A+bK| \tag{5.26}$$

of the closed-loop system (5.24) is a n th order polynomial in s containing the feedback gains $k_1, k_2, \cdots, k_n$, and the vector

$$K=[k_1 \quad k_2 \quad \cdots \quad k_n] \tag{5.27}$$

is the state feedback gain vector. Moreover, the expected characteristic polynomial of the closed-loop system can be determined by (5.18). Hence, the required elements $k_1, k_2, \cdots, k_n$ can be obtained by matching coefficients in the characteristic polynomial (5.26) and in the expected characteristic polynomial (5.18). This forces the characteristic polynomial to be the expected form by introducing the feedback control, and the closed-loop poles to the expected locations.

Example 5.2 Consider the LTI system described by

$$\dot{X}=\begin{bmatrix}0 & 0 & 0\\ 1 & -6 & 0\\ 0 & 1 & -12\end{bmatrix}X+\begin{bmatrix}1\\ 0\\ 0\end{bmatrix}u$$

$$y=[0 \quad 0 \quad 1]X$$

Design the state feedback gain vector K that assigns the set of closed-loop eigenvalues as $\{-2, -1+\mathrm{j}, -1-\mathrm{j}\}$.

Solution The expected characteristic polynomial of the system is

$$\alpha^*(s)=(s+2)(s+1-\mathrm{j})(s+1+\mathrm{j})=s^3+4s^2+6s+4=s^3+a_1^*s^2+a_2^*s+a_3^*$$

and the coefficients of it are

$$a_1^*=4, a_2^*=6, a_3^*=4$$

Suppose the state feedback gain vector is

$$\boldsymbol{K}=[k_1 \quad k_2 \quad k_3]$$

By introducing the state feedback control $u(t)=-\boldsymbol{KX}(t)+r(t)$, the characteristic polynomial

$$\alpha(s)=|s\boldsymbol{I}-\boldsymbol{A}+\boldsymbol{bK}|=\left|s\boldsymbol{I}-\begin{bmatrix}0 & 0 & 0\\ 1 & -6 & 0\\ 0 & 1 & -12\end{bmatrix}+\begin{bmatrix}1\\0\\0\end{bmatrix}[k_1 \quad k_2 \quad k_3]\right|$$

$$=\begin{vmatrix}s+k_1 & k_2 & k_3\\ -1 & s+6 & 0\\ 0 & -1 & s+12\end{vmatrix}$$

$$=s^3+(k_1+18)s^2+(18k_1+k_2+72)s+(72k_1+12k_2+k_3)$$

Equating the coefficients with like powers of s in $\alpha(s)$ and $\alpha^*(s)$ yields the feedback gains

$$k_1=-14, k_2=186, k_3=-1220$$

Obviously, the results are the same as the solutions of Example 5.1.

5.1.4 Zeros of the Closed-Loop System

Since the system (5.21) is a controllable canonical form, the transfer function of it can be calculated as

$$\bar{g}(s)=\frac{b_1s^{n-1}+\cdots+b_{n-1}s+b_n}{s^n+a_1^*s^{n-1}+a_2^*s^{n-2}+\cdots+a_{n-1}^*s+a_n^*}$$

$$=\bar{\boldsymbol{c}}(s\boldsymbol{I}-\bar{\boldsymbol{A}}+\bar{\boldsymbol{b}}\bar{\boldsymbol{K}})^{-1}\bar{\boldsymbol{b}}=\boldsymbol{cP}(s\boldsymbol{I}-\boldsymbol{P}^{-1}\boldsymbol{AP}+\boldsymbol{P}^{-1}\boldsymbol{bKP})^{-1}\boldsymbol{P}^{-1}\boldsymbol{b}$$

$$=\boldsymbol{cPP}^{-1}(s\boldsymbol{I}-\boldsymbol{A}+\boldsymbol{bK})^{-1}\boldsymbol{PP}^{-1}\boldsymbol{b}=\boldsymbol{c}(s\boldsymbol{I}-\boldsymbol{A}+\boldsymbol{bK})^{-1}\boldsymbol{b}$$

It means that the zeros of the closed-loop system (5.24) will not to be changed by introducing the state feedback control into the open-loop system (5.9) in the case $\boldsymbol{d}=\boldsymbol{0}$.

Example 5.3 Consider the LTI system described by

$$G(s)=\frac{1}{s(s-2)}$$

(1) Design a state feedback gain vector $\boldsymbol{K}$ that assigns the set of closed-loop eigenvalues as $\{-1,-1\}$.

(2) Determine the transfer function of the closed-loop system.

Solution (1) Rewrite the transfer function as

$$G(s)=\frac{1}{s(s-2)}=\frac{1}{s^2-2s}$$

The coefficients of the transfer function are noted as

$$b_1=0,b_2=1;a_1=-2,a_2=0$$

Then the state space description of the system in controllable canonical form can be obtained as

$$\begin{bmatrix}\dot{x}_1\\ \dot{x}_2\end{bmatrix}=\begin{bmatrix}0 & 1\\ -a_2 & -a_1\end{bmatrix}\begin{bmatrix}x_1\\ x_2\end{bmatrix}+\begin{bmatrix}0\\ 1\end{bmatrix}u=\begin{bmatrix}0 & 1\\ 0 & 2\end{bmatrix}\begin{bmatrix}x_1\\ x_2\end{bmatrix}+\begin{bmatrix}0\\ 1\end{bmatrix}u$$

$$y=[b_2 \quad b_1]\begin{bmatrix}x_1\\ x_2\end{bmatrix}=[1 \quad 0]\begin{bmatrix}x_1\\ x_2\end{bmatrix}$$

and the characteristic polynomial of the system is

$$\alpha(s)=s^2-2s=s^2+a_1s+a_2$$

Furthermore, the expected characteristic polynomial of the system can be determined as

$$\alpha^*(s)=(s+1)^2=s^2+2s+1=s^2+a_1^*s+a_2^*$$

Where $a_1^*=2;a_2^*=1$.

So, for the controllable canonical form realization of the system, the state feedback gain vector can be constructed directly as

$$\boldsymbol{K}=[k_1 \quad k_2]=[a_2^*-a_2 \quad a_1^*-a_1]=[1 \quad 4]$$

By introducing the state feedback control $u(t)=-\boldsymbol{KX}(t)+r(t)$, the poles of the closed-loop system can be placed at the expected locations, $-1,-1$, and the closed-loop system is shown in Figure 5.3.

(2) Since the transfer function of the system is strictly proper, i. e. $\boldsymbol{d}=\boldsymbol{0}$, then, introducing the state feedback control had no influence on zeros of the system, but placed the poles at the expected locations $-1,-1$ from the original locations 0, 2.

So, the transfer function of the closed-loop system can be determined directly as

$$G'(s)=\frac{1}{(s+1)^2}$$

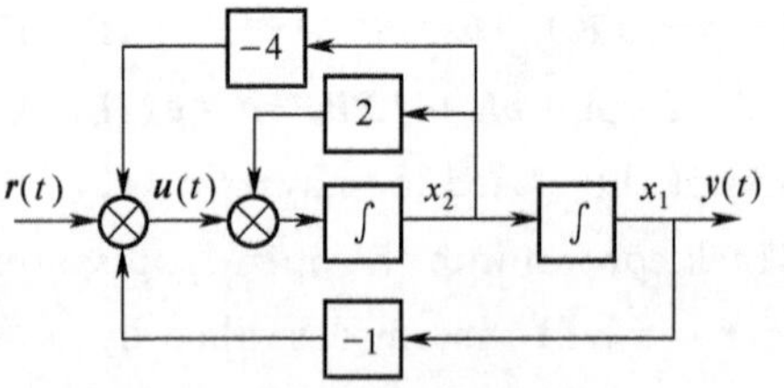

Figure 5.3 Closed-Loop System

5.2 Design of the State Observer

State feedback can be used to improve the system properties under the assumption that all the state variables are available. In most cases, not all the state variables can be measured. The cost of the required sensors may be prohibitive, or it may be physically impossible to measure all the state variables. Hence, in order to apply state feedback to optimize a system, a reasonable substitute for the state vector often has to be found.

In this section, we demonstrate how to reconstruct all the state variables by using the available inputs and outputs. Such a device that gives an estimation of the state vector is called a **state observer** or a **state estimator**.

5.2.1 Full-Order State Observer

One method of estimating the state is to reconstruct a full-order model of the original estimated system such as (5.9). The state reconstruct procedure is shown in Figure 5.4.

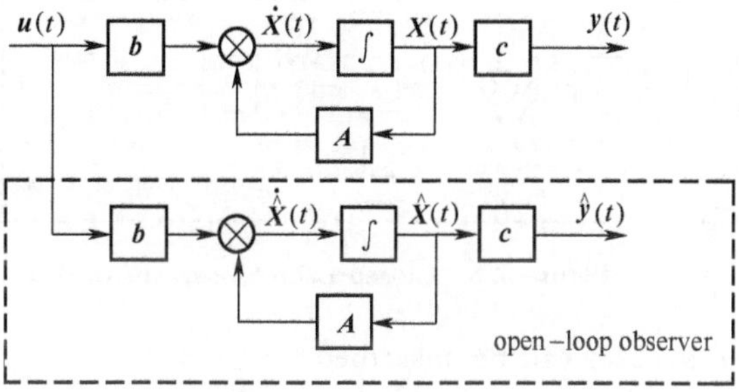

Figure 5.4 Open-Loop Observer

Obviously, the model is the open-loop form, and the observer shown above is called the open-loop state observer.

The reconstructed system can be described by (5.28)

$$\dot{\hat{\boldsymbol{X}}}=\boldsymbol{A}\hat{\boldsymbol{X}}+\boldsymbol{b}u$$
$$\hat{y}=\boldsymbol{c}\hat{\boldsymbol{X}} \tag{5.28}$$

Where $\hat{\boldsymbol{X}}$ is the estimate of the actual state $\boldsymbol{X}$. To study the dynamics of this estimate procedure, we define the error in the estimate to be

$$\tilde{\boldsymbol{X}}=\boldsymbol{X}-\hat{\boldsymbol{X}} \tag{5.29}$$

We can confirm the estimates can be used in place of the actual state variables when

$$\lim_{t\to\infty}\tilde{\boldsymbol{X}}=\lim_{t\to\infty}\boldsymbol{X}-\hat{\boldsymbol{X}}=\boldsymbol{0} \tag{5.30}$$

Substituting (5.9) and (5.28) into (5.29), we can obtain

$$\dot{\tilde{\boldsymbol{X}}}=\dot{\boldsymbol{X}}-\dot{\hat{\boldsymbol{X}}}=\boldsymbol{A}(\boldsymbol{X}-\hat{\boldsymbol{X}})=\boldsymbol{A}\tilde{\boldsymbol{X}} \tag{5.31}$$

The solution of (5.31) is

$$\tilde{\boldsymbol{X}}(t)=\mathrm{e}^{\boldsymbol{A}(t-t_0)}\tilde{\boldsymbol{X}}(t_0)=\mathrm{e}^{\boldsymbol{A}(t-t_0)}[\boldsymbol{X}(t_0)-\hat{\boldsymbol{X}}(t_0)] \tag{5.32}$$

We know $\boldsymbol{A}$, $\boldsymbol{b}$ and $u(t)$. Hence, this observer will be satisfied for use if we can obtain the correct initial condition $\boldsymbol{X}(t_0)$ and set $\hat{\boldsymbol{X}}(t_0)$ equal to it. However, the information about $\boldsymbol{X}(t_0)$ is unknown. Thus, if we made a poor estimate for the initial condition, the estimated state would have a continually growing error or an error that goes to zero too slowly to be used. For this, the open-loop observer shown in Figure 5. 4 is not practical.

Feeding back the difference between the measured and estimated outputs and correcting the model continuously with this error signal, we can obtain the closed-loop form observer shown in Figure 5. 5.

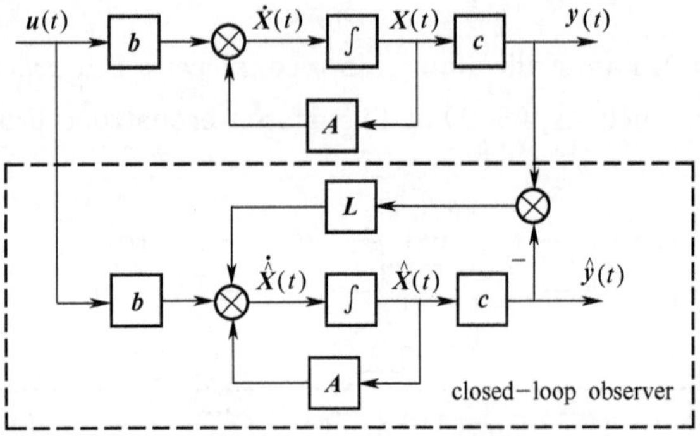

Figure 5. 5 Closed-Loop Observer

The reconstructed system can be described by (5. 33)

$$\dot{\hat{\boldsymbol{X}}}=(\boldsymbol{A}-\boldsymbol{Lc})\hat{\boldsymbol{X}}+\boldsymbol{L}y+\boldsymbol{b}u \qquad (5.33)$$
$$\hat{y}=\boldsymbol{c}\hat{\boldsymbol{X}}$$

Where $\hat{\boldsymbol{X}}$ is the estimate of the actual state $\boldsymbol{X}$, and the feedback gain vector

$$\boldsymbol{L}=[l_1 \quad l_2 \quad \cdots \quad l_n] \qquad (5.34)$$

will be chosen to achieve satisfactory error characteristics.

Obviously, the dimension of the closed-loop observer (5. 33) is equal to it of the estimated original system (5. 9), so the observer (5. 33) is called a **full-order observer.**

Substituting (5. 9) and (5. 33) into (5. 29), we can obtain

$$\dot{\tilde{\boldsymbol{X}}}=\dot{\boldsymbol{X}}-\dot{\hat{\boldsymbol{X}}}=(\boldsymbol{A}-\boldsymbol{Lc})\tilde{\boldsymbol{X}} \qquad (5.35)$$

The solution of (5. 35) is

$$\tilde{\boldsymbol{X}}(t)=\mathrm{e}^{(\boldsymbol{A}-\boldsymbol{Lc})(t-t_0)}\tilde{\boldsymbol{X}}(t_0)=\mathrm{e}^{(\boldsymbol{A}-\boldsymbol{Lc})(t-t_0)}[\boldsymbol{X}(t_0)-\hat{\boldsymbol{X}}(t_0)] \qquad (5.36)$$

If we choose $\boldsymbol{L}$ so that $\boldsymbol{A}-\boldsymbol{Lc}$ has asymptotically stable and reasonably fast eigenvalues, then the state error $\tilde{\boldsymbol{X}}(t)$ will decay to zeros and remain there. It is in dependent of the input $u(t)$ and the initial condition $\boldsymbol{X}(t_0)$. A natural question would be: Under what condition can the eigenvalues of $\boldsymbol{A}-\boldsymbol{Lc}$ be placed arbitrarily?

5. 2. 2 Design of the Full-Order State Observer

Theorem 5. 3 The eigenvalues of ($\boldsymbol{A}$-$\boldsymbol{Lc}$) can be placed arbitrarily iff the estimated original system (5. 9) is observable.

Proof. Consider a SISO LTI system, which is completely observable, such as (5.9). The dual system of it can be described by

$$\begin{aligned} \dot{\boldsymbol{Z}}(t) &= \boldsymbol{A}^{\mathrm{T}}\boldsymbol{Z}(t) + \boldsymbol{c}^{\mathrm{T}} v(t) \\ w(t) &= \boldsymbol{b}^{\mathrm{T}}\boldsymbol{Z}(t) \end{aligned} \tag{5.37}$$

Since the system (5.9) is completely observable, then its dual system (5.37) is completely controllable, and the eigenvalues of it can be placed arbitrarily by introducing the state feedback control

$$v(t) = -\boldsymbol{K}\boldsymbol{Z}(t) + \eta(t) \tag{5.38}$$

Furthermore, the closed-loop system by introducing the state feedback control (5.38) into (5.37) can be described by

$$\begin{aligned} \dot{\boldsymbol{Z}}(t) &= (\boldsymbol{A}^{\mathrm{T}} - \boldsymbol{c}^{\mathrm{T}}\boldsymbol{K})\boldsymbol{Z}(t) + \boldsymbol{c}^{\mathrm{T}}\eta(t) \\ w(t) &= \boldsymbol{b}^{\mathrm{T}}\boldsymbol{Z}(t) \end{aligned} \tag{5.39}$$

It means the eigenvalues of $(\boldsymbol{A}^{\mathrm{T}} - \boldsymbol{c}^{\mathrm{T}}\boldsymbol{K})$ can be placed arbitrarily at the expected locations $s_1^*, s_2^*, \cdots, s_n^*$.

Moreover, since

$$|s\boldsymbol{I} - (\boldsymbol{A}^{\mathrm{T}} - \boldsymbol{c}^{\mathrm{T}}\boldsymbol{K})| = |s\boldsymbol{I} - (\boldsymbol{A} - \boldsymbol{K}^{\mathrm{T}}\boldsymbol{c})|$$

then, it can be obtained that the eigenvalues of $(\boldsymbol{A} - \boldsymbol{L}\boldsymbol{c})$ can also be placed arbitrarily at the expected locations $s_1^*, s_2^*, \cdots s_n^*$, when the feedback gain vector of the closed-loop observer (5.33) is chosen as

$$\boldsymbol{L} = [l_1 \quad l_2 \quad \cdots \quad l_n]^{\mathrm{T}} = \boldsymbol{K}^{\mathrm{T}} \tag{5.40}$$

The procedure above is called the poles placement for the full-order observer, and $s_1^*, s_2^*, \cdots s_n^*$ is the expected poles of the observer.

Example 5.4 Consider the LTI system described by

$$\begin{aligned} \dot{\boldsymbol{X}} &= \begin{bmatrix} -1 & -2 & -2 \\ 0 & -1 & 1 \\ 1 & 0 & -1 \end{bmatrix} \boldsymbol{X} + \begin{bmatrix} 2 \\ 0 \\ 1 \end{bmatrix} u \\ y &= [1 \quad 1 \quad 0]\boldsymbol{X} \end{aligned}$$

Design a full-order observer and place the poles of it at $s_1^* = -3, s_2^* = -3, s_3^* = -4$.

Solution The observability matrix of the system is

$$\boldsymbol{Q}_o = \begin{bmatrix} \boldsymbol{c} \\ \boldsymbol{c}\boldsymbol{A} \\ \boldsymbol{c}\boldsymbol{A}^2 \end{bmatrix} = \begin{bmatrix} 1 & 1 & 0 \\ -1 & -3 & -1 \\ 0 & 5 & 0 \end{bmatrix}$$

Since rank $\boldsymbol{Q}_o = 3$, then the system is completely observable, and a full-order observer can be constructed with the expected poles.

The dual system of the original system is

$$\begin{aligned} \dot{\boldsymbol{Z}}(t) &= \boldsymbol{A}^{\mathrm{T}}\boldsymbol{Z}(t) + \boldsymbol{c}^{\mathrm{T}} v(t) = \begin{bmatrix} -1 & 0 & 1 \\ -2 & -1 & 0 \\ -2 & 1 & -1 \end{bmatrix} \boldsymbol{Z}(t) + \begin{bmatrix} 1 \\ 1 \\ 0 \end{bmatrix} v(t) \\ w(t) &= \boldsymbol{b}^{\mathrm{T}}\boldsymbol{Z}(t) = [2 \quad 0 \quad 1]\boldsymbol{Z}(t) \end{aligned}$$

and it must be completely controllable.

The characteristic polynomial of it is

$$\alpha_0(s)=|sI-A^{\mathrm{T}}|=|sI-A|=\begin{vmatrix} s+1 & 0 & -1 \\ 2 & s+1 & 0 \\ 2 & -1 & s+1 \end{vmatrix}$$

$$=s^3+3s^2+5s+5=s^3+a_1s^2+a_2s+a_3$$

and the coefficients are

$$a_1=3, a_2=5, a_3=5$$

Furthermore, letting the expected poles of the dual system are equal to the expected poles of the observer, the expected characteristic polynomial of the dual system can be calculated as

$$\alpha^*(s)=(s+3)(s+3)(s+4)=s^3+10s^2+33s+36=s^3+a_1^*s^2+a_2^*s+a_3^*$$

The coefficients are

$$a_1^*=10, a_2^*=33, a_3^*=36$$

Let

$$\overline{K}=[a_3^*-a_3 \quad a_2^*-a_2 \quad a_1^*-a_1]=[31 \quad 28 \quad 7]$$

Since the dual system is completely controllable, there exists a nonsingular transformation $Z=P\overline{Z}$, which transforms the dual system into the controllable canonical form.

A nonsingular transformation matrix can be constructed as

$$P=[(A^{\mathrm{T}})^2c^{\mathrm{T}} \quad A^{\mathrm{T}}c^{\mathrm{T}} \quad c^{\mathrm{T}}]\begin{bmatrix} 1 & 0 & 0 \\ \alpha_1 & 1 & 0 \\ \alpha_2 & \alpha_1 & 1 \end{bmatrix}$$

$$=\begin{bmatrix} 0 & -1 & 1 \\ 5 & -3 & 1 \\ 0 & -1 & 0 \end{bmatrix}\begin{bmatrix} 1 & 0 & 0 \\ 3 & 1 & 0 \\ 5 & 3 & 1 \end{bmatrix}=\begin{bmatrix} 2 & 2 & 1 \\ 1 & 0 & 1 \\ -3 & -1 & 0 \end{bmatrix}$$

and its inverse matrix is

$$P^{-1}=\frac{1}{5}\begin{bmatrix} -1 & 1 & -2 \\ 3 & -3 & 1 \\ 1 & 4 & 2 \end{bmatrix}$$

So, the state feedback gain vector for the dual system can be constructed as

$$K=\overline{K}P^{-1}=[31 \quad 28 \quad 7]\cdot\frac{1}{5}\cdot\begin{bmatrix} -1 & 1 & -2 \\ 3 & -3 & 1 \\ 1 & 4 & 2 \end{bmatrix}=[12 \quad -5 \quad -4]$$

By introducing the state feedback control $v(t)=-KZ(t)+\eta(t)$, the poles of the closed-loop dual system can be placed at the expected locations, $-3, -3, -4$.

Moreover, if the feedback gain vector of the closed-loop observer is chosen as

$$L=K^{\mathrm{T}}=[12 \quad -5 \quad -4]^{\mathrm{T}}$$

the eigenvalues of it can also be placed at the expected locations $-3, -3, -4$.

The closed-loop observer can be constructed as

$$\dot{\hat{\boldsymbol{X}}}(t)=(\boldsymbol{A}-\boldsymbol{Lc})\hat{\boldsymbol{X}}(t)+\boldsymbol{b}u(t)+\boldsymbol{L}y(t)$$

$$=\begin{bmatrix}-13 & -14 & -2\\ 5 & 4 & 1\\ 5 & 4 & -1\end{bmatrix}\hat{\boldsymbol{X}}(t)+\begin{bmatrix}2\\0\\1\end{bmatrix}u(t)+\begin{bmatrix}12\\-5\\-4\end{bmatrix}y(t)$$

$$\hat{y}(t)=\boldsymbol{c}\hat{\boldsymbol{X}}(t)=[1\quad 1\quad 0]\hat{\boldsymbol{X}}(t)$$

In fact, the characteristic polynomial

$$\alpha(s)=|s\boldsymbol{I}-\boldsymbol{A}+\boldsymbol{Lc}| \tag{5.41}$$

of the closed-loop observer (5.33) is a nth order polynomial in s containing the feedback gains $l_1, l_2, \cdots, l_n$, and the vector

$$\boldsymbol{L}=[l_1\quad l_2\quad \cdots\quad l_n]^{\mathrm{T}} \tag{5.42}$$

is the state feedback gain vector of the closed-loop observer. Moreover, the expected characteristic polynomial of the closed-loop observer can be determined as

$$\alpha^*(s)=\prod_{i=1}^{n}(s-s_i^*)=s^n+a_1^* s^{n-1}+\cdots+a_{n-1}^* s+a_n^* \tag{5.43}$$

Hence, the required elements, $l_1, l_2, \cdots, l_n$, can be obtained by matching coefficients in the characteristic polynomial (5.41) and in the expected characteristic polynomial (5.43).

Example 5.5 Consider the LTI system described by

$$\dot{\boldsymbol{X}}=\begin{bmatrix}-1 & 1\\ 0 & -2\end{bmatrix}\boldsymbol{X}+\begin{bmatrix}0\\1\end{bmatrix}u$$

$$y=[2\quad 0]\boldsymbol{X}$$

Design a full-order observer and place the poles of it at $s_1^*=-10, s_2^*=-10$.

Solution The observability matrix of the system is

$$\boldsymbol{Q}_o=\begin{bmatrix}\boldsymbol{c}\\ \boldsymbol{cA}\end{bmatrix}=\begin{bmatrix}2 & 0\\ -2 & 2\end{bmatrix}$$

Since rank $\boldsymbol{Q}_o=2$, then the system is completely observable, and a full-order observer can be constructed with the expected poles.

Let the feedback gain vector of the full-order observer is

$$\boldsymbol{L}=\begin{bmatrix}l_1\\ l_2\end{bmatrix}$$

Then, it can be calculated that

$$(\boldsymbol{A}-\boldsymbol{Lc})=\begin{bmatrix}-1 & 1\\ 0 & -2\end{bmatrix}-\begin{bmatrix}l_1\\ l_2\end{bmatrix}[2\quad 0]=\begin{bmatrix}-1-2l_1 & 1\\ -2l_2 & -2\end{bmatrix}$$

The characteristic polynomial of the closed-loop observer is

$$\alpha(s)=|s\boldsymbol{I}-\boldsymbol{A}+\boldsymbol{Lc}|=\begin{vmatrix}s+1+2l_1 & -1\\ 2l_2 & s+2\end{vmatrix}$$

$$=s^2+(2l_1+3)s+(4l_1+2l_2+2)$$

The expected characteristic polynomial of the closed-loop observer can be determined by

$$\alpha^*(s)=(s+10)^2=s^2+20s+100$$

Equating the coefficients with like powers of s in $\alpha(s)$ and $\alpha^*(s)$ yields the feedback gains

$$l_1=8.5, l_2=32$$

The closed-loop observer can be constructed as

$$\dot{\hat{\boldsymbol{X}}}(t)=(\boldsymbol{A}-\boldsymbol{L}\boldsymbol{c})\hat{\boldsymbol{X}}(t)+\boldsymbol{b}u(t)+\boldsymbol{L}y(t)$$
$$=\begin{bmatrix}-18 & 1\\ -64 & -2\end{bmatrix}\hat{\boldsymbol{X}}(t)+\begin{bmatrix}0\\1\end{bmatrix}u(t)+\begin{bmatrix}8.5\\32\end{bmatrix}y(t)$$
$$\hat{y}(t)=\boldsymbol{c}\hat{\boldsymbol{X}}(t)=[2\quad 0]\hat{\boldsymbol{X}}(t)$$

The state space description above can also be rewritten in another form such as

$$\dot{\hat{\boldsymbol{X}}}(t)=\boldsymbol{A}\hat{\boldsymbol{X}}(t)+\boldsymbol{b}u(t)+\boldsymbol{L}[y(t)-\hat{y}(t)]$$
$$=\begin{bmatrix}-1 & 1\\ 0 & -2\end{bmatrix}\hat{\boldsymbol{X}}(t)+\begin{bmatrix}0\\1\end{bmatrix}u(t)+\begin{bmatrix}8.5\\32\end{bmatrix}[y(t)-\hat{y}(t)]$$
$$\hat{y}(t)=\boldsymbol{c}\hat{\boldsymbol{X}}(t)=[2\quad 0]\hat{\boldsymbol{X}}(t)$$

The closed-loop observer is shown in Figure 5.6.

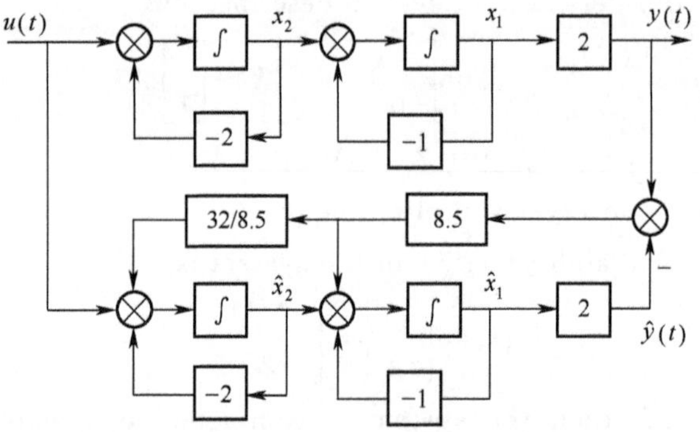

Figure 5.6 The Closed-Loop Observer

The discussion of state estimators so far has ignored information about the state that is provided directly by the output signal. In fact, we should be able to use the output information, and construct an observer only for the state variables that are not directly known from the output. Supposing rank $\boldsymbol{c}=q$, the minimal dimension of the observer may be $n-q$, and the observer is called a **reduced-order observer.** In the textbook, we will not study the reduced-order observer in detail.

5.3 Feedback System with the State Observer

It is clear that if a system is controllable and the states are available for feedback, then the poles of the system can be placed arbitrarily by introducing a constant feedback control. However, in most practical applications, the states are not completely accessible and all the designer knows are the output y and input u. Hence, the estimation of the states from the given output information y and input u is necessary to satisfy some specific design objectives. The closed-loop system by introducing a feedback control is shown in Figure 5.7, in which the state is estimated by a full-order observer.

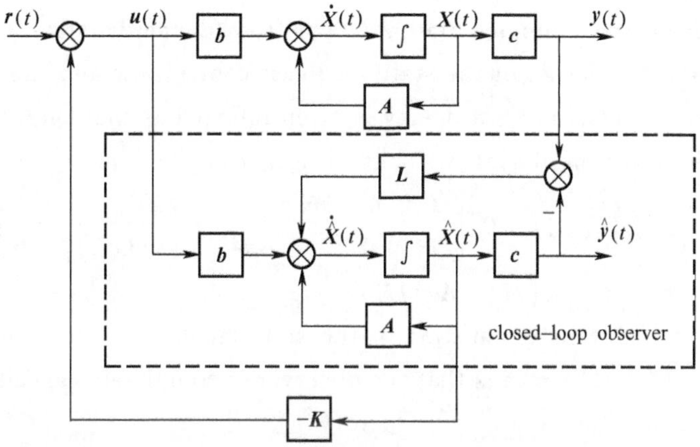

Figure 5.7 Closed-Loop System via Estimated State Feedback

Since the feedback control law is designed for the actual state rather than the estimated state, then we will discuss what effect has on the system dynamics if the estimated state $\hat{\boldsymbol{X}}(t)$ instead of the actual state $\boldsymbol{X}(t)$

Obviously, the dimension of the closed-loop system shown in Figure 5.7 is $2n$, i.e. the system has n actual state variables and n estimated state variables. The system can be described by

$$\begin{bmatrix}\dot{\boldsymbol{X}}(t)\\ \dot{\hat{\boldsymbol{X}}}(t)\end{bmatrix}=\begin{bmatrix}\boldsymbol{A} & -\boldsymbol{bK}\\ \boldsymbol{Lc} & \boldsymbol{A}-\boldsymbol{Lc}-\boldsymbol{bK}\end{bmatrix}\begin{bmatrix}\boldsymbol{X}(t)\\ \hat{\boldsymbol{X}}(t)\end{bmatrix}+\begin{bmatrix}\boldsymbol{b}\\ \boldsymbol{b}\end{bmatrix}r(t) \tag{5.44}$$

$$y(t)=[\boldsymbol{c}\quad \boldsymbol{0}]\begin{bmatrix}\boldsymbol{X}(t)\\ \hat{\boldsymbol{X}}(t)\end{bmatrix}$$

Introducing a nonsingular transformation

$$\begin{bmatrix}\boldsymbol{X}(t)\\ \hat{\boldsymbol{X}}(t)\end{bmatrix}=\begin{bmatrix}\boldsymbol{I}_n & \boldsymbol{0}\\ \boldsymbol{I}_n & -\boldsymbol{I}_n\end{bmatrix}\begin{bmatrix}\overline{\boldsymbol{X}}(t)\\ \overline{\hat{\boldsymbol{X}}}(t)\end{bmatrix} \tag{5.45}$$

the system (5.44) can be transformed into (5.46).

$$\begin{bmatrix}\dot{\overline{\boldsymbol{X}}}(t)\\ \dot{\overline{\boldsymbol{X}}}(t)-\dot{\hat{\boldsymbol{X}}}(t)\end{bmatrix}=\begin{bmatrix}\boldsymbol{A}-\boldsymbol{bK} & \boldsymbol{bK}\\ \boldsymbol{0} & \boldsymbol{A}-\boldsymbol{Lc}\end{bmatrix}\begin{bmatrix}\overline{\boldsymbol{X}}(t)\\ \overline{\boldsymbol{X}}(t)-\hat{\boldsymbol{X}}(t)\end{bmatrix}+\begin{bmatrix}\boldsymbol{b}\\ \boldsymbol{0}\end{bmatrix}r(t)$$

$$y(t)=[\boldsymbol{c}\quad \boldsymbol{0}]\begin{bmatrix}\overline{\boldsymbol{X}}(t)\\ \overline{\boldsymbol{X}}(t)-\hat{\boldsymbol{X}}(t)\end{bmatrix} \tag{5.46}$$

Obviously, the characteristic polynomial of the system (5.46) can be calculated as

$$\begin{vmatrix} sI-(A-bK) & -bK \\ 0 & sI-(A-Lc) \end{vmatrix} = |sI-(A-bK)| \cdot |sI-(A-Lc)| \quad (5.47)$$

Since a change of state by the nonsingular transformation does not change the eigenvalues, we conclude from (5.47) that eigenvalues, or poles, of the closed-loop system shown in Figure 5.7 are the union of the control poles and the observer poles, which are determined by $\boldsymbol{A}-\boldsymbol{bK}$ and $\boldsymbol{A}-\boldsymbol{Lc}$ separately. It is clear that there is no difference in state feedback from the estimated state $\hat{\boldsymbol{X}}(t)$ or from the actual state $\boldsymbol{X}(t)$.

Consequently, the design of the state feedback control law and the observer can be carried out independently. This property is often referred as the **separation principle.**

The transfer function of system (5.46) can be calculated as

$$\begin{aligned} g(s) &= [\boldsymbol{c} \quad \boldsymbol{0}] \begin{bmatrix} s\boldsymbol{I}-(\boldsymbol{A}-\boldsymbol{bK}) & -\boldsymbol{bK} \\ \boldsymbol{0} & s\boldsymbol{I}-(\boldsymbol{A}-\boldsymbol{Lc}) \end{bmatrix}^{-1} \begin{bmatrix} \boldsymbol{b} \\ \boldsymbol{0} \end{bmatrix} \\ &= \boldsymbol{c}[s\boldsymbol{I}-(\boldsymbol{A}-\boldsymbol{bK})]^{-1}\boldsymbol{b} \end{aligned} \quad (5.48)$$

It is equal to the transfer function of the state feedback system without using the observer such as (5.7). It means that the observer is completely cancelled and does not appear in the transfer function from $r(t)$ to $y(t)$.

In order to make the dynamic of the state error converge to zero as well as much faster than the closed-loop dynamic, the expected poles of observer are often chosen as

$$\mathrm{Re}[\lambda_i^*(\boldsymbol{A}-\boldsymbol{Lc})] \leqslant (2\sim3)\mathrm{Re}[\lambda_i^*(\boldsymbol{A}-\boldsymbol{bK})] < 0 \quad (5.49)$$

Where $\lambda_i^*(\boldsymbol{A}-\boldsymbol{bK})$ are the expected poles of the close-loop system by introducing the state feedback control.

Problems

5.1 Consider the LTI system described by

$$\dot{\boldsymbol{X}} = \begin{bmatrix} 0 & 1 \\ -2 & -4 \end{bmatrix} \boldsymbol{X} + \begin{bmatrix} 0 \\ 1 \end{bmatrix} u$$

$$y = [1 \quad 2]\boldsymbol{X}$$

(1) Design a state feedback control law that assigns the set of closed-loop eigenvalues as $\{-3, -6\}$.

(2) Determine the transfer function of the closed-loop system by introducing the state feedback control law above into the original system.

5.2 Consider the LTI system described by

$$\dot{\boldsymbol{X}} = \begin{bmatrix} -1 & -2 & -2 \\ 0 & -1 & 1 \\ 1 & 0 & -1 \end{bmatrix} \boldsymbol{X} + \begin{bmatrix} 2 \\ 0 \\ 1 \end{bmatrix} u$$

$$y = [1 \quad 1 \quad 0]\boldsymbol{X}$$

(1) Design a state feedback control law that assigns the set of closed-loop eigenval-

ues as $\{-1, -2, -2\}$.

(2) Draw the state simulation diagram of the closed-loop system after state feedback.

5.3 Consider the LTI system described by

$$G(s)=\frac{(s-1)(s+2)}{(s+1)(s+3)(s-2)}$$

Is it possible to change the transfer function to

$$G'(s)=\frac{s-1}{(s+2)(s+3)}$$

by a state feedback control ? If possible, try to do it.

5.4 Consider the 2 dimension system

$$\dot{\boldsymbol{X}}=\begin{bmatrix}0 & 1\\ 2 & 1\end{bmatrix}\boldsymbol{X}+\begin{bmatrix}1\\ 0\end{bmatrix}u$$

$$y=\begin{bmatrix}1 & 0\end{bmatrix}\boldsymbol{X}$$

Design a full-order observer and place the poles of it at $s_{1,2}^{*}=-1\pm \mathrm{j}$.

5.5 Consider the 3 dimension system

$$\dot{\boldsymbol{X}}=\begin{bmatrix}0 & 1 & 0\\ 0 & 0 & 1\\ -1 & -2 & -3\end{bmatrix}\boldsymbol{X}+\begin{bmatrix}0\\ 0\\ 4\end{bmatrix}u$$

$$y=\begin{bmatrix}2 & -4 & 0\end{bmatrix}\boldsymbol{X}$$

(1) Design a full-order observer and place the poles of it at $s_{1,2}^{*}=-1\pm 2\mathrm{j}$ and $s_3^{*}=-10$.

(2) Draw the state simulation diagram of the full-order observer.

5.6 Consider the LTI system described by

$$G(s)=\frac{1}{s(s-2)}$$

(1) Design a state feedback control law that place the poles of the closed-loop system at $s_{1,2}^{*}=-1\pm 2\mathrm{j}$.

(2) Suppose the states, which are used to feedback above, are impossible to be measured. Design a full-order observer to estimate them and place the poles of it at $s_{1,2}^{*}=-1$.

(3) Draw the state simulation diagram of the closed-loop system via estimated states, and determine the transfer function of it.

5.7 A DC motor control system has the form shown in Figure P5.7.

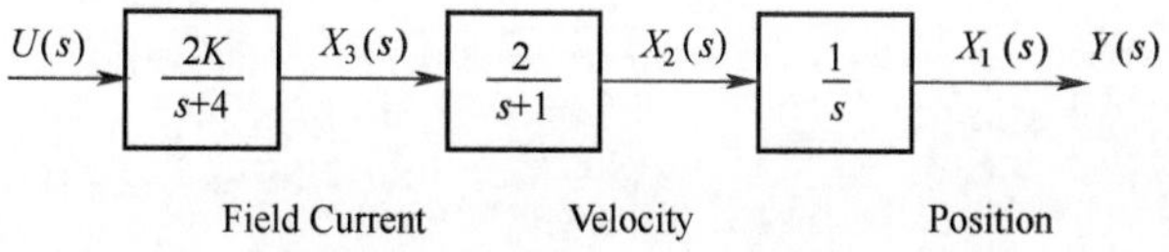

Figure P5.7 Field-Controlled DC motor

All the state variables noted on Figure P5.7 are impossible to be measured. Design a scheme to make the system has a steady-state error equal to zero for a step input and response with a percent overshoot less than 3%.

Harry Nyquist (1889—1976) was an important contributor to information theory.

As an engineer at Bell Laboratories, Nyquist did important work on thermal noise ("*Johnson-Nyquist noise*"), the stability of feedback amplifiers, telegraphy, facsimile, television, and other important communications problems. With Herbert E. Ives, he helped to develop AT&T's first facsimile machines that were made public in 1924. In 1932, he published a classical paper on stability of feedback amplifiers. *The Nyquist stability criterion* can now be found in all textbooks on feedback control theory.

His early theoretical work on determining the bandwidth requirements for transmitting information laid the foundations for later advances by Claude Shannon, which led to the development of information theory. In particular, Nyquist determined that the number of independent pulses that could be put through a telegraph channel per unit time is limited to twice the bandwidth of the channel, and published his results in the papers *Certain factors affecting telegraph speed* (1924) and *Certain topics in Telegraph Transmission Theory* (1928). This rule is essentially a dual of what is now known as *the Nyquist-Shannon sampling theorem*.

Chapter6 Discrete Time Control System

6.1 State Space Description of Discrete Time System

If the system variables (namely input variables、state variables and output variables) only have value in separate time, the system is called the **discrete time system** or **discrete system.** This kind of system may be the nature existence, for example dynamic system which only has value in discrete time in society, economy, project domains and so on, may also be discrete system that is obtained by sampling continuous time system. In modern control theory, methods of system description, system analysis and synthesis are very similar to continuous time system. Thus, we can expand some basic concepts and methods of continuous time system to discrete time system.

6.1.1 State Space Description of Discrete Time System

Similar to continuous time system, in the state space description, discrete system also has concepts and expressions of state variables, state vector, state equation and output equation. What is different, the discrete system takes a set of first order difference equations as the state equation. Therefore, the state space description of discrete time linear system is given by

$$\boldsymbol{X}(k+1)=\boldsymbol{G}(k)\boldsymbol{X}(k)+\boldsymbol{H}(k)\boldsymbol{u}(k) \tag{6.1}$$

$$\boldsymbol{y}(k)=\boldsymbol{C}(k)\boldsymbol{X}(k)+\boldsymbol{D}(k)\boldsymbol{u}(k) \tag{6.2}$$

The variables $x(k)$, $u(k)$, $y(k)$ represent values at time instant $t=kT$, $(k=0,1,2,\cdots)$, where T is sampling period, for convenience, T is omitted. Difference equation (6.1) describes the relationship between state variables at time instant $(k+1)T$ and state variables and input variable at time instant kT, is called a **state equation.** Equation (6.2) gives the relation of output variables at time instant kT with state variables and input variable at time instant kT, is called a **output equation.** Both state equation (6.1) and output equation (6.2) are called a **state space description** of the discrete time system.

In the state space description (6.1) and (6.2), $\boldsymbol{X}(k)$ is the **state vector**, and is a n dimension column vector. $\boldsymbol{u}(k)$ is the input vector, and is a p dimension column vector. $\boldsymbol{y}(k)$ is the output vector, and is a q dimension column vector. $\boldsymbol{G}(k)$ is called the **system matrix** or **coefficient matrix**, and is a $n\times n$ matrix. $\boldsymbol{H}(k)$ is called the **input matrix**, and is a $n\times p$ real matrix. $\boldsymbol{C}(k)$ is called the **output matrix**, and is a $q\times n$ real matrix. $\boldsymbol{D}(k)$ is called the **forward matrix**, and is a $q\times p$ real matrix.

Since $\boldsymbol{G}(k)$、$\boldsymbol{H}(k)$、$\boldsymbol{C}(k)$ and $\boldsymbol{D}(k)$ change with the passage of time, the system described by equation (6.1) and (6.2) is called a **discrete time linear time-varying (LTV) system.**

If $\boldsymbol{G}(k)$、$\boldsymbol{H}(k)$、$\boldsymbol{C}(k)$ and $\boldsymbol{D}(k)$ are independent of time instant kT, that is, they are real constant matrixes, then the equation (6.1) and (6.2) reduce to

$$\boldsymbol{X}(k+1)=\boldsymbol{G}\boldsymbol{X}(k)+\boldsymbol{H}\boldsymbol{u}(k) \tag{6.3}$$

$$\boldsymbol{y}(k)=\boldsymbol{C}\boldsymbol{X}(k)+\boldsymbol{D}\boldsymbol{u}(k) \tag{6.4}$$

The system with the description (6.3) and (6.4) is called a **discrete time linear time-invariant (LTI) system.**

The block diagram form of state space description (6.3) and (6.4) is shown in Figure 6.1. Where z^{-1} denotes a unit time delay.

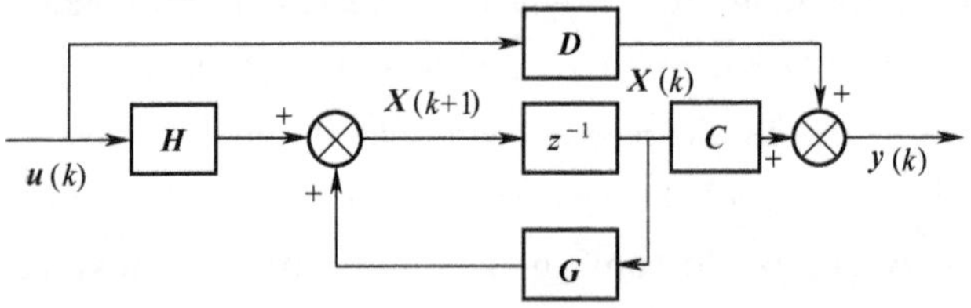

Figure 6.1 Block Diagram of Discrete Time Linear Time-Invariant System

6.1.2 Obtaining State Space Description from Difference Equation or Impulse Transfer Function

1. Obtaining State Space Description from Difference Equation

Obtaining state space description from difference equation is similar to obtaining state space description from differential equation in continuous-time system. Suppose the difference equation of a SISO n-order discrete time LTI system is

$$\begin{aligned} &y(k+n)+a_1y(k+n-1)+\cdots+a_{n-1}y(k+1)+a_ny(k) \\ &=b_0u(k+m)+b_1u(k+m-1)+\cdots+b_{m-1}u(k+1)+b_mu(k) \end{aligned} \tag{6.5}$$

In this subsection, we will discuss how to obtain the state space description of system from its difference equation description, such as (6.5), mainly. Two cases will be considered, that is, the case $m=0$ and the case $m\leqslant n$.

Case 1: $m=0$

In this case, the difference equation (6.5) reduces to

$$y(k+n)+a_1y(k+n-1)+\cdots+a_{n-1}y(k+1)+a_ny(k)=b_0u(k) \tag{6.6}$$

The state variables can be selected as the phase variables, which are defined by

$$\begin{cases} x_1(k)=y(k) \\ x_2(k)=y(k+1) \\ \vdots \\ x_n(k)=y(k+n-1) \end{cases} \tag{6.7}$$

Differencing (6.7), we get a set of first-order difference equations shown as

$$\begin{cases} x_1(k+1)=y(k+1)=x_2(k) \\ x_2(k+1)=y(k+2)=x_3(k) \\ \quad\vdots \\ x_{n-1}(k+1)=y(k+n-1)=x_n(k) \\ x_n(k+1)=y(k+n)=-a_n y(k)-a_{n-1}y(k+1)-\cdots-a_1 y(k+n-1)+b_0 u(k) \\ \qquad =-a_n x_1(k)-a_{n-1}x_2(k)-\cdots-a_1 x_n(k)+b_0 u(k) \end{cases} \tag{6.8}$$

The difference equations (6.8) can be written in matrix form as

$$\begin{bmatrix} x_1(k+1) \\ x_2(k+1) \\ \vdots \\ x_{n-1}(k+1) \\ x_n(k+1) \end{bmatrix} = \begin{bmatrix} 0 & 1 & 0 & \cdots & 0 \\ 0 & 0 & 1 & \cdots & 0 \\ \vdots & \vdots & \vdots & \ddots & \vdots \\ 0 & 0 & 0 & \cdots & 1 \\ -a_n & -a_{n-1} & -a_{n-2} & \cdots & -a_1 \end{bmatrix} \begin{bmatrix} x_1(k) \\ x_2(k) \\ \vdots \\ x_{n-1}(k) \\ x_n(k) \end{bmatrix} + \begin{bmatrix} 0 \\ 0 \\ \vdots \\ 0 \\ b_0 \end{bmatrix} u(k) \tag{6.9}$$

The output equation of the system is given by

$$y(k)=[1 \quad 0 \quad 0 \quad \cdots \quad 0] \begin{bmatrix} x_1(k) \\ x_2(k) \\ \vdots \\ x_{n-1}(k) \\ x_n(k) \end{bmatrix} \tag{6.10}$$

The state space description (6.9) and (6.10) can be written directly from the original difference equation. Obviously, the system matrix $\boldsymbol{G}$ is a **companion matrix.**

Example 6.1 Find a state space description for the system described by the following difference equation.

$$y(k+3)+5y(k+2)+11y(k+1)+6y(k)=5u(k)$$

Solution For the system

$$n=3;\quad a_1=5,\quad a_2=11,\quad a_3=6;\quad b_0=5$$

The state variables are chosen as

$$x_1(k)=y(k), x_2(k)=y(k+1)=x_1(k+1), x_3(k)=y(k+2)=x_2(k+1)$$

Thus, the state equation of the system is

$$\begin{bmatrix} x_1(k+1) \\ x_2(k+1) \\ x_3(k+1) \end{bmatrix} = \begin{bmatrix} 0 & 1 & 0 \\ 0 & 0 & 1 \\ -a_3 & -a_2 & -a_1 \end{bmatrix} \begin{bmatrix} x_1(k) \\ x_2(k) \\ x_3(k) \end{bmatrix} + \begin{bmatrix} 0 \\ 0 \\ b_0 \end{bmatrix} u(k)$$

$$= \begin{bmatrix} 0 & 1 & 0 \\ 0 & 0 & 1 \\ -6 & -11 & -5 \end{bmatrix} \begin{bmatrix} x_1(k) \\ x_2(k) \\ x_3(k) \end{bmatrix} + \begin{bmatrix} 0 \\ 0 \\ 5 \end{bmatrix} u(k)$$

and the output equation is

$$y(k)=\begin{bmatrix}1 & 0 & 0\end{bmatrix}\begin{bmatrix}x_1(k)\\x_2(k)\\x_3(k)\end{bmatrix}$$

Case 2: $m \leqslant n$

In this case, the difference equation description of a SISO n-order discrete time LTI system is the general form shown as (6.5).

The state variables can be defined by

$$\begin{cases}x_1(k)=y(k)-\beta_0 u(k)\\x_2(k)=y(k+1)-\beta_0 u(k+1)-\beta_1 u(k)\\\vdots\\x_{n-1}(k)=y(k+n-2)-\beta_0 u(k+n-2)-\cdots-\beta_{n-3}u(k+1)-\beta_{n-2}u(k)\\x_n(k)=y(k+n-1)-\beta_0 u(k+n-1)-\cdots-\beta_{n-2}u(k+1)-\beta_{n-1}u(k)\end{cases} \tag{6.11}$$

Firstly, we will determine the parameters $\beta_i\,(i=0,1,2,\cdots,n)$ from the coefficients of (6.5) as

$$\begin{cases}\beta_0=b_0\\\beta_1=b_1-a_1\beta_0\\\beta_2=b_2-a_1\beta_1-a_2\beta_0\\\vdots\\\beta_{n-1}=b_{n-1}-a_1\beta_{n-2}-a_2\beta_{n-3}-\cdots-a_{n-2}\beta_1-a_{n-1}\beta_0\\\beta_n=b_n-a_1\beta_{n-1}-a_2\beta_{n-2}-\cdots-a_{n-2}\beta_2-a_{n-1}\beta_1-a_n\beta_0\end{cases} \tag{6.12}$$

i. e

$$\begin{cases}b_0=\beta_0\\b_1=\beta_1+a_1\beta_0\\b_2=\beta_2+a_1\beta_1+a_2\beta_0\\\vdots\\b_{n-1}=\beta_{n-1}+a_1\beta_{n-2}+a_2\beta_{n-3}+\cdots+a_{n-2}\beta_1+a_{n-1}\beta_0\\b_n=\beta_n+a_1\beta_{n-1}+a_2\beta_{n-2}+\cdots+a_{n-2}\beta_2+a_{n-1}\beta_1+a_n\beta_0\end{cases} \tag{6.13}$$

From (6.12) and (6.13) we can obtain (6.14) and (6.15).

$$\begin{cases}x_1(k+1)=x_2(k)+\beta_1 u(k)\\x_2(k+1)=x_3(k)+\beta_2 u(k)\\x_3(k+1)=x_4(k)+\beta_3 u(k)\\\vdots\\x_{n-1}(k+1)=x_n(k)+\beta_{n-1}u(k)\\x_n(k+1)=(-a_n x_1(k)-a_{n-1}x_2(k)-\cdots-a_2 x_{n-1}(k)-a_1 x_n(k))+\beta_n u(k)\end{cases} \tag{6.14}$$

and

$$y(k)=x_1(k)+\beta_0 u(k) \tag{6.15}$$

By expressing (6.14) and (6.15) in matrix form, we can obtain the state space description of system as (6.16) and (6.17).

$$\begin{bmatrix} x_1(k+1) \\ x_2(k+1) \\ x_3(k+1) \\ \vdots \\ x_{n-1}(k+1) \\ x_n(k+1) \end{bmatrix} = \begin{bmatrix} 0 & 1 & 0 & \cdots & 0 & 0 \\ 0 & 0 & 1 & \cdots & 0 & 0 \\ 0 & 0 & 0 & \ddots & 0 & 0 \\ \vdots & \vdots & \vdots & \ddots & \ddots & \vdots \\ 0 & 0 & 0 & \cdots & 0 & 1 \\ -a_n & -a_{n-1} & -a_{n-2} & \cdots & -a_2 & -a_1 \end{bmatrix} \begin{bmatrix} x_1(k) \\ x_2(k) \\ x_3(k) \\ \vdots \\ x_{n-1}(k) \\ x_n(k) \end{bmatrix} + \begin{bmatrix} \beta_1 \\ \beta_2 \\ \beta_3 \\ \vdots \\ \beta_{n-1} \\ \beta_n \end{bmatrix} u(k) \tag{6.16}$$

$$y(k) = [1 \quad 0 \quad 0 \quad \cdots \quad 0 \quad 0] \begin{bmatrix} x_1(k) \\ x_2(k) \\ x_3(k) \\ \vdots \\ x_{n-1}(k) \\ x_n(k) \end{bmatrix} + [\beta_0] u(k) \tag{6.17}$$

Example 6.2 Find a state space description for the system described by the following difference equation.

$$y(k+3)+2y(k+2)+3y(k+1)+y(k)=u(k+1)+5u(k)$$

Solution For the system

$$a_1=2, a_2=3, a_3=1; \quad b_0=b_1=0, b_2=1, b_3=5$$

The parameters $\beta_i\ (i=0,1,2,3)$ are calculated as

$$\begin{cases} \beta_0=b_0=0 \\ \beta_1=b_1-a_1\beta_0=0 \\ \beta_2=b_2-a_1\beta_1-a_2\beta_0=b_2=1 \\ \beta_3=b_3-a_1\beta_2-a_2\beta_1-a_3\beta_0=5-2\times1=3 \end{cases}$$

The state variables can be defined by

$$\begin{cases} x_1(k)=y(k)-\beta_0 u(k) \\ x_2(k)=y(k+1)-\beta_0 u(k+1)-\beta_1 u(k) \\ x_3(k)=y(k+2)-\beta_0 u(k+2)-\beta_1 u(k+1)-\beta_2 u(k) \end{cases}$$

The state space description of system can be obtained as

$$\boldsymbol{X}(k+1)=\begin{bmatrix} 0 & 1 & 0 \\ 0 & 0 & 1 \\ -a_3 & -a_2 & -a_1 \end{bmatrix}\boldsymbol{X}(k)+\begin{bmatrix} \beta_1 \\ \beta_2 \\ \beta_3 \end{bmatrix}u(k)$$

$$=\begin{bmatrix} 0 & 1 & 0 \\ 0 & 0 & 1 \\ -1 & -3 & -2 \end{bmatrix}\boldsymbol{X}(k)+\begin{bmatrix} 0 \\ 1 \\ 3 \end{bmatrix}u(k)$$

$$y(k)=[1 \quad 0 \quad 0]\boldsymbol{X}(k)+[\beta_0]u(k)$$

$$=[1 \quad 0 \quad 0]\boldsymbol{X}(k)$$

Where $\boldsymbol{X}(k)=[x_1(k) \quad x_2(k) \quad x_3(k)]^T$.

Example 6.3 Find a state space description for the system described by the following difference equation.

$$y(k+3)+16y(k+2)+194y(k+1)+640y(k)=4u(k+3)+160u(k+1)+720u(k)$$

Solution For the system

$$a_1=16, a_2=194, a_3=640;$$
$$b_0=4, b_1=0, b_2=160, b_3=720$$

The parameters $\beta_i\ (i=0,1,2,3)$ are calculated as

$$\begin{cases}\beta_0=b_0=4\\ \beta_1=b_1-a_1\beta_0=0-16\times4=-64\\ \beta_2=b_2-a_1\beta_1-a_2\beta_0=160-16\times(-64)-194\times4=408\\ \beta_3=b_3-a_1\beta_2-a_2\beta_1-a_3\beta_0=720-16\times408-194\times(-64)-640\times4=4048\end{cases}$$

The state variables can be defined by

$$\begin{cases}x_1(k)=y(k)-\beta_0u(k)\\ x_2(k)=y(k+1)-\beta_0u(k+1)-\beta_1u(k)\\ x_3(k)=y(k+2)-\beta_0u(k+2)-\beta_1u(k+1)-\beta_2u(k)\end{cases}$$

The state space description of system can be obtained as

$$\boldsymbol{X}(k+1)=\begin{bmatrix}0 & 1 & 0\\ 0 & 0 & 1\\ -a_3 & -a_2 & -a_1\end{bmatrix}\boldsymbol{X}(k)+\begin{bmatrix}\beta_1\\ \beta_2\\ \beta_3\end{bmatrix}u(k)$$
$$=\begin{bmatrix}0 & 1 & 0\\ 0 & 0 & 1\\ -640 & -194 & -16\end{bmatrix}\boldsymbol{X}(k)+\begin{bmatrix}-64\\ 408\\ 4048\end{bmatrix}u(k)$$

$$y(k)=[1\quad 0\quad 0]\boldsymbol{X}(k)+[\beta_0]u(k)=[1\quad 0\quad 0]\boldsymbol{X}(k)+4u(k)$$

Where $\boldsymbol{X}(k)=[x_1(k)\quad x_2(k)\quad x_3(k)]^{\mathrm{T}}$.

2. Obtaining State Space Description from Impulse Transfer Function

Suppose the system difference equation of a SISO n-order discrete time LTI system is

$$y(k+n)+a_1y(k+n-1)+\cdots+a_{n-1}y(k+1)+a_ny(k)$$
$$=b_0u(k+n)+b_1u(k+n-1)+\cdots+b_{n-1}u(k+1)+b_nu(k) \tag{6.18}$$

We have the z transform $Z[y(k+i)]=z^iy(z)$ if the zero initial conditions hold true. By taking the z transform of (6.18), the **impulse transfer function** of a proper rational system is obtained as

$$G(z)=\frac{Y(z)}{U(z)}=\frac{b_0z^n+b_1z^{n-1}+b_2z^{n-2}+\cdots+b_{n-1}z+b_n}{z^n+a_1z^{n-1}+\cdots+a_{n-1}z+a_n}$$
$$=b_0+\frac{(b_1-b_0a_1)z^{n-1}+(b_2-b_0a_2)z^{n-2}+\cdots+(b_{n-1}-b_0a_{n-1})z+(b_n-b_0a_n)}{z^n+a_1z^{n-1}+\cdots+a_{n-1}z+a_n}$$
$$=b_0+G'(z) \tag{6.19}$$

and can be represented with the block diagram which is shown in Figure 6.2.

Where $G'(z)$ is a rational fraction of z and

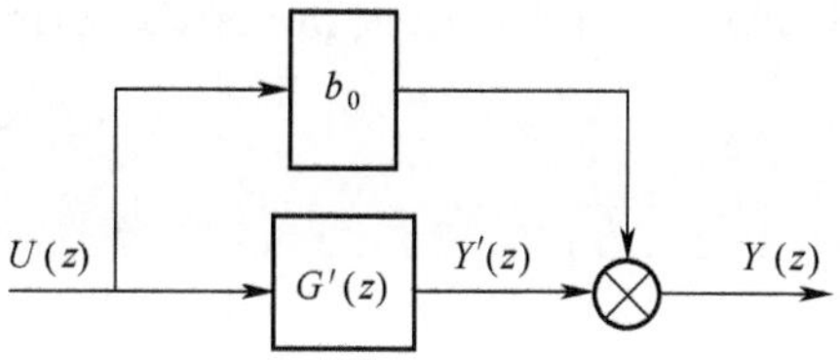

Figure 6.2 Block Diagram

$$G'(z)=\frac{(b_1-b_0a_1)z^{n-1}+(b_2-b_0a_2)z^{n-2}+\cdots+(b_{n-1}-b_0a_{n-1})z+(b_n-b_0a_n)}{z^n+a_1z^{n-1}+\cdots+a_{n-1}z+a_n} \tag{6.20}$$

By introducing an intermediate variable $V(z)$, the equation (6.21) can be derived from (6.20).

$$G'(z)=\frac{N(z)}{D(z)}=\frac{Y'(z)}{U(z)}=\frac{Y'(z)/V(z)}{U(z)/V(z)} \tag{6.21}$$

Where $N(z)$ and $D(z)$ are numerator and denominator of (6.20), respectively, thus we have

$$\frac{U(z)}{V(z)}=D(z)=z^n+a_1z^{n-1}+\cdots+a_{n-1}z+a_n \tag{6.22}$$

Rewriting (6.22) the following equation is obtained as

$$z^nV(z)=-a_1z^{n-1}V(z)-\cdots-a_{n-1}zV(z)-a_nV(z)+U(z) \tag{6.23}$$

The state variables can be selected as

$$\begin{cases} x_1(z)=V(z) \\ x_2(z)=zV(z)=zx_1(z) \\ x_3(z)=z^2V(z)=zx_2(z) \\ \vdots \\ x_n(z)=z^{n-1}V(z)=zx_{n-1}(z) \end{cases} \tag{6.24}$$

Multiplying (6.24) by z and substituting (6.23) into its last equation the following equations are obtained as

$$\begin{cases} zx_1(z)=zV(z)=x_2(z) \\ zx_2(z)=z^2V(z)=x_3(z) \\ \quad\vdots \\ zx_{n-1}(z)=z^{n-1}V(z)=x_n(z) \\ zx_n(z)=z^nV(z)=-a_1z^{n-1}V(z)-\cdots-a_{n-1}zV(z)-a_nV(z)+U(z) \\ \qquad =-a_1x_n(z)-\cdots-a_{n-1}x_2(z)-a_nx_1(z)+U(z) \end{cases} \tag{6.25}$$

By taking the inverse z transform of (6.25), we get a set of first-order difference equations shown as

$$\begin{cases} x_1(k+1)=x_2(k) \\ x_2(k+1)=x_3(k) \\ \quad\vdots \\ x_{n-1}(k+1)=x_n(k) \\ x_n(k+1)=-a_nx_1(k)-a_{n-1}x_2(k)-\cdots-a_1x_n(k)+u(k) \end{cases} \tag{6.26}$$

The state equation (6.26) may be written in matrix form as

$$\begin{bmatrix} x_1(k+1) \\ x_2(k+1) \\ \vdots \\ x_{n-1}(k+1) \\ x_n(k+1) \end{bmatrix} = \boldsymbol{GX}(k)+\boldsymbol{h}u(k)$$

$$= \begin{bmatrix} 0 & 1 & 0 & \cdots & 0 \\ 0 & 0 & 1 & \cdots & 0 \\ \vdots & \vdots & \vdots & \ddots & \vdots \\ 0 & 0 & 0 & \cdots & 1 \\ -a_n & -a_{n-1} & -a_{n-2} & \cdots & -a_1 \end{bmatrix} \begin{bmatrix} x_1(k) \\ x_2(k) \\ \vdots \\ x_{n-1}(k) \\ x_n(k) \end{bmatrix} + \begin{bmatrix} 0 \\ 0 \\ \vdots \\ 0 \\ 1 \end{bmatrix} u(k) \tag{6.27}$$

According to equation (6.21), we have

$$\frac{Y'(z)}{V(z)} = N(z)$$
$$= (b_1 - b_0 a_1) z^{n-1} + (b_2 - b_0 a_2) z^{n-2} + \cdots + (b_{n-1} - b_0 a_{n-1}) z + (b_n - b_0 a_n) \tag{6.28}$$

Rewriting (6.28) and substituting (6.24) into (6.28), we get equation (6.29).

$$Y'(z) = (b_1 - b_0 a_1) z^{n-1} V(z) + (b_2 - b_0 a_2) z^{n-2} V(z) + \cdots + (b_{n-1} - b_0 a_{n-1}) z V(z) + (b_n - b_0 a_n) V(z)$$
$$= (b_1 - b_0 a_1) x_n(z) + (b_2 - b_0 a_2) x_{n-1}(z) + \cdots + (b_{n-1} - b_0 a_{n-1}) x_2(z) + (b_n - b_0 a_n) x_1(z) \tag{6.29}$$

By taking the inverse z transform of (6.29) and assuming zero initial conditions hold true, the output equation can be obtained as

$$y'(k) = (b_1 - b_0 a_1) x_n(k) + (b_2 - b_0 a_2) x_{n-1}(k) + \cdots + (b_{n-1} - b_0 a_{n-1}) x_2(k) + (b_n - b_0 a_n) x_1(k) \tag{6.30}$$

According to (6.19), we have

$$y(k) = y'(k) + b_0 u(k) \tag{6.31}$$

Substituting (6.30) into (6.31), we get the output equation (6.32).

$$y'(k) = (b_1 - b_0 a_1) x_n(k) + (b_2 - b_0 a_2) x_{n-1}(k) + \cdots + (b_{n-1} - b_0 a_{n-1}) x_2(k) + (b_n - b_0 a_n) x_1(k) + b_0 u(k) \tag{6.32}$$

The output equation (6.32) may be written in matrix form as

$$y'(k) = \boldsymbol{cX}(k) + du(k)$$

$$= [b_n - b_0 a_n \quad b_{n-1} - b_0 a_{n-1} \quad \cdots \quad b_2 - b_0 a_2 \quad b_1 - b_0 a_1] \begin{bmatrix} x_1(k) \\ x_2(k) \\ \vdots \\ x_{n-1}(k) \\ x_n(k) \end{bmatrix} + [b_0] u(k) \tag{6.33}$$

Obviously, the state space description, such as (6.27) and (6.33), is the **controllable canonical form.**

Particularly, $d=\mathbf{0}$ when $b_0=0$, in other words, when the system is a strictly proper system, the state space description of the discrete time LTI system reduces to

$$\begin{cases} \boldsymbol{X}(k+1)=\boldsymbol{G}\boldsymbol{X}(k)+\boldsymbol{h}u(k) \\ y(k)=\boldsymbol{c}\boldsymbol{X}(k) \end{cases} \tag{6.34}$$

Where $$\boldsymbol{G}=\begin{bmatrix} 0 & 1 & 0 & \cdots & 0 \\ 0 & 0 & 1 & \cdots & 0 \\ \vdots & \vdots & \vdots & \ddots & \vdots \\ 0 & 0 & 0 & \cdots & 1 \\ -a_n & -a_{n-1} & -a_{n-2} & \cdots & -a_1 \end{bmatrix}; \quad \boldsymbol{h}=\begin{bmatrix} 0 \\ 0 \\ \vdots \\ 0 \\ 1 \end{bmatrix};$$

$$\boldsymbol{c}=[b_n \quad b_{n-1} \quad \cdots \quad b_2 \quad b_1].$$

Example 6.4 Consider a system described by the transfer function

$$G(z)=\frac{z^2+z+3}{z^3+9z^2+24z+20}$$

Find a state space description by the method of direct decomposition.

Solution By introducing an intermediate variable $V(z)$, the state variables can be selected as the phase variables, which are defined by

$$\begin{cases} x_1(z)=V(z) \\ x_2(z)=zV(z)=zx_1(z) \\ x_3(z)=z^2V(z)=zx_2(z) \\ \vdots \\ x_n(z)=z^{n-1}V(z)=zx_{n-1}(z) \end{cases}$$

For the strictly proper system

$$a_1=9, a_2=24, a_3=20; \quad b_0=0, b_1=1, b_2=1, b_3=3$$

$$\boldsymbol{X}(k+1)=\begin{bmatrix} 0 & 1 & 0 \\ 0 & 0 & 1 \\ -a_3 & -a_2 & -a_1 \end{bmatrix}\boldsymbol{X}(k)+\begin{bmatrix} 0 \\ 0 \\ 1 \end{bmatrix}u(k)$$

$$=\begin{bmatrix} 0 & 1 & 0 \\ 0 & 0 & 1 \\ -20 & -24 & -9 \end{bmatrix}\boldsymbol{X}(k)+\begin{bmatrix} 0 \\ 0 \\ 1 \end{bmatrix}u(k)$$

$$y(k)=[b_3 \quad b_2 \quad b_1]\boldsymbol{X}(k)=[3 \quad 1 \quad 1]\boldsymbol{X}(k)$$

From the discussion above we can see that constructing the state space description of the discrete time system is very similar to the continuous time system. It can be obtained from physical mechanism、difference equation or impulse transfer function. We only introduce the methods above, and the other state space descriptions of discrete time system can be easily obtained on the basis of the continuous time system.

6.1.3 Obtaining Impulse Transfer Function Matrix from State Space Description

Considering the discrete time LTI system, its state space description is

$$\boldsymbol{X}(k+1)=\boldsymbol{G}\boldsymbol{X}(k)+\boldsymbol{H}\boldsymbol{u}(k) \tag{6.35}$$

$$\boldsymbol{y}(k)=\boldsymbol{C}\boldsymbol{X}(k)+\boldsymbol{D}\boldsymbol{u}(k) \tag{6.36}$$

Where $\boldsymbol{X}(k)$ denotes the n dimension state; $\boldsymbol{u}(k)$ denotes the p dimension input; $\boldsymbol{y}(k)$ denotes the q dimension output.

Definition 6.1 If the z transforms of $\boldsymbol{y}(k)$ and $\boldsymbol{u}(k)$ are $\boldsymbol{Y}(z)$ and $\boldsymbol{U}(z)$ respectively, namely

$$\boldsymbol{Y}(z)=Z[\boldsymbol{y}(k)], \boldsymbol{U}(z)=Z[\boldsymbol{u}(k)] \tag{6.37}$$

assuming the zero initial conditions hold true, the **impulse transfer function matrix** of discrete time LTI system described by (6.35) and (6.36) can be defined by a $q\times p$ rational fraction matrix $\boldsymbol{G}(z)$ and it satisfies

$$\boldsymbol{Y}(z)=\boldsymbol{G}(z)\boldsymbol{U}(z) \tag{6.38}$$

Where z is a complex variable and $\boldsymbol{G}(z)$ can be calculated as

$$\boldsymbol{G}(z)=\boldsymbol{C}(z\boldsymbol{I}-\boldsymbol{G})^{-1}\boldsymbol{H}+\boldsymbol{D} \tag{6.39}$$

Then we will discuss how to obtain the transfer function matrix from the state space description.

Consider the discrete time LTI system such as (6.35) and (6.36), taking the z transform and assuming $\boldsymbol{X}(0)=\boldsymbol{X}_0$, we will obtain

$$z\boldsymbol{X}(z)-z\boldsymbol{X}(0)=\boldsymbol{G}\boldsymbol{X}(z)+\boldsymbol{H}\boldsymbol{U}(z) \tag{6.40}$$

$$\boldsymbol{Y}(z)=\boldsymbol{C}\boldsymbol{X}(z)+\boldsymbol{D}\boldsymbol{U}(z) \tag{6.41}$$

According to the definition of impulse transfer function matrix and assuming the initial state of system $\boldsymbol{X}_0=\boldsymbol{0}$, and rewriting (6.40), we have

$$\boldsymbol{X}(z)=(z\boldsymbol{I}-\boldsymbol{G})^{-1}\boldsymbol{H}\boldsymbol{U}(z) \tag{6.42}$$

Substituting (6.42) into (6.41), the equation (6.43) is obtained.

$$\begin{aligned}\boldsymbol{Y}(z)&=\boldsymbol{C}\boldsymbol{X}(z)+\boldsymbol{D}\boldsymbol{U}(z)\\&=[\boldsymbol{C}(z\boldsymbol{I}-\boldsymbol{G})^{-1}\boldsymbol{H}+\boldsymbol{D}]\boldsymbol{U}(z)=\boldsymbol{G}(z)\boldsymbol{U}(z)\end{aligned} \tag{6.43}$$

Then, the impulse transfer function matrix can be calculated as

$$\boldsymbol{G}(z)=\frac{\boldsymbol{Y}(z)}{\boldsymbol{U}(z)}=\boldsymbol{C}(z\boldsymbol{I}-\boldsymbol{G})^{-1}\boldsymbol{H}+\boldsymbol{D} \tag{6.44}$$

Example 6.5 A system is described by

$$\boldsymbol{X}(k+1)=\begin{bmatrix}0 & 1\\-2 & -3\end{bmatrix}\boldsymbol{X}(k)+\begin{bmatrix}1 & 0\\1 & 1\end{bmatrix}\boldsymbol{u}(k)$$

$$\boldsymbol{y}(k)=\begin{bmatrix}1 & 0\\1 & 1\end{bmatrix}\boldsymbol{X}(k)$$

Find the impulse transfer function matrix of it.

Solution From the state space description $\{\boldsymbol{G},\boldsymbol{H},\boldsymbol{C}\}$ of the system, we can obtain

$$(z\boldsymbol{I}-\boldsymbol{G})=\begin{bmatrix} z & -1 \\ 2 & z+3 \end{bmatrix}$$

Its inverse matrix is

$$(z\boldsymbol{I}-\boldsymbol{G})^{-1}=\frac{1}{z(z+3)+2}\begin{bmatrix} z+3 & 1 \\ -2 & z \end{bmatrix}=\frac{1}{z^2+3z+2}\begin{bmatrix} z+3 & 1 \\ -2 & z \end{bmatrix}$$

Then, the impulse transfer function matrix of the system can be obtained as

$$\begin{aligned} \boldsymbol{G}(z)&=\boldsymbol{C}(z\boldsymbol{I}-\boldsymbol{G})^{-1}\boldsymbol{H}+\boldsymbol{D} \\ &=\begin{bmatrix} 1 & 0 \\ 1 & 1 \end{bmatrix}\left(\frac{1}{z^2+3z+2}\right)\begin{bmatrix} z+3 & 1 \\ -2 & z \end{bmatrix}\begin{bmatrix} 1 & 0 \\ 1 & 1 \end{bmatrix} \\ &=\begin{bmatrix} \dfrac{z+4}{(z+1)(z+2)} & \dfrac{1}{(z+1)(z+2)} \\ \dfrac{2}{z+2} & \dfrac{1}{z+2} \end{bmatrix} \end{aligned}$$

6.2 State Equation Solution of Discrete Time LTI System

Usually, there are two methods for solving the state equation of discrete system, iterative method and z transform method. The discussion in this section is restricted to discrete time LTI system described by the following state equation.

$$\boldsymbol{X}(k+1)=\boldsymbol{G}\boldsymbol{X}(k)+\boldsymbol{H}\boldsymbol{u}(k),\boldsymbol{X}(0)=\boldsymbol{X}_0 \quad k=0,1,2,\cdots \tag{6.45}$$

6.2.1 Iterative Method

Consider the discrete time LTI system described by (6.45), and assume the initial state $\boldsymbol{X}(0)=\boldsymbol{X}_0$. If the inputs at every sampling instant of time are $\boldsymbol{u}(0),\boldsymbol{u}(1),\boldsymbol{u}(2),\cdots$, and the final time instant is noted as a positive integer l, then the system state response can be calculated by the following steps.

Step1: When $k=1$, since $\boldsymbol{X}(0)$ and $\boldsymbol{u}(0)$ are known, then we can obtain the state at time instant $k=1$ from (6.45)

$$\boldsymbol{X}(1)=\boldsymbol{G}\boldsymbol{X}(0)+\boldsymbol{H}\boldsymbol{u}(0)$$

Step2: When $k=1$, according to $\boldsymbol{X}(1)$ obtained above and input variable $\boldsymbol{u}(1)$, we can obtain the state at time instant $k=2$ from (6.45)

$$\boldsymbol{X}(2)=\boldsymbol{G}\boldsymbol{X}(1)+\boldsymbol{H}\boldsymbol{u}(1)=\boldsymbol{G}^2\boldsymbol{X}(0)+\boldsymbol{G}\boldsymbol{H}\boldsymbol{u}(0)+\boldsymbol{H}\boldsymbol{u}(1)$$

Step3: When $k=2$, according to $\boldsymbol{X}(2)$ obtained above and input variable $\boldsymbol{u}(2)$, we can also obtain the state at time instant $k=3$ from (6.45)

$$\boldsymbol{X}(3)=\boldsymbol{G}\boldsymbol{X}(2)+\boldsymbol{H}\boldsymbol{u}(2)=\boldsymbol{G}^3\boldsymbol{X}(0)+\boldsymbol{G}^2\boldsymbol{H}\boldsymbol{u}(0)+\boldsymbol{G}\boldsymbol{H}\boldsymbol{u}(1)+\boldsymbol{H}\boldsymbol{u}(2)$$

$$\vdots$$

By applying the induction method, we can get the solution formula (6.46) of the system (6.45) as

$$\boldsymbol{X}(k)=\boldsymbol{G}^{k}\boldsymbol{X}(0)+\sum_{i=0}^{k-1}\boldsymbol{G}^{k-i-1}\boldsymbol{Hu}(i) \tag{6.46}$$

Equation (6.46) is called a **state response** of the discrete time LTI system.

Example 6.6 Considering the discrete time LTI system

$$\boldsymbol{X}(k+1)=\begin{bmatrix}0 & 1\\ -0.21 & -1\end{bmatrix}\boldsymbol{X}(k)+\begin{bmatrix}1\\1\end{bmatrix}u(k)$$

Where

$$\boldsymbol{X}(0)=[1\quad -1]^{\mathrm{T}},\quad u(k)=1$$

Please determine the state variables $\boldsymbol{X}(k)$, when $k=1,2,3,4$.

Solution According to the iterative method, when $k=0$, we have

$$\boldsymbol{X}(1)=\boldsymbol{GX}(0)+\boldsymbol{h}u(0)=\begin{bmatrix}0 & 1\\ -0.21 & -1\end{bmatrix}\begin{bmatrix}1\\-1\end{bmatrix}+\begin{bmatrix}1\\1\end{bmatrix}\times 1=\begin{bmatrix}0\\1.79\end{bmatrix}$$

When $k=1$, we have

$$\boldsymbol{X}(2)=\boldsymbol{GX}(1)+\boldsymbol{h}u(1)=\begin{bmatrix}0 & 1\\ -0.21 & -1\end{bmatrix}\begin{bmatrix}0\\1.79\end{bmatrix}+\begin{bmatrix}1\\1\end{bmatrix}\times 1=\begin{bmatrix}2.79\\-0.79\end{bmatrix}$$

When $k=2$, we have

$$\boldsymbol{X}(3)=\boldsymbol{GX}(2)+\boldsymbol{h}u(2)=\begin{bmatrix}0 & 1\\ -0.21 & -1\end{bmatrix}\begin{bmatrix}2.79\\-0.79\end{bmatrix}+\begin{bmatrix}1\\1\end{bmatrix}\times 1=\begin{bmatrix}0.21\\1.204\end{bmatrix}$$

When $k=3$, we have

$$\boldsymbol{X}(4)=\boldsymbol{GX}(3)+\boldsymbol{h}u(3)=\begin{bmatrix}0 & 1\\ -0.21 & -1\end{bmatrix}\begin{bmatrix}0.21\\1.204\end{bmatrix}+\begin{bmatrix}1\\1\end{bmatrix}\times 1=\begin{bmatrix}2.204\\-0.248\end{bmatrix}$$

$$\vdots$$

We need to indicate that the state solutions at any time instant $t=kT(k=1,2,3,\cdots)$ can be obtained by using the iterative method, but the solution in enclosed form such as equation (6.46) can not be obtained. However, the z transform method can get it, and we will discuss it later.

Based on the definition of the state transition matrix of continuous time LTI system, we will give the definition of state transition matrix of discrete time LTI system.

Definition 6.2 If a $n\times n$ matrix $\boldsymbol{\Phi}(k)$ is the solution matrix of matrix difference equation

$$\boldsymbol{\Phi}(k+1)=\boldsymbol{G\Phi}(k) \tag{6.47}$$

and its initial value is $\boldsymbol{\Phi}(0)=\boldsymbol{I}$, it is called the **state transition matrix** of the discrete time LTI system described by (6.45).

Theorem 6.1 The state transition matrix of the discrete time LTI system described by (6.45), namely the solution matrix $\boldsymbol{\Phi}(k)$ of matrix equation (6.47) is

$$\boldsymbol{\Phi}(k)=\boldsymbol{GG}\cdots\boldsymbol{G}=\boldsymbol{G}^{k} \tag{6.48}$$

Proof. Consider the discrete time LTI system described by (6.45), when $k=k+1$

and $k=0$, equations (6.49) and (6.50) can be derived from (6.48) directly, respectively

$$\boldsymbol{\Phi}(k+1)=\boldsymbol{G}\cdot[\boldsymbol{G}\boldsymbol{G}\cdots\boldsymbol{G}]=\boldsymbol{G}\cdot\boldsymbol{G}^k=\boldsymbol{G}\cdot\boldsymbol{\Phi}(k) \tag{6.49}$$

and

$$\boldsymbol{\Phi}(0)=\boldsymbol{G}^0=\boldsymbol{I} \tag{6.50}$$

This illustrates that, $\boldsymbol{\Phi}(k)$ determined by equation (6.48) is the solution matrix of (6.47) and satisfies the initial condition. Then it is the state transition matrix of the discrete time LTI system (6.45).

By applying the definition of state transition matrix, the state response (6.46) of the discrete time LTI system can be obtained as

$$\boldsymbol{X}(k)=\boldsymbol{\Phi}(k)\boldsymbol{X}_0+\sum_{i=0}^{k-1}\boldsymbol{\Phi}(k-i-1)\boldsymbol{H}\boldsymbol{u}(i) \tag{6.51}$$

or

$$\boldsymbol{X}(k)=\boldsymbol{\Phi}(k)\boldsymbol{X}_0+\sum_{i=0}^{k-1}\boldsymbol{\Phi}(i)\boldsymbol{H}\boldsymbol{u}(k-i-1) \tag{6.52}$$

Furthermore, substituting the state transition matrix (6.48) into (6.52), we have

$$\boldsymbol{X}(k)=\boldsymbol{G}^k\boldsymbol{X}_0+\sum_{i=0}^{k-1}\boldsymbol{G}^i\boldsymbol{H}\boldsymbol{u}(k-i-1) \tag{6.53}$$

The state response of the discrete time LTI system is composed of two parts, the zero input response $\boldsymbol{X}_{0u}(k)$ and the zero initial state response $\boldsymbol{X}_{0X}(k)$, that is

$$\boldsymbol{X}(k)=\boldsymbol{X}_{0u}(k)+\boldsymbol{X}_{0X}(k)$$

Where

$$\begin{aligned}&\boldsymbol{X}_{0u}(k)=\boldsymbol{\Phi}(k)\boldsymbol{X}_0\\&\boldsymbol{X}_{0X}(k)=\sum_{i=0}^{k-1}\boldsymbol{\Phi}(k-i-1)\boldsymbol{H}\boldsymbol{u}(i)=\sum_{i=0}^{k-1}\boldsymbol{\Phi}(i)\boldsymbol{H}\boldsymbol{u}(k-i-1)\end{aligned} \tag{6.54}$$

6.2.2 *z* Transform Method

Consider the discrete time LTI system described by (6.45), the z transform method can be used to solve the state equation.

By taking the z transform of (6.45), we have

$$z\boldsymbol{X}(z)-z\boldsymbol{X}(0)=\boldsymbol{G}\boldsymbol{X}(z)+\boldsymbol{H}\boldsymbol{U}(z) \tag{6.55}$$

Rewriting (6.55), we get

$$\boldsymbol{X}(z)=(z\boldsymbol{I}-\boldsymbol{G})^{-1}z\boldsymbol{X}(0)+(z\boldsymbol{I}-\boldsymbol{G})^{-1}\boldsymbol{H}\boldsymbol{U}(z) \tag{6.56}$$

By taking the inverse z transform of (6.56), we obtain

$$\begin{aligned}\boldsymbol{X}(k)&=Z^{-1}[z\boldsymbol{I}-\boldsymbol{G})^{-1}z]\boldsymbol{X}(0)+Z^{-1}[(z\boldsymbol{I}-\boldsymbol{G})^{-1}\boldsymbol{H}\boldsymbol{U}(z)]\\&=\boldsymbol{\Phi}(k)\boldsymbol{X}_0+\sum_{i=0}^{k-1}\boldsymbol{\Phi}(k-i-1)\boldsymbol{H}\boldsymbol{u}(i)\end{aligned} \tag{6.57}$$

Therefore

$$\boldsymbol{\Phi}(k)=Z^{-1}[(z\boldsymbol{I}-\boldsymbol{G})^{-1}z]=\boldsymbol{G}^k \qquad (6.58)$$

$$\sum_{i=0}^{k-1}\boldsymbol{\Phi}(k-i-1)\boldsymbol{H}\boldsymbol{u}(i)=Z^{-1}[(z\boldsymbol{I}-\boldsymbol{G})^{-1}\boldsymbol{H}\boldsymbol{U}(z)]=\sum_{i=0}^{k}\boldsymbol{G}^{k-i-1}\boldsymbol{H}\boldsymbol{u}(i) \qquad (6.59)$$

6.2.3 Calculation of the State Transition Matrix

Solving the state transition matrix of discrete time LTI system is very similar to the calculation of continuous time system.

Method 1: Direct Method

Calculating the state transition matrix $\boldsymbol{\Phi}(k)$ directly by its definition, i. e

$$\boldsymbol{\Phi}(k)=\boldsymbol{G}^k$$

This method is easy to be realized by computer, but difficult to obtain the closed type expression of $\boldsymbol{\Phi}(k)$.

Method 2: z Transform Method

We have proved that the state transition matrix $\boldsymbol{\Phi}(k)$ can be calculated as

$$\boldsymbol{\Phi}(k)=Z^{-1}[(z\boldsymbol{I}-\boldsymbol{G})^{-1}z]$$

Method 3: Similarly Transformation Method

Case 1: when the eigenvalues of discrete time LTI system are distinct, system matrix $\boldsymbol{G}$ can be transferred to diagonal matrix by similarly transformation, namely

$$\boldsymbol{P}^{-1}\boldsymbol{G}\boldsymbol{P}=\boldsymbol{\Lambda} \qquad (6.60)$$

then, the state transition matrix $\boldsymbol{\Phi}(k)$ of discrete time system can be calculated as

$$\boldsymbol{\Phi}(k)=\boldsymbol{G}^k=\boldsymbol{P}\boldsymbol{\Lambda}^k\boldsymbol{P}^{-1} \qquad (6.61)$$

Where $\boldsymbol{\Lambda}$ is the diagonal matrix. If the eigenvalues of characteristic equation $|\lambda\boldsymbol{I}-\boldsymbol{G}|=0$ are $\lambda_1, \lambda_2, \cdots, \lambda_n$, we have

$$\boldsymbol{\Lambda}=\begin{bmatrix}\lambda_1 & 0 & \cdots & 0\\ 0 & \lambda_2 & \ddots & \vdots\\ \vdots & \ddots & \ddots & 0\\ 0 & \cdots & 0 & \lambda_n\end{bmatrix} \quad \boldsymbol{\Lambda}^k=\begin{bmatrix}\lambda_1^k & 0 & \cdots & 0\\ 0 & \lambda_2^k & \ddots & \vdots\\ \vdots & \ddots & \ddots & 0\\ 0 & \cdots & 0 & \lambda_n^k\end{bmatrix} \qquad (6.62)$$

and

$$\boldsymbol{\Phi}(k)=\boldsymbol{P}\begin{bmatrix}\lambda_1^k & 0 & \cdots & 0\\ 0 & \lambda_2^k & \ddots & \vdots\\ \vdots & \ddots & \ddots & 0\\ 0 & \cdots & 0 & \lambda_n^k\end{bmatrix}\boldsymbol{P}^{-1} \qquad (6.63)$$

Where $\boldsymbol{P}$ is a nonsingular matrix, which transforms a general system matrix $\boldsymbol{G}$ into a diagonal matrix.

Case 2: If the characteristic equation $|\lambda\boldsymbol{I}-\boldsymbol{G}|=0$ of the discrete system has multiplicity eigenvalues roots, system matrix $\boldsymbol{G}$ can be transferred to the Jordan matrix by the similarly transformation, namely

$$P^{-1}GP=J \tag{6.64}$$

the state transition matrix $\boldsymbol{\Phi}(k)$ of the discrete time system can be calculated as

$$\boldsymbol{\Phi}(k)=G^k=PJ^kP^{-1} \tag{6.65}$$

Where J denotes the Jordan matrix; P is a nonsingular matrix, which transforms the general system matrix G into the Jordan matrix. The calculations of them were discussed in Chapter 1.

Method 4: Cayley-Hamilton Theorem Method

Applying Cayley-Hamilton Theorem, system matrix G is a matrix root of the characteristic polynomial, thus representing G^k as a polynomial of G^i, we obtain

$$\boldsymbol{\Phi}(k)=G^k=\alpha_0(k)I+\alpha_1(k)G+\alpha_2(k)G^2+\cdots+\alpha_{n-1}(k)G^{n-1} \tag{6.66}$$

Where $\alpha_0(k),\alpha_1(k),\cdots,\alpha_{n-1}(k)$ are undetermined coefficients, which can be obtained by consulting the calculation of continuous system.

Example 6.7 Consider the discrete time LTI system, its state equation is

$$X(k+1)=\begin{bmatrix}0 & 1\\ -0.16 & -1\end{bmatrix}X(k)+\begin{bmatrix}0\\0\end{bmatrix}u(k)$$

and its initial state $X(0)=[1\quad -1]^{\mathrm{T}}$, please determine its state transition matrix by applying the four methods introduced above. Furthermore, determine the state solution of the discrete system when $u(k)=1$.

Solution From the state equation, the system matrix G and input matrix H are

$$G=\begin{bmatrix}0 & 1\\ -0.16 & -1\end{bmatrix}, h=\begin{bmatrix}0\\0\end{bmatrix}, X(0)=\begin{bmatrix}1\\-1\end{bmatrix}$$

Method 1: Direct Method

The state response of the discrete time LTI system is

$$X(k)=\boldsymbol{\Phi}(k)X_0+\sum_{i=0}^{k-1}\boldsymbol{\Phi}(k-i-1)hu(i), k=0,1,2,\cdots$$

Where

$$\boldsymbol{\Phi}(k)=G^k=\begin{bmatrix}0 & 1\\ -0.16 & -1\end{bmatrix}^k$$

when $k=1$, $\boldsymbol{\Phi}(1)=G^1=\begin{bmatrix}0 & 1\\ -0.16 & -1\end{bmatrix}$

when $k=2$, $\boldsymbol{\Phi}(2)=G^2=\begin{bmatrix}0 & 1\\ -0.16 & -1\end{bmatrix}^2=\begin{bmatrix}-0.16 & -1\\ 0.16 & 0.84\end{bmatrix}$

when $k=3$, $\boldsymbol{\Phi}(3)=G^3=\begin{bmatrix}0 & 1\\ -0.16 & -1\end{bmatrix}^3=\begin{bmatrix}0.16 & 0.84\\ -0.13 & -0.68\end{bmatrix}$

⋮

Thus, the state response is

$$X(1)=\boldsymbol{\Phi}(1)X(0)=\begin{bmatrix}-1\\0.84\end{bmatrix}$$

$$X(2)=\Phi(2)X(0)=\begin{bmatrix}0.84\\-0.68\end{bmatrix}$$

$$X(3)=\Phi(3)X(0)=\begin{bmatrix}-0.68\\0.55\end{bmatrix}$$

$$\vdots$$

Method 2: z Transform Method

$$(zI-G)^{-1}=\begin{bmatrix}z & -1\\0.16 & z+1\end{bmatrix}^{-1}=\begin{bmatrix}\frac{z+1}{(z+0.2)(z+0.8)} & \frac{1}{(z+0.2)(z+0.8)}\\ \frac{-0.16}{(z+0.2)(z+0.8)} & \frac{z}{(z+0.2)(z+0.8)}\end{bmatrix}$$

$$\Phi(k)=Z^{-1}[(zI-G)^{-1}z]$$

$$=\begin{bmatrix}\frac{4}{3}(-0.2)^k-\frac{1}{3}(-0.8)^k & \frac{5}{3}(-0.2)^k-\frac{5}{3}(-0.8)^k\\ -\frac{0.8}{3}(-0.2)^k+\frac{0.8}{3}(-0.8)^k & -\frac{1}{3}(-0.2)^k+\frac{4}{3}(-0.8)^k\end{bmatrix}$$

Method 3: Similarly Transformation Method

(1) The characteristic equation of the discrete time LTI system is

$$|\lambda I-G|=\begin{vmatrix}\lambda & -1\\0.16 & \lambda+1\end{vmatrix}=(\lambda+0.2)(\lambda+0.8)=0$$

Its eigenvalues are $\lambda_1=-0.2, \lambda_2=-0.8$.

(2) Since G is a companion matrix, and the eigenvalues are distinct, then the similarly transformation matrix P can be chosen directly as

$$P=\begin{bmatrix}1 & 1\\\lambda_1 & \lambda_2\end{bmatrix}=\begin{bmatrix}1 & 1\\-0.2 & -0.8\end{bmatrix}$$

and its inverse matrix

$$P^{-1}=\frac{1}{3}\begin{bmatrix}4 & 5\\-1 & -5\end{bmatrix}$$

After similarly transformation the diagonal matrix can be obtained as

$$\Lambda=P^{-1}GP=\begin{bmatrix}\lambda_1 & 0\\0 & \lambda_2\end{bmatrix}=\begin{bmatrix}-0.2 & 0\\0 & -0.8\end{bmatrix}$$

(3) Then, the state transition matrix can be calculated as

$$\Phi(k)=G^k=P\Lambda^kP^{-1}$$

$$=\begin{bmatrix}1 & 1\\-0.2 & -0.8\end{bmatrix}\cdot\begin{bmatrix}(-0.2)^k & 0\\0 & (-0.8)^k\end{bmatrix}\cdot\frac{1}{3}\begin{bmatrix}4 & 5\\-1 & -5\end{bmatrix}$$

$$=\begin{bmatrix}\frac{4}{3}(-0.2)^k-\frac{1}{3}(-0.8)^k & \frac{5}{3}(-0.2)^k-\frac{5}{3}(-0.8)^k\\ -\frac{0.8}{3}(-0.2)^k+\frac{0.8}{3}(-0.8)^k & -\frac{1}{3}(-0.2)^k+\frac{4}{3}(-0.8)^k\end{bmatrix}$$

Method 4: Cayley-Hamilton Theorem Method

Its eigenvalues were $\lambda_1=-0.2, \lambda_2=-0.8$.

The undetermined coefficients $\alpha_i(t)$ can be obtained by

$$\begin{cases}(\lambda_1)^k = \alpha_0(k) + \alpha_1(k)\lambda_1 \\ (\lambda_2)^k = \alpha_0(k) + \alpha_1(k)\lambda_2\end{cases}$$

Substituting $\lambda_1 = -0.2$ and $\lambda_2 = -0.8$ into the equation above, we have

$$\begin{cases}(-0.2)^k = \alpha_0(k) - 0.2\alpha_1(k) \\ (-0.8)^k = \alpha_0(k) - 0.8\alpha_1(k)\end{cases}$$

From the equations above, we get

$$\begin{cases}\alpha_0(k) = \frac{4}{3}(-0.2)^k - \frac{1}{3}(-0.8)^k \\ \alpha_1(k) = \frac{5}{3}(-0.2)^k - \frac{5}{3}(-0.8)^k\end{cases}$$

Then, according to (6.66) we get the state transition matrix as

$$\begin{aligned}\boldsymbol{\Phi}(k) &= \alpha_0(k)\boldsymbol{I} + \alpha_1(k)\boldsymbol{G} \\ &= \left(\frac{4}{3}(-0.2)^k - \frac{1}{3}(-0.8)^k\right)\begin{bmatrix}1 & 0 \\ 0 & 1\end{bmatrix} + \left(\frac{5}{3}(-0.2)^k - \frac{5}{3}(-0.8)^k\right)\begin{bmatrix}0 & 1 \\ -0.16 & -1\end{bmatrix} \\ &= \begin{bmatrix}\frac{4}{3}(-0.2)^k - \frac{1}{3}(-0.8)^k & \frac{5}{3}(-0.2)^k - \frac{5}{3}(-0.8)^k \\ -\frac{0.8}{3}(-0.2)^k + \frac{0.8}{3}(-0.8)k & -\frac{1}{3}(-0.2)^k + \frac{4}{3}(-0.8)^k\end{bmatrix}\end{aligned}$$

We can see that the three methods above obtain the same results of state transition matrix. Then the state response of the discrete time LTI system is

$$\begin{aligned}\boldsymbol{X}(k) &= \boldsymbol{\Phi}(k)\boldsymbol{X}_0 \\ &= \begin{bmatrix}\frac{4}{3}(-0.2)^k - \frac{1}{3}(-0.8)^k & \frac{5}{3}(-0.2)^k - \frac{5}{3}(-0.8)^k \\ -\frac{0.8}{3}(-0.2)^k + \frac{0.8}{3}(-0.8)^k & -\frac{1}{3}(-0.2)^k + \frac{4}{3}(-0.8)^k\end{bmatrix}\begin{bmatrix}1 \\ -1\end{bmatrix} \\ &= \begin{bmatrix}-\frac{1}{3}(-0.2)^k + \frac{4}{3}(-0.8)^k \\ \frac{0.2}{3}(-0.2)^k - \frac{3.2}{3}(-0.8)^k\end{bmatrix}\end{aligned}$$

We need to indicate that in this example the input matrix $\boldsymbol{h}$ is a zero matrix. If $\boldsymbol{h}$ is not a zero matrix, the state response will contain the progression items and be much more complicate.

6.3 Data-Sampled Control System

Digital computers and microprocessors are used in many control systems. The size and cost advantages associated with the microcomputer make its use as a controller economical and practical. Using such a digital processor with a plant that operates in the continuous-time domain requires the input signal be discretized by sampling. The sample device may be an inherent or extrinsically device of the system. If there is no inherent sample device, an analog-to-digital (A/D) converter must be incorporated in a digital or data-sampled control system. And the output of controller must be transformed into the anolog signal from a discrete signal by a digital-to-analog (D/A) converter.

6.3.1 Realization Method

A functional block diagram of such a system is shown in Figure 6.3. In which the controlled object is a continuous-time system. The control device is composed of digital-to-analog (D/A) converter, computer and analog-to-digital (A/D) converter. In order to connect the two parts, sample device and hold device are introduced in. The sample device transfers the continuous signal $\boldsymbol{y}(t)$ to the discrete sequence $\boldsymbol{y}(k)$. The hold device converts the discrete sequence $\boldsymbol{u}(k)$ into a continuous signal $\boldsymbol{u}(t)$. If we regard the hold device-continuous time system-sample device as a entirety, and denote $\boldsymbol{X}(k)$ as the discretized state variable, there forms the discrete time system whose variables are $\boldsymbol{X}(k)$, $\boldsymbol{u}(k)$ and $\boldsymbol{y}(k)$. Its state space description is the discretization model of the continuous time system, and the control device would face the discretization model. The block diagram of such a system is shown in Figure 6.4.

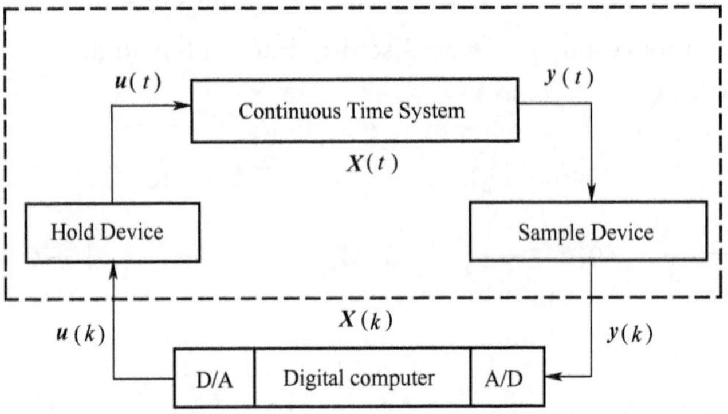

Figure 6.3 Block Diagram of Discretization Continuous Time System

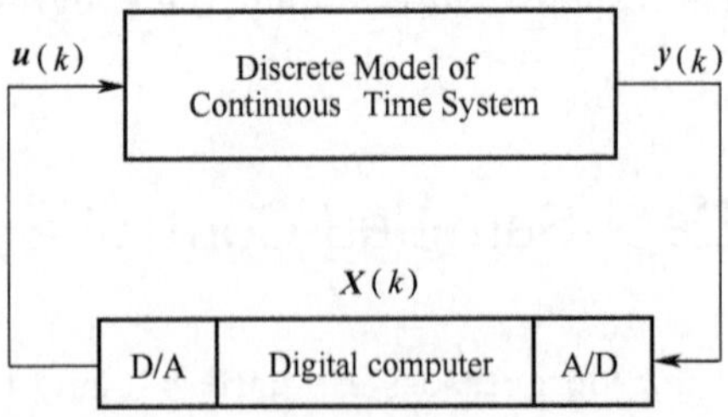

Figure 6.4 Discrete Model of Continuous Time System

The essential problems of discretizing the continuous time system are based on some sampling method and holding operator, obtaining the discrete time state space description from the continuous time system, and constructing the relational expression of their coefficient matrixes.

6.3.2 Three Basic Assumptions

In order to obtain the discretization description of the continuous time system, and ensure that it can be restored, three basic assumptions are introduced firstly.

1. Assumption about the Sampling Operator

The sampling device utilizes the fixed-interval sampling operation and its period is constant T, the sampling instant is $t_k - kT$, $k - 0,1,2,\cdots$. The sampling distance Δ is far less than the sampling period T, therefore it can be look upon zero. Notations $y(t)$ and $y(k)$ denote the input and output signal of the sampling device, respectively. Then we assume the relational expression of them is

$$y(k)=\begin{cases} y(t), & t_k = kT \\ 0, & t_k \neq kT \end{cases} \tag{6.67}$$

Where $k=0,1,2,\cdots$. Such a sampling operation is shown in Figure 6.5. $y_i(t)$ and $y_i(k)$ are the i th component of vector $y(t)$ and $y(k)$, respectively, $i=1,2,\cdots,q$.

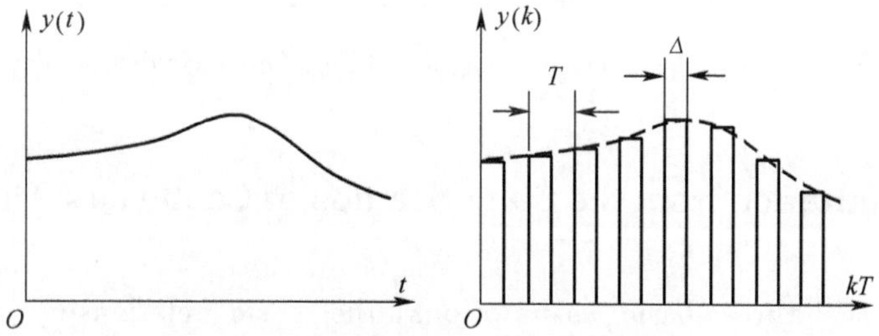

Figure 6.5 Schematic Diagram of Sampling

2. Assumption about the Sampling Period T

The sampling period should satisfy the conditions that are determined by the Nyquist-Shannon sampling Theorem. Suppose the amplitude spectrum of continuous signal $y_i(t)$ is shown in Figure 6.6. It is ordinate symmetry, and ω_c is the superior limit frequency. The Nyquist-Shannon sampling theorem indicates: in order to restore the original continuous signal $y_i(t)$ completely from the discretized signal $y_i(k)$, the sampling frequency ω_s should satisfy the following inequality

$$\omega_s > 2\omega_c \tag{6.68}$$

Considering $\omega_s = 2\pi/T$, then (6.68) can be rewritten to inequality (6.69).

$$T < \pi/\omega_c \tag{6.69}$$

In fact, considering some other properties, the sampling period T can be further determined by equation (6.70).

$$T=\frac{1}{10\sim 20}\times\frac{\pi}{\omega_c} \tag{6.70}$$

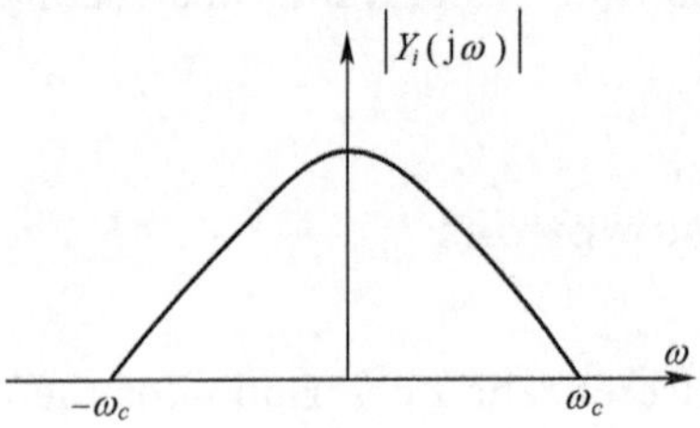

Figure 6.6 The Amplitude Spectrum of Continuous Signal

3. Assumption about the Holding Operation

The zero-order hold is generally used. A schematic diagram of zero-order holding mode is shown in Figure 6.7. It has the following features: the output $u_j(t)$ is equal to the discrete sequence $u_j(k)$ at the sampling instant, it remains constant between two sampling instants and equal to the previous sampling instant value.

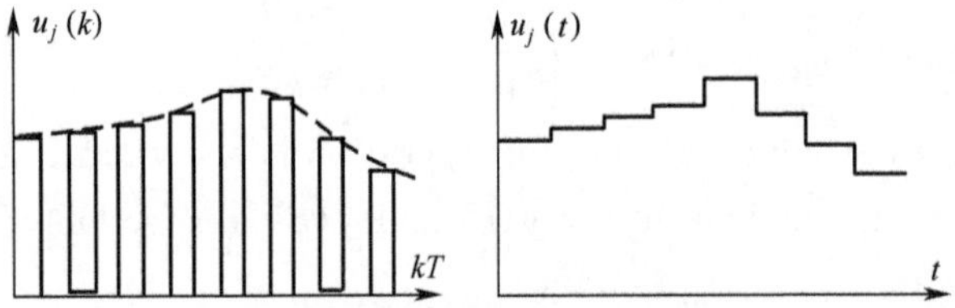

Figure 6.7 The Schematic Diagram of Zero-Order Holding Mode

6.3.3 Discretization from the State Solution of Continuous Time System

Based on the three basic assumptions, the basic relationship and associated properties of discretization problem for continuous time LTI system are represented as the following conclusions.

Theorem 6.2 Consider the continuous time LTI system

$$\begin{cases}\dot{\boldsymbol{X}}(t)=\boldsymbol{AX}(t)+\boldsymbol{Bu}(t),\boldsymbol{X}(t_0)=\boldsymbol{X}_0 \quad t\geqslant 0\\ \boldsymbol{y}(t)=\boldsymbol{CX}(t)+\boldsymbol{Du}(t)\end{cases} \tag{6.71}$$

the discretized model of (6.71) based on the basic assumptions is

$$\begin{cases}\boldsymbol{X}(k+1)=\boldsymbol{GX}(k)+\boldsymbol{Hu}(k),\boldsymbol{X}(0)=\boldsymbol{X}_0 \quad k=0,1,\cdots\\ \boldsymbol{y}(k)=\boldsymbol{CX}(k)+\boldsymbol{Du}(k)\end{cases} \tag{6.72}$$

The variables have the following relations between (6.71) and (6.72) as

$$\boldsymbol{X}(k)=[\boldsymbol{X}(t)]_{t=kT},\boldsymbol{u}(k)=[\boldsymbol{u}(t)]_{t=kT},\boldsymbol{y}(k)=[\boldsymbol{y}(t)]_{t=kT} \tag{6.73}$$

and the coefficient matrixes have the following relations:

$$\boldsymbol{G}=\mathrm{e}^{\boldsymbol{A}T},\boldsymbol{H}=\int_0^T \mathrm{e}^{\boldsymbol{A}t}\,\mathrm{d}t\boldsymbol{B};$$

$$\boldsymbol{C}=[\boldsymbol{C}(t)]_{t=kT},\boldsymbol{D}=[\boldsymbol{D}(t)]_{t=kT} \tag{6.74}$$

Where T is the sampling period.

Proof. According to Chapter 2 we know that, the state response of continuous time LTI system (6.71) is

$$\boldsymbol{X}(t)=\mathrm{e}^{\boldsymbol{A}(t-t_0)}\boldsymbol{X}(t_0)+\int_{t_0}^{t}\mathrm{e}^{\boldsymbol{A}(t-\tau)}\boldsymbol{B}\boldsymbol{u}(\tau)\mathrm{d}\tau \tag{6.75}$$

When $t_0=kT, t=(k+1)T$, we can obtain (6.76) from (6.75).

$$\boldsymbol{X}((k+1)T)=\mathrm{e}^{\boldsymbol{A}T}\boldsymbol{X}(kT)+\int_{kT}^{(k+1)T}\mathrm{e}^{\boldsymbol{A}[(k+1)T-\tau]}\boldsymbol{B}\boldsymbol{u}(\tau)\mathrm{d}\tau \tag{6.76}$$

Where T is sampling period, $\tau=kT+\xi(0\leqslant\xi\leqslant T)$. Considering the assumption about zero-order hold, it can be obtained that

$$\boldsymbol{u}(\tau)=\boldsymbol{u}(kT+\xi)=\boldsymbol{u}(kT)\quad k=0,1,2,\cdots \tag{6.77}$$

Shifting out $\boldsymbol{u}(\tau)$ from the integral expression (6.76), and substituting $\tau=kT+\xi$ into (6.76), it can be derived that

$$\begin{aligned}\boldsymbol{X}((k+1)T)&=\mathrm{e}^{\boldsymbol{A}T}\boldsymbol{X}(kT)+\int_{0}^{T}\mathrm{e}^{\boldsymbol{A}[(k+1)T-(kT+\xi)]}\boldsymbol{B}\mathrm{d}(kT+\xi)\boldsymbol{u}(kT)\\&=\mathrm{e}^{\boldsymbol{A}T}\boldsymbol{X}(kT)+\int_{0}^{T}\mathrm{e}^{\boldsymbol{A}(T-\xi)}\boldsymbol{B}\mathrm{d}\xi\,\boldsymbol{u}(kT)\end{aligned} \tag{6.78}$$

Because of the sampling mode using equal intervals, that is T is fixed, then $\mathrm{e}^{\boldsymbol{A}T}$ and $\int_0^T\mathrm{e}^{\boldsymbol{A}(T-\xi)}\boldsymbol{B}\mathrm{d}\xi$ is constant, and can be represented as

$$\boldsymbol{G}=\mathrm{e}^{\boldsymbol{A}T} \tag{6.79}$$

$$\boldsymbol{H}=\int_0^{\mathrm{T}}\mathrm{e}^{\boldsymbol{A}(T-\xi)}\boldsymbol{B}\mathrm{d}\xi \tag{6.80}$$

when $t=T-\xi$ and substituting it into (6.80), then (6.81) is obtained as

$$\boldsymbol{H}=\int_{T-0}^{T-T}\mathrm{e}^{\boldsymbol{A}t}\boldsymbol{B}\mathrm{d}(T-t)=-\int_{T}^{0}\mathrm{e}^{\boldsymbol{A}t}\boldsymbol{B}\mathrm{d}t=\int_0^T\mathrm{e}^{\boldsymbol{A}t}\,\mathrm{d}t\boldsymbol{B} \tag{6.81}$$

For convenience, omitting T, then (6.78) can be rewritten as

$$\boldsymbol{X}(k+1)=\boldsymbol{G}\boldsymbol{X}(k)+\boldsymbol{H}\boldsymbol{u}(k) \tag{6.82}$$

Furthermore, when $t=kT$, the discrete expression of output equation can be derived as

$$\boldsymbol{y}(k)=\boldsymbol{C}\boldsymbol{X}(k)+\boldsymbol{D}\boldsymbol{u}(k) \tag{6.83}$$

Example 6.8 Consider the continuous time LTI system described by

$$\begin{cases}\dot{\boldsymbol{X}}(t)=\begin{bmatrix}0 & 1\\-2 & -3\end{bmatrix}\boldsymbol{X}(t)+\begin{bmatrix}0\\1\end{bmatrix}u(t)\\ y(t)=[1\quad 1]\boldsymbol{X}(t)\end{cases}$$

Please discretize the system firstly, and the sampling period $T=0.02\mathrm{s}$, then determine the state and output response value at every sampling point, when the initial state is $\boldsymbol{X}(0)=[1\quad -1]^{\mathrm{T}}$, and the input is unit step signal.

Solution The state transition matrix of the original continuous time system has been calculated in Chapter 2 as

$$\boldsymbol{\Phi}(t)=\mathrm{e}^{\boldsymbol{A}t}=\begin{bmatrix}2\mathrm{e}^{-t}-\mathrm{e}^{-2t} & \mathrm{e}^{-t}-\mathrm{e}^{-2t}\\-2\mathrm{e}^{-t}+2\mathrm{e}^{-2t} & -\mathrm{e}^{-t}+2\mathrm{e}^{-2t}\end{bmatrix}$$

The system matrix of the discretized system is obtained from (6.74).

$$G=\Phi(k)=e^{AT}\mid_{T=0.02}=\begin{bmatrix}2e^{-0.02}-e^{-2\times 0.02} & e^{-0.02}-e^{-2\times 0.02}\\ -2e^{-0.02}+2e^{-2\times 0.02} & -e^{-0.02}+2e^{-2\times 0.02}\end{bmatrix}$$

$$=\begin{bmatrix}0.9996 & 0.0194\\ -0.0388 & 0.9414\end{bmatrix}$$

and the input matrix is

$$h=\int_0^T e^{At}\,dt\,b=\int_0^{0.02}\begin{bmatrix}2e^{-t}-e^{-2t} & e^{-t}-e^{-2t}\\ -2e^{-t}+2e^{-2t} & -e^{-t}+2e^{-2t}\end{bmatrix}\begin{bmatrix}0\\1\end{bmatrix}dt$$

$$=\int_0^{0.02}\begin{bmatrix}e^{-t}-e^{-2t}\\ -e^{-t}+2e^{-2t}\end{bmatrix}dt=\begin{bmatrix}-e^{-t}+\frac{1}{2}e^{-2t}\\ e^{-t}-e^{-2t}\end{bmatrix}\Bigg|_0^{0.02}$$

$$=\begin{bmatrix}\frac{1}{2}-e^{-0.02}-\frac{1}{2}e^{-0.04}\\ e^{-0.02}-e^{-0.04}\end{bmatrix}=\begin{bmatrix}0.0002\\ 0.0194\end{bmatrix}$$

By taking the place of continuous time t by sampling instant kT, the discretized output equation is obtained. Then the state space description of the discretized system is determined by

$$\begin{cases}X(k+1)=\begin{bmatrix}0.9996 & 0.0194\\ -0.0388 & 0.9414\end{bmatrix}X(k)+\begin{bmatrix}0.0002\\ 0.0194\end{bmatrix}u(k)\\ y(k)=[1\quad 1]X(k)\end{cases}$$

From $X(0)=[1\quad -1]^T$ and $u(k)=1\quad (k=0,1,2,\cdots)$, the state values at every sampling point can be get by using the recursion method.

$$X(1)=GX(0)+hu(0)=\begin{bmatrix}0.9996 & 0.0194\\ -0.0388 & 0.9414\end{bmatrix}\begin{bmatrix}1\\ -1\end{bmatrix}+\begin{bmatrix}0.0002\\ 0.0194\end{bmatrix}=\begin{bmatrix}0.9804\\ -0.9608\end{bmatrix}$$

$$X(2)=GX(1)+hu(1)=\begin{bmatrix}0.9996 & 0.0194\\ -0.0388 & 0.9414\end{bmatrix}\begin{bmatrix}0.9804\\ -0.9608\end{bmatrix}+\begin{bmatrix}0.0002\\ 0.0194\end{bmatrix}=\begin{bmatrix}0.9616\\ -0.9231\end{bmatrix}$$

$$X(3)=GX(2)+hu(2)=\begin{bmatrix}0.9996 & 0.0194\\ -0.0388 & 0.9414\end{bmatrix}\begin{bmatrix}0.9616\\ -0.9231\end{bmatrix}+\begin{bmatrix}0.0002\\ 0.0194\end{bmatrix}=\begin{bmatrix}0.9435\\ -0.8869\end{bmatrix}$$

$$X(4)=GX(3)+hu(3)=\begin{bmatrix}0.9996 & 0.0194\\ -0.0388 & 0.9414\end{bmatrix}\begin{bmatrix}0.9435\\ -0.8869\end{bmatrix}+\begin{bmatrix}0.0002\\ 0.0194\end{bmatrix}=\begin{bmatrix}0.9261\\ -0.8521\end{bmatrix}$$

$\cdots$

By applying the output equation, $y(k)=x_1(k)+x_2(k)$. Then the output values at every sampling point are

$$y(0)=0,\ y(1)=0.0196,\ y(2)=0.0385,\ y(3)=0.0556,\cdots$$

6.3.4 Approximate Discretization

According to the definition of derivative, $\dot{X}(kT)$ can be approximately calculated as

$$\dot{X}(kT)\approx\frac{1}{T}[X((k+1)T)-X(kT)] \tag{6.84}$$

Where T is the sampling period.

Substituting (6. 84) into the state equation of linear time LTI system (6. 85)

$$\dot{\boldsymbol{X}}(t)=\boldsymbol{AX}(t)+\boldsymbol{Bu}(t) \tag{6.85}$$

and letting $t=kT$, we get

$$\frac{1}{T}[\boldsymbol{X}((k+1)T)-\boldsymbol{X}(kT)]=\boldsymbol{AX}(kT)+\boldsymbol{Bu}(kT)$$

Then, (6. 86) can be derived

$$\boldsymbol{X}((k+1)T)=(\boldsymbol{I}+T\boldsymbol{A})\boldsymbol{X}(kT)+T\boldsymbol{Bu}(kT) \tag{6.86}$$

Denoting

$$\boldsymbol{G}=\boldsymbol{I}+T\boldsymbol{A} \tag{6.87}$$

$$\boldsymbol{H}=T\boldsymbol{B} \tag{6.88}$$

Then the approximate discretization equation of LTI system (6. 85) is

$$\boldsymbol{X}(k+1)=\boldsymbol{GX}(k)+\boldsymbol{Hu}(k)$$

It is not difficult to see that the smaller the sampling period T, the higher the approximate level. In general, the higher approximate precision can be get, when the sampling period T is set as about one tenth of the smallest time constant of system.

Example 6. 9 Use the approximate method to discretize the following system

$$\begin{cases}\dot{\boldsymbol{X}}(t)=\begin{bmatrix}0 & 1\\ -2 & -3\end{bmatrix}\boldsymbol{X}(t)+\begin{bmatrix}0\\ 1\end{bmatrix}u(t)\\ y(t)=[1 \quad 1]\boldsymbol{X}(t)\end{cases}$$

Assume the sampling period $T=0.02$s.

Solution Since the system is a LTI system, the system matrix and input matrix of the discretization system are obtained from (6. 87) and (6. 88).

$$\boldsymbol{G}=\boldsymbol{I}+T\boldsymbol{A}=\begin{bmatrix}1 & 0\\ 0 & 1\end{bmatrix}+\begin{bmatrix}0 & T\\ -2T & -3T\end{bmatrix}=\begin{bmatrix}1 & 0.02\\ -0.04 & 0.94\end{bmatrix}$$

$$\boldsymbol{h}=T\boldsymbol{b}=\begin{bmatrix}0\\ 0.02\end{bmatrix}$$

Then, the discretized state equation is

$$\boldsymbol{X}(k+1)=\begin{bmatrix}1 & 0.02\\ -0.04 & 0.94\end{bmatrix}\boldsymbol{X}(k)+\begin{bmatrix}0\\ 0.02\end{bmatrix}u(k)$$

Comparing with Example 6. 8 we know that, the precision of approximate discretization method is lower than the precision of discretizing state response directly. For some continuous system, one-order hold or higher order hold maybe used to improve the discretization precision. In addition, reducing the sampling period can also raise the precision.

6. 4 Discrete Time System Stability Analysis and Criteria

6. 4. 1 Lyapunov Stability of Discrete Time System

Consider the discrete time system described by difference equation (6. 89).

$$X(k+1)=f(X(k),k) \tag{6.89}$$

Where $k \in I=\{h+i \mid i=0,1,2,\cdots;h \geqslant 0\}$, $X \in R^n$; $f: R^n \times I \to R^n$. We need to suppose that, for any initial state $X_0 \in R^n$ and any initial instant $h \geqslant 0$, there exists a unique solution $X(k;X_0,h)$ to equation (6.89), and $X(h;X_0,h)=X_0$. We also assume that, for any $k \in I$, the necessary and sufficient condition of $f(X,k)=X$ holding true is $X=\mathbf{0}$. Therefore, system (6.89) has a unique equilibrium point $X_e=\mathbf{0}$.

The definitions of Stability about the equilibrium point $X_e=\mathbf{0}$ of a discrete time system (6.89) are given as follows.

Definition 6.3 The equilibrium point $X_e=\mathbf{0}$ of the discrete time system (6.89) is said to be **Lyapunov stable**, if for any given $\varepsilon>0$, and any non-negative integer h, there exists a real $\delta=\delta(\varepsilon,h)>0$ such that if $\|X_0\|<\delta$, then the zero-input state solution $X(k;X_0,h)$ with initial state X_0 satisfies inequality $\|X(k;X_0,h)\|<\varepsilon$ for any $k \geqslant h$.

In the above definition, δ relies on ε and h. If δ is independent of h, i.e. $\delta=\delta(\varepsilon)$, then the equilibrium point $X_e=\mathbf{0}$ is said to be **uniformly stable.**

Definition 6.4 The equilibrium point $X_e=\mathbf{0}$ of the discrete time system (6.89) is said to be **asymptotically stable**, if

(1) it is stable.

(2) there exists a number $\eta(h)>0$ such that whenever $\|X_0\|<\eta$ the zero-input state solution satisfies $\lim_{k\to\infty} X(k;X_0,h)=0$.

Definition 6.5 The equilibrium point $X_e=\mathbf{0}$ of the discrete time system (6.89) is said to be **uniformly asymptotically stable**, if

(1) it is uniformly stable.

(2) for any given $\varepsilon>0$ and any non negative integer h, there exists a real δ_0 which is independent of ε and h, and a real $T(\varepsilon)>0$ that is independent of h such that if $\|X_0\|<\delta_0$, then the zero-input state solution $X(k;X_0,h)$ satisfies inequality $\|X(k;X_0,h)\|<\varepsilon$ for any $k \geqslant h+T(\varepsilon)$.

Definition 6.6 The equilibrium point $X_e=\mathbf{0}$ of the discrete time system (6.89) is said to be **unstable**, if the conditions of Definition 6.3 are not true.

About the stable range, we have the following definitions.

Definition 6.7 The solution of the discrete time system (6.89) is said to be **uniformly bounded**, if for any given $\alpha>0$ and non-negative integer h, there exists a real $\beta=\beta(\alpha)>0$ which is independent of h, makes the inequality $\|X(k;X_0,h)\|<\beta$ hold true for any $k \geqslant h$, when $\|X_0\|<\alpha$.

Definition 6.8 The equilibrium point $X_e=\mathbf{0}$ of the discrete time system (6.89) is said to be **globally stable**, if it is stable, and any zero-input state solution of (6.89) intends to zero, when $k\to\infty$.

Definition 6.9 The equilibrium point $X_e=\mathbf{0}$ of the discrete time system (6.89) is said to be **globally asymptotically stable**, if

(1) it is uniformly stable.

(2) the state solution of (6.89) is uniformly bounded.

(3) for any given $\alpha>0$, ε and $h\in\boldsymbol{R}^+$, there exists a real $T(\varepsilon,\alpha)$ that is independent of h, makes the inequality $\|\boldsymbol{X}(k;\boldsymbol{X}_0,h)\|<\varepsilon$ hold true for any $k\geqslant h+T(\varepsilon,\alpha)$, when $\|\boldsymbol{X}_0\|<\alpha$.

6.4.2 Lyapunov Stability Theorem of Discrete Time System

For convenience, we only discuss the **discrete time time-invariant (TI) system.**

Consider a **discrete time nonlinear time-invariant (TI) system**, its autonomous state equation is

$$\boldsymbol{X}(k+1)=f[\boldsymbol{X}(k)],\boldsymbol{X}(0)=\boldsymbol{X}_0 \quad k=0,1,2,\cdots \tag{6.90}$$

Where $\boldsymbol{X}\in\boldsymbol{R}^n$ is the state, $f(0)=0$, i.e. the origin of state space $\boldsymbol{X}_e=\boldsymbol{0}$ is the equilibrium point. Similarly to continuous time system, the **main Lyapunov stability theorem** for discrete time nonlinear TI system (6.90) can be given out.

Theorem 6.3 The origin equilibrium point $\boldsymbol{X}_e=\boldsymbol{0}$ of the discrete time nonlinear TI system (6.90) is global asymptotically stable if there exists a **scalar function** $V(\boldsymbol{X}(k))$ such that for any $\boldsymbol{X}(k)\in\boldsymbol{R}^n$ satisfies:

(1) $V(\boldsymbol{X}(k))$ is positive definite.

(2) denote $\Delta V(\boldsymbol{X}(k))=V(\boldsymbol{X}(k+1))-V(\boldsymbol{X}(k))$, and $\Delta V(\boldsymbol{X}(k))$ is negative definite.

(3) $V(\boldsymbol{X}(k))\to\infty$ as $\|\boldsymbol{X}(k)\|\to\infty$.

In practical applications of Theorem 6.3, because of condition (2) is conservative, a failure estimate maybe caused for some problems. Thus, a new main Lyapunov stability theorem can be obtain through extending the condition.

Theorem 6.4 The origin equilibrium point $\boldsymbol{X}_e=\boldsymbol{0}$ of the discrete time nonlinear TI system (6.90) is global asymptotically stable if there exists a scalar function $V(\boldsymbol{X}(k))$ such that for any $\boldsymbol{X}(k)\in\boldsymbol{R}^n$ satisfies:

(1) $V(\boldsymbol{X}(k))$ is positive definite.

(2) denote $\Delta V(\boldsymbol{X}(k))=V(\boldsymbol{X}(k+1))-V(\boldsymbol{X}(k))$, and $\Delta V(\boldsymbol{X}(k))$ is negative semidefinite.

(3) $\Delta V(\boldsymbol{X}(k))$ is not almost zero for all free motions determined by any initial state $\boldsymbol{X}(0)\in\mathbf{R}^n$.

(4) $V(\boldsymbol{X}(k))\to\infty$ as $\|\boldsymbol{X}(k)\|\to\infty$.

Based on the main Lyapunov stability theorems above, we can deduce another stability criterion that is simple in form and easy to use.

For the discrete time nonlinear TI system described by (6.90), assume $\boldsymbol{X}_e=\boldsymbol{0}$ is the system equilibrium point, then it is said to be global asymptotically stable if $f(\boldsymbol{X}(k))$ is convergent for any $\boldsymbol{X}(k)\neq\boldsymbol{0}$, namely

$$\| f(\boldsymbol{X}(k)) \| < \| \boldsymbol{X}(k) \| \tag{6.91}$$

Proof. For the discrete time nonlinear TI system described by (6.90), chose the Lyapunov function as

$$V(\boldsymbol{X}(k)) = \| \boldsymbol{X}(k) \| \tag{6.92}$$

Obviously, $V(\boldsymbol{X}(k))$ is positive definite, then we can deduce equation (6.93).

$$\begin{aligned}\Delta V(\boldsymbol{X}(k)) &= V(\boldsymbol{X}(k+1)) - V(\boldsymbol{X}(k)) \\ &= \| \boldsymbol{X}(k+1) \| - \| \boldsymbol{X}(k) \| = \| f(\boldsymbol{X}(k)) \| - \| \boldsymbol{X}(k) \|\end{aligned} \tag{6.93}$$

From equation (6.91), we know that $\Delta V(\boldsymbol{X}(k))$ is negative, and $V(\boldsymbol{X}(k))\to\infty$ as $\| \boldsymbol{X}(k) \| \to\infty$. According to Theorem 6.3, the equilibrium point $\boldsymbol{X}_e=\boldsymbol{0}$ is global asymptotically stable.

6.4.3 Stability Criteria of Discrete Time LTI System

Consider the discrete time LTI system, its autonomous state equation is

$$\boldsymbol{X}(k+1)=\boldsymbol{G}\boldsymbol{X}(k), \boldsymbol{X}(0)=\boldsymbol{X}_0 \quad k=0,1,2,\cdots \tag{6.94}$$

Where, $\boldsymbol{X}$ is a n dimension state, the state solutions of $\boldsymbol{G}\boldsymbol{X}_e=\boldsymbol{0}$ are the equilibrium points of the system. If system matrix $\boldsymbol{G}$ is singular then there exists non-zero equilibrium points except for origin equilibrium point $\boldsymbol{X}_e=\boldsymbol{0}$. If system matrix $\boldsymbol{G}$ is nonsingular then origin equilibrium point $\boldsymbol{X}_e=\boldsymbol{0}$ is the uniquely equilibrium point. Corresponding with the continuous time system, system (6.94) has the following stability determinant theorems.

Theorem 6.5 Consider the discrete time LTI autonomous system (6.94), the origin equilibrium point is Lyapunov stable if and only if, all the amplitudes of eigenvalues $\lambda_i(\boldsymbol{G})(i=1,2,\cdots,n)$ of $\boldsymbol{G}$ satisfies inequality $|\lambda_i(\boldsymbol{G})|\leqslant 1$. If the amplitudes of some eigenvalues are equal to one, they must be the single roots of the smallest polynomial of $\boldsymbol{G}$.

Theorem 6.6 Consider the discrete time LTI autonomous system (6.94), the origin equilibrium point is Lyapunov asymptotically stable if and only if, all the amplitudes of eigenvalues $\lambda_i(\boldsymbol{G})(i=1,2,\cdots,n)$ of $\boldsymbol{G}$ satisfies inequality $|\lambda_i(\boldsymbol{G})|<1$.

In the two criterias above, the stability is determined by the eigenvalues of $\boldsymbol{G}$, thus they are called **eigenvalue criteria** also.

Theorem 6.7 A necessary and sufficient condition for the discrete time LTI autonomous system (6.94) to be asymptotically stable is that, for any symmetric positive definite matrix $\boldsymbol{Q}$, there exists a $n\times n$ symmetric positive definite matrix $\boldsymbol{P}$ that satisfies the discrete type Lyapunov equation

$$\boldsymbol{G}^{\mathrm{T}}\boldsymbol{P}\boldsymbol{G}-\boldsymbol{P}=-\boldsymbol{Q} \tag{6.95}$$

Moreover, if the system is stable, $\boldsymbol{P}$ is the unique solution to (6.95) and the quadratic form scalar function

$$V(\boldsymbol{X}(k))=\boldsymbol{X}^{\mathrm{T}}(k)\boldsymbol{P}\boldsymbol{X}(k) \tag{6.96}$$

is the Lyapunov function of system (6.94).

If we have

$$\lim_{\|\boldsymbol{X}(k)\|\to\infty} V(\boldsymbol{X}(k))=\infty \tag{6.97}$$

Then the equilibrium point $\boldsymbol{X}_e=\boldsymbol{0}$ is globally asymptotically stable.

Proof. Assume the Lyapunov function is

$$V(\boldsymbol{X}(k))=\boldsymbol{X}^{\mathrm{T}}(k)\boldsymbol{P}\boldsymbol{X}(k)$$

Where $\boldsymbol{P}$ is a real symmetric positive definite matrix. Then $V(\boldsymbol{X}(k))$ is positive definite. Its change rate is

$$\begin{aligned}\Delta V(\boldsymbol{X}(k),k)&=V(\boldsymbol{X}(k+1))-V(\boldsymbol{X}(k))=\boldsymbol{X}^{\mathrm{T}}(k+1)\boldsymbol{P}\boldsymbol{X}(k+1)-\boldsymbol{X}^{\mathrm{T}}(k)\boldsymbol{P}\boldsymbol{X}(k)\\&=[\boldsymbol{G}\boldsymbol{X}(k)]^{\mathrm{T}}\boldsymbol{P}\boldsymbol{G}\boldsymbol{X}(k)-\boldsymbol{X}^{\mathrm{T}}(k)\boldsymbol{P}\boldsymbol{X}(k)=\boldsymbol{X}^{\mathrm{T}}(k)(\boldsymbol{G}^{\mathrm{T}}\boldsymbol{P}\boldsymbol{G}-\boldsymbol{P})\boldsymbol{X}(k)\\&=\boldsymbol{X}^{\mathrm{T}}(k)(-\boldsymbol{Q})\boldsymbol{X}(k)\end{aligned}$$

Since $V(\boldsymbol{X}(k))$ is positive definite, in accordance with the conditions of asymptotically stable that

$$\Delta V(\boldsymbol{X}(k))=\boldsymbol{X}^{\mathrm{T}}(k)(-\boldsymbol{Q})\boldsymbol{X}(k)$$

is negative definite. That is

$$\boldsymbol{Q}=-(\boldsymbol{G}^{\mathrm{T}}\boldsymbol{P}\boldsymbol{G}-\boldsymbol{P}) \tag{6.98}$$

should be positive definite.

Therefore, for the selected $\boldsymbol{P}>\boldsymbol{0}$, the sufficient and necessary condition for system (6.94) to be asymptotically stable is $\boldsymbol{Q}>\boldsymbol{0}$.

On the other hand, give a real symmetric positive definite matrix $\boldsymbol{Q}$ firstly, then solute matrix $\boldsymbol{P}$ from matrix equation (6.95). In order to ensure that the system equilibrium point $\boldsymbol{X}_e=\boldsymbol{0}$ is asymptotically stable, the matrix $\boldsymbol{P}$ must be a real symmetric positive definite matrix. Consequently, the matrix $\boldsymbol{P}$ is a real symmetric positive definite matrix is the necessary condition.

Example 6.10 Consider the discrete time LTI system described by the following state equation

$$\boldsymbol{X}(k+1)=\begin{bmatrix}0 & 1\\ \frac{1}{2} & 0\end{bmatrix}\boldsymbol{X}(k)$$

Determine the stability of its equilibrium point.

Solution Since the system matrix $\boldsymbol{G}=\begin{bmatrix}0 & 1\\ \frac{1}{2} & 0\end{bmatrix}$ is nonsingular, $\boldsymbol{X}_e=\boldsymbol{0}$ is the unique equilibrium point. According to the Lyapunov equation

$$\boldsymbol{G}^{\mathrm{T}}\boldsymbol{P}\boldsymbol{G}-\boldsymbol{P}=-\boldsymbol{Q}$$

Let $\boldsymbol{Q}=\boldsymbol{I}$, then $\boldsymbol{G}^{\mathrm{T}}\boldsymbol{P}\boldsymbol{G}-\boldsymbol{P}=-\boldsymbol{I}$

Letting $\boldsymbol{P}=\begin{bmatrix}p_{11} & p_{12}\\ p_{12} & p_{22}\end{bmatrix}$ and substituting $\boldsymbol{P}$ into the Lyapunov equation, we can obtain

$$\begin{bmatrix} 0 & \frac{1}{2} \\ 1 & 0 \end{bmatrix}\begin{bmatrix} p_{11} & p_{12} \\ p_{12} & p_{22} \end{bmatrix}\begin{bmatrix} 0 & 1 \\ \frac{1}{2} & 0 \end{bmatrix}-\begin{bmatrix} p_{11} & p_{12} \\ p_{12} & p_{22} \end{bmatrix}=\begin{bmatrix} -1 & 0 \\ 0 & -1 \end{bmatrix}$$

Its solution matrix is

$$\boldsymbol{P}=\begin{bmatrix} \frac{5}{3} & 0 \\ 0 & \frac{8}{3} \end{bmatrix}$$

Every leading order principal minor determinant of $\boldsymbol{P}$ are

$$\Delta_1=p_{11}=\frac{5}{3}>0,\Delta_2=\begin{vmatrix} p_{11} & p_{12} \\ p_{12} & p_{22} \end{vmatrix}=\begin{vmatrix} \frac{5}{3} & 0 \\ 0 & \frac{8}{3} \end{vmatrix}=\frac{40}{9}>0$$

According to the Sylvester criterion, $\boldsymbol{P}$ is positive definite. Therefore, the system equilibrium point $\boldsymbol{X}_e=\boldsymbol{0}$ is asymptotically stable, and the quadratic form function

$$V(\boldsymbol{X}(k))=\boldsymbol{X}^{\mathrm{T}}(k)\boldsymbol{P}\boldsymbol{X}(k)=[x_1(k) \quad x_2(k)]\begin{bmatrix} \frac{5}{3} & 0 \\ 0 & \frac{8}{3} \end{bmatrix}\begin{bmatrix} x_1(k) \\ x_2(k) \end{bmatrix}$$

$$=\frac{5}{3}x_1^2(k)+\frac{8}{3}x_2^2(k)>0$$

is a Lyapunov function of the system.

In addition $V(\boldsymbol{X}(k))\to\infty$ as $\|\boldsymbol{X}(k)\|\to\infty$, thus the equilibrium point is globally asymptotically stable.

Example 6.11 Consider the discrete-time LTI system

$$\boldsymbol{X}(k+1)=\begin{bmatrix} \lambda_1 & 0 \\ 0 & \lambda_2 \end{bmatrix}\boldsymbol{X}(k) \qquad \text{where}, \lambda_1\lambda_2\neq 1$$

Determine the conditions of that the equilibrium point $\boldsymbol{X}_e=\boldsymbol{0}$ is asymptotically stable.

Solution According to Theorem 6.7, the Lyapunov equation is

$$\boldsymbol{G}^{\mathrm{T}}\boldsymbol{P}\boldsymbol{G}-\boldsymbol{P}=-\boldsymbol{Q}$$

Let $\boldsymbol{Q}=\boldsymbol{I}$ and $\boldsymbol{P}=\begin{bmatrix} p_{11} & p_{12} \\ p_{12} & p_{22} \end{bmatrix}$, substitute $\boldsymbol{P}$ into the Lyapunov equation, it is obtained that

$$\begin{bmatrix} \lambda_1 & 0 \\ 0 & \lambda_2 \end{bmatrix}\begin{bmatrix} p_{11} & p_{12} \\ p_{12} & p_{22} \end{bmatrix}\begin{bmatrix} \lambda_1 & 0 \\ 0 & \lambda_2 \end{bmatrix}-\begin{bmatrix} p_{11} & p_{12} \\ p_{12} & p_{22} \end{bmatrix}=\begin{bmatrix} -1 & 0 \\ 0 & -1 \end{bmatrix}$$

$$\begin{bmatrix} p_{11}(1-\lambda_1^2) & p_{12}(1-\lambda_1\lambda_2) \\ p_{12}(1-\lambda_1\lambda_2) & p_{22}(1-\lambda_2^2) \end{bmatrix}=\begin{bmatrix} 1 & 0 \\ 0 & 1 \end{bmatrix}$$

Then we can obtained

$$p_{11}=\frac{1}{1-\lambda_1^2}, p_{12}=0, p_{22}=\frac{1}{1-\lambda_2^2}$$

In order to ensure that $\boldsymbol{P}$ is positive definite, according to the Sylvester criterion, we have

$$p_{11}>0, p_{11}p_{22}-p_{12}p_{12}>0$$

Substituting p_{11}、p_{12} and p_{22} into the inequalities above, the conditions of that the system is asymptotically stable can be obtained

$$|\lambda_1|<1, |\lambda_2|<1$$

The quadratic form function

$$V(\boldsymbol{X}(k))=\boldsymbol{X}^{\mathrm{T}}(k)\boldsymbol{P}\boldsymbol{X}(k)=[x_1(k) \quad x_2(k)]\begin{bmatrix} \dfrac{1}{1-\lambda_1^2} & 0 \\ 0 & \dfrac{1}{1-\lambda_2^2} \end{bmatrix}\begin{bmatrix} x_1(k) \\ x_2(k) \end{bmatrix}$$

$$=\frac{1}{1-\lambda_1^2}x_1^2(k)+\frac{1}{1-\lambda_2^2}x_2^2(k)>0$$

is a Lyapunov function, and $V(\boldsymbol{X}(k))\to\infty$ when $\|\boldsymbol{X}(k)\|\to\infty$. Furthermore, from the state equation we know that λ_1 and λ_2 are eigenvalues of the system. Thus, the equilibrium point $\boldsymbol{X}_e=\boldsymbol{0}$ is globally asymptotically stable if and only of all the eigenvalues are located in the unit circle. This conclusion is consistent with the stability criterion for sampled control system in classical control theory and Theorem 6. 6.

6. 5 Controllability and Observability of Discrete Time LTI System

6. 5. 1 Controllability

Consider the state space description of the discrete time LTI system described by

$$\begin{cases} \boldsymbol{X}(k+1)=\boldsymbol{G}\boldsymbol{X}(k)+\boldsymbol{H}\boldsymbol{u}(k) \\ \boldsymbol{y}(k)=\boldsymbol{C}\boldsymbol{X}(k)+\boldsymbol{D}\boldsymbol{u}(k) \end{cases} \tag{6.99}$$

Where $\boldsymbol{X}(k)$ is the n dimension state vector; $\boldsymbol{u}(k)$ is the p dimension input vector; $\boldsymbol{y}(k)$ is the q dimension output vector; $\boldsymbol{G}$ is a $n\times n$ system matrix; $\boldsymbol{H}$ is a $n\times p$ input matrix; $\boldsymbol{C}$ is a $q\times n$ output matrix; $\boldsymbol{D}$ is a $q\times p$ matrix.

Definition 6. 10 Consider the discrete time LTI system described by (6. 99), the non-zero initial state $\boldsymbol{X}(h)=\boldsymbol{X}_0$ is said to be **controllable**, if it is possible to find a available input (or sequence in discrete case) $\boldsymbol{u}(k), k>h$ such that the system state is transmitted to origin $\boldsymbol{X}(l)=\boldsymbol{0}$ at time instant l. Otherwise, the state $\boldsymbol{X}_0$ is said to be **uncontrollable.**

If any non-zero state is controllable, the system is said to be **completely controllable** or, in simply, controllable.

Definition 6. 11 Consider the discrete time LTI system described by (6. 99), the non-zero final state $\boldsymbol{X}(l)=\boldsymbol{X}_l$ is said to be **reachable**, if it is possible to find a available input (or sequence in discrete case) $\boldsymbol{u}(k)$, $k>h$ such that the system state is transmitted to final state $\boldsymbol{X}(l)=\boldsymbol{X}_l$ from $\boldsymbol{X}(h)=\boldsymbol{0}$. Otherwise, the state $\boldsymbol{X}_l$ is said to be **unreachable.**

If any non-zero state is reachable, the system is said to be **completely reachable** or, in simply, reachable.

Consider the discrete time LTI system described by (6.99), the state solution of system (6.99) has been calculated in Section 6.2 as

$$\boldsymbol{X}(k)=\boldsymbol{G}^k\boldsymbol{X}_0+\sum_{i=0}^{k-1}\boldsymbol{G}^{k-i-1}\boldsymbol{H}\boldsymbol{u}(i) \tag{6.100}$$

Assume that any non-zero initial state $\boldsymbol{X}(0)=\boldsymbol{X}_0$ can be transmitted to final state $\boldsymbol{X}(l)=\boldsymbol{0}$ over l beats, then (6.100) can be rewritten as

$$\boldsymbol{X}(l)=\boldsymbol{G}^l\boldsymbol{X}_0+\sum_{i=0}^{l-1}\boldsymbol{G}^{l-i-1}\boldsymbol{H}\boldsymbol{u}(i)=\boldsymbol{0} \tag{6.101}$$

that is

$$\begin{aligned}-\boldsymbol{G}^l\boldsymbol{X}_0&=\sum_{i=0}^{l-1}\boldsymbol{G}^{l-i-1}\boldsymbol{H}\boldsymbol{u}(i)\\&=\boldsymbol{H}\boldsymbol{u}(l-1)+\boldsymbol{G}\boldsymbol{H}\boldsymbol{u}(l-2)+\cdots+\boldsymbol{G}^{l-2}\boldsymbol{H}\boldsymbol{u}(1)+\boldsymbol{G}^{l-1}\boldsymbol{H}\boldsymbol{u}(0)\end{aligned} \tag{6.102}$$

It can be expressed in vector form that

$$-\boldsymbol{G}^l\boldsymbol{X}_0=[\boldsymbol{H}\quad \boldsymbol{G}\boldsymbol{H}\quad \cdots\quad \boldsymbol{G}^{l-1}\boldsymbol{H}]\begin{bmatrix}\boldsymbol{u}(l-1)\\ \boldsymbol{u}(l-2)\\ \vdots\\ \boldsymbol{u}(0)\end{bmatrix} \tag{6.103}$$

For any non-zero initial state $\boldsymbol{X}_0$, a control sequence $\boldsymbol{u}(0)$, $\boldsymbol{u}(1)$, …, $\boldsymbol{u}(l-1)$ can be get from (6.103), namely there exists a available input $\boldsymbol{u}(k)$, such that the system is controllable. The problem is transferred to the calculation of $l\times p$ unknown quantities from n non-homogeneous linear algebraic equations. According to linear algebraic theory, equation (6.104) must be satisfied

$$\begin{aligned}&\operatorname{rank}[\boldsymbol{H}\quad \boldsymbol{G}\boldsymbol{H}\quad \cdots\quad \boldsymbol{G}^{l-1}\boldsymbol{H}]\\&=\operatorname{rank}[\boldsymbol{H}\quad \boldsymbol{G}\boldsymbol{H}\quad \cdots\quad \boldsymbol{G}^{l-1}\boldsymbol{H}\quad \vdots\quad -\boldsymbol{G}^l\boldsymbol{X}_0]=n\end{aligned} \tag{6.104}$$

In SI case, because the matrix$[\boldsymbol{h}\quad \boldsymbol{G}\boldsymbol{h}\cdots\boldsymbol{G}^{l-1}\boldsymbol{h}]$is a $n\times l$ matrix, and the matrix$[\boldsymbol{G}\quad \boldsymbol{G}\boldsymbol{h}\cdots\boldsymbol{G}^{l-1}h\ \vdots -\boldsymbol{G}^lX_0]$is a $n\times(l+1)$ matrix, there must satisfy $l\geqslant n$. While in MI case, the matrix$[\boldsymbol{H}\quad \boldsymbol{G}\boldsymbol{H}\cdots\boldsymbol{G}^{l-1}\boldsymbol{H}]$is a $n\times lp$ matrix, and the matrix$[\boldsymbol{G}\quad \boldsymbol{G}\boldsymbol{H}\quad \cdots\boldsymbol{G}^{l-1}\boldsymbol{H}\ \vdots\quad -\boldsymbol{G}^l\boldsymbol{X}_0]$is a $n\times(lp+1)$matrix. Then the control sequence vector $\boldsymbol{u}(k)$ is not unique. This subsection is restricted to the SI case.

According to linear algebraic theory, $\boldsymbol{G}^l$ is nonsingular as the system matrix $\boldsymbol{G}$ is nonsingular. In this case, holding equation (6.104) true is not only the necessary condition but also the sufficient condition for solving a control sequence $\boldsymbol{u}(0)$, $\boldsymbol{u}(1)$, …, $\boldsymbol{u}(l-1)$ from (6.103). Hence, equation (6.104) is the sufficient and necessary condition of that system (6.99) is completely controllable.

Similarly, G^l is singular as the system matrix G is singular. Thus, the n components of $-G^l X_0$ are linearly dependent. In this case, holding equation (6.104) true is the sufficient condition but not the necessary condition for solving a control sequence $u(0)$、$u(1)$、…、$u(l-1)$ from (6.103). An extreme case, $-G^l X_0 = 0$ as $G^l = 0$, no matter matrix $[h \quad GH \quad \cdots \quad G^{l-1}h]$ is full row rank, there must exists a control sequence $u(0)=u(1)=\cdots=u(l-1)=0$ such that equation (6.103) holds true. Hence, (6.104) is the sufficient condition of that system (6.105) is completely controllable when G is singular.

Consider a single input discrete time LTI system described by

$$X(k+1) = GX(k) + hu(k), \quad k=0,1,2,\cdots \tag{6.105}$$

From the discussions above, the controllability criteria of system (6.105) can be concluded as follows.

Theorem 6.8 Rank criteria of Controllability

Consider a SI discrete time LTI system described by (6.105), if the system matrix G is nonsingular, the system or state is completely controllable iff

$$\text{rank}Q_c = \text{rank}[h \quad Gh \quad \cdots \quad G^{l-1}h] = n \tag{6.106}$$

Where $Q_c = [h \quad Gh \quad \cdots G^{l-1}h]$ is called the **controllability matrix** of the system.

If the system matrix G is singular, the condition above is sufficient but not necessary.

Theorem 6.9 Rank criteria of Reachability

Consider a SI discrete time LTI system described by (6.105), the system or state is completely reachable iff

$$\text{rank}Q_c = \text{rank}[h \quad Gh \quad \cdots \quad G^{l-1}h] = n \tag{6.107}$$

Proof. According to equation (6.101) and the reachability definition we can obtain

$$X(l) = [h \quad Gh \quad \cdots \quad G^{l-1}h]\begin{bmatrix} u(l-1) \\ u(l-2) \\ \vdots \\ u(0) \end{bmatrix} \tag{6.108}$$

For any non-zero final state $X(l)=X_l$, if a control sequence $u(0)$、$u(1)$、…、$u(l-1)$ can be calculated from equation (6.107), then the system is reachable. Obviously, the sufficient and necessary condition is

$$\text{rank}[h \quad Gh \quad \cdots \quad G^{l-1}h] = n$$

It can be seen that controllability is equivalent to reachability as the system matrix G is nonsingular, but this is not true while G is singular.

Example 6.12 Consider the discrete time LTI system described by

$$X(k+1) = \begin{bmatrix} 3 & 2 \\ 6 & 4 \end{bmatrix} X(k) + \begin{bmatrix} 1 \\ 2 \end{bmatrix} u(k)$$

Please try to determine its controllability and reachability.

Solution

(1) Reachability. Since $|G|=0$, G is singular.

Substituting $G=\begin{bmatrix} 3 & 2 \\ 6 & 4 \end{bmatrix}$ and $h=\begin{bmatrix} 1 \\ 2 \end{bmatrix}$ into the controllability matrix (6.105), we get

$$Q_c = [\boldsymbol{h} \quad \boldsymbol{Gh}] = \begin{bmatrix} 1 & 7 \\ 2 & 14 \end{bmatrix}$$

Its rank is $1 < 2 = n$, thus the system state is not completely reachable.

(2) Controllability. Since the system matrix G is singular, holding rank$Q_c = n$ true is nothing but a sufficient condition. We can not determine the controllability.

In fact, according to

$$\mathbf{0} = \boldsymbol{X}(1) = \begin{bmatrix} 3 & 2 \\ 6 & 4 \end{bmatrix} \boldsymbol{X}(0) + \begin{bmatrix} 1 \\ 2 \end{bmatrix} u(0)$$

a unique equation is obtained

$$3X_1(0) + 2X_2(0) + u(0) = 0$$

Rewriting it, we get

$$u(0) = -3X_1(0) - 2X_2(0)$$

Consequently, a control input $\boldsymbol{u}(0) = -3X_1(0) - 2X_2(0)$ can be constructed such that the system state is transmitted to origin $\boldsymbol{X}(1) = \mathbf{0}$ from any non-zero initial state $(X_1(0) \neq 0, X_2(0) \neq 0)$. According to the definition of controllability, the system state is completely controllable and it is one beat controllable.

This example illustrates that there exists some cases, the system is controllable but not reachable when the system matrix $\boldsymbol{G}$ is singular.

Theorem 6.10 Minimum Beat Control

Consider a SI discrete time LTI system described by (6.105), and suppose $\boldsymbol{G}$ is nonsingular. If the system is completely controllable, there exists a set of input control sequence

$$\begin{bmatrix} \boldsymbol{u}(0) \\ \boldsymbol{u}(1) \\ \vdots \\ \boldsymbol{u}(n-2) \\ \boldsymbol{u}(n-1) \end{bmatrix} = -[\boldsymbol{G}^{-1}\boldsymbol{h} \quad \boldsymbol{G}^{-2}\boldsymbol{h} \quad \cdots \quad \boldsymbol{G}^{-(n-1)}\boldsymbol{h} \quad \boldsymbol{G}^{-n}\boldsymbol{h}]^{-1}\boldsymbol{X}_0 \tag{6.109}$$

such that the system state can be transmitted to origin $\boldsymbol{X}(l) = \mathbf{0}$ from any non-zero initial state $\boldsymbol{X}(0) = \boldsymbol{X}_0$ by using less than n beats. This set of control input sequence is called **minimum beat control.**

Proof. Substituting $k = n$ into the state response equation (6.100) of system (6.105), we have

$$\begin{aligned} \boldsymbol{X}(n) &= \boldsymbol{G}^n\boldsymbol{X}_0 + [\boldsymbol{G}^{n-1}\boldsymbol{hu}(0) \quad \boldsymbol{G}^{n-2}\boldsymbol{hu}(1) \quad \cdots \quad \boldsymbol{Ghu}(n-2) \quad \boldsymbol{hu}(n-1)] \\ &= \boldsymbol{G}^n\boldsymbol{X}_0 + \boldsymbol{G}^n[\boldsymbol{G}^{-1}\boldsymbol{hu}(0) \quad \boldsymbol{G}^{-2}\boldsymbol{hu}(1) \quad \cdots \quad \boldsymbol{G}^{-(n-1)}\boldsymbol{hu}(n-2) \quad \boldsymbol{G}^{-n}\boldsymbol{hu}(n-1)] \\ &= \boldsymbol{G}^n\boldsymbol{X}_0 - \boldsymbol{G}^n[\boldsymbol{G}^{-1}\boldsymbol{h} \quad \boldsymbol{G}^{-2}\boldsymbol{h} \quad \cdots \quad \boldsymbol{G}^{-(n-1)}\boldsymbol{h} \quad \boldsymbol{G}^{-n}\boldsymbol{h}] \begin{bmatrix} \boldsymbol{u}(0) \\ \boldsymbol{u}(1) \\ \vdots \\ \boldsymbol{u}(n-2) \\ \boldsymbol{u}(n-1) \end{bmatrix} \end{aligned} \tag{6.110}$$

Since the system is completely controllable, and $\boldsymbol{G}$ is nonsingular, equation (6.111) can be derived as

$$[\boldsymbol{G}^{n-1}\boldsymbol{h} \quad \boldsymbol{G}^{n-2}\boldsymbol{h} \quad \cdots \quad \boldsymbol{Gh} \quad \boldsymbol{h}]=\boldsymbol{G}^{n}[\boldsymbol{G}^{-1}\boldsymbol{h} \quad \boldsymbol{G}^{-2}\boldsymbol{h} \quad \cdots \quad \boldsymbol{G}^{-(n-1)}\boldsymbol{h} \quad \boldsymbol{G}^{-n}\boldsymbol{h}] \tag{6.111}$$

Obviously

$$[\boldsymbol{G}^{-1}\boldsymbol{h} \quad \boldsymbol{G}^{-2}\boldsymbol{h} \quad \cdots \quad \boldsymbol{G}^{-(n-1)}\boldsymbol{h} \quad \boldsymbol{G}^{-n}\boldsymbol{h}] \tag{6.112}$$

is nonsingular.

It indicates that the control input given by equation (6.109) is constructible. Substituting (6.109) into (6.110), equation (6.113) can be got.

$$\begin{aligned}\boldsymbol{X}(n)&=\boldsymbol{G}^{n}\boldsymbol{X}_0-\boldsymbol{G}^{n}[\boldsymbol{G}^{-1}\boldsymbol{h} \quad \boldsymbol{G}^{-2}\boldsymbol{h} \quad \cdots \quad \boldsymbol{G}^{-(n-1)}\boldsymbol{h} \quad \boldsymbol{G}^{-n}\boldsymbol{h}]\\ &\quad [\boldsymbol{G}^{-1}\boldsymbol{h} \quad \boldsymbol{G}^{-2}\boldsymbol{h} \quad \cdots \quad \boldsymbol{G}^{-(n-1)}\boldsymbol{h} \quad \boldsymbol{G}^{-n}\boldsymbol{h}]^{-1}\boldsymbol{X}_0\\ &=\boldsymbol{G}^{n}\boldsymbol{X}_0-\boldsymbol{G}^{n}\boldsymbol{X}_0=\boldsymbol{0}\end{aligned} \tag{6.113}$$

Equation (6.113) shows that the non-zero initial state $\boldsymbol{X}_0$ is transmitted to origin $\boldsymbol{X}(l)=\boldsymbol{0}$ by using $l=n$ beats, this is a minimum beat control.

Example 6.13 Determine the controllability of the following system described by

$$\boldsymbol{X}(k+1)=\begin{bmatrix}-2 & 2 & -1\\ 0 & -2 & 0\\ 1 & -4 & 0\end{bmatrix}\boldsymbol{X}(k)+\begin{bmatrix}0\\1\\0\end{bmatrix}u(k)$$

If the system is completely controllable, try to find a minimum beat control $u(k)$, $k=0$, 1, 2. Suppose the initial state $\boldsymbol{X}_0=[1 \quad 2 \quad 1]^T$.

Solution Since the determinant of system matrix $\boldsymbol{G}$ is

$$\det\begin{bmatrix}-2 & 2 & -1\\ 0 & -2 & 0\\ 1 & -4 & 0\end{bmatrix}=-2\neq 0$$

the system matrix $\boldsymbol{G}$ is nonsingular.

The controllability matrix of the system is

$$\begin{aligned}rank\boldsymbol{Q}_c&=rank[\boldsymbol{h} \quad \boldsymbol{Gh} \quad \boldsymbol{G}^2\boldsymbol{h}]\\ &=rank\begin{bmatrix}0 & 2 & -12\\ 1 & -2 & 4\\ 0 & -4 & 10\end{bmatrix}=3=n\end{aligned}$$

According to the rank criteria of controllability, the system is completely controllable.

Thus, there exists a minimum beat control and can be calculated as

$$\begin{aligned}\begin{bmatrix}u(0)\\u(1)\\u(2)\end{bmatrix}&=-[\boldsymbol{G}^{-1}\boldsymbol{h} \quad \boldsymbol{G}^{-2}\boldsymbol{h} \quad \boldsymbol{G}^{-3}\boldsymbol{h}]^{-1}\boldsymbol{X}_0\\ &=-\begin{bmatrix}-2 & -2 & -7\\ -0.5 & 0.25 & -0.125\\ -3 & -6.5 & -15.75\end{bmatrix}^{-1}\begin{bmatrix}1\\2\\1\end{bmatrix}=\begin{bmatrix}7.2143\\4.7857\\-3.2857\end{bmatrix}\end{aligned}$$

Theorem 6.11 The state of a n-dimension discrete time SI LTI system would never be transmitted to zero state if it can not be transmitted to zero state through n beats.

Proof. For a SI system, if its system state can not be transmitted to zero state by using n beats, then rank of controllability matrix Q_c satisfies

$$\operatorname{rank}\boldsymbol{Q}_c=\operatorname{rank}[\boldsymbol{h}\quad \boldsymbol{G}\boldsymbol{h}\quad \cdots \quad \boldsymbol{G}^{n-1}\boldsymbol{h}]<n \tag{6.114}$$

It indicates that n column vectors of matrix Q_c are linearly independent. We can assume that

$$\boldsymbol{G}^{n-1}\boldsymbol{h}=a_0\boldsymbol{h}+a_1\boldsymbol{G}\boldsymbol{h}+\cdots a_{n-2}\boldsymbol{G}^{n-2}\boldsymbol{h} \tag{6.115}$$

then

$$\begin{aligned}\boldsymbol{G}^n\boldsymbol{h}&=\boldsymbol{G}(\boldsymbol{G}^{n-1}\boldsymbol{h})=\boldsymbol{G}[a_0\boldsymbol{h}+a_1\boldsymbol{G}\boldsymbol{h}+\cdots+a_{n-2}\boldsymbol{G}^{n-2}\boldsymbol{h}]\\&=a_0\boldsymbol{G}\boldsymbol{h}+a_1\boldsymbol{G}^2\boldsymbol{h}+\cdots+a_{n-2}\boldsymbol{G}^{n-1}\boldsymbol{h}\end{aligned} \tag{6.116}$$

Substitute (6.115) into (6.116), and (6.117) can be obtained as

$$\boldsymbol{G}^n\boldsymbol{h}=a_{n-2}a_0\boldsymbol{h}+(a_0+a_{n-2}a_1)\boldsymbol{G}\boldsymbol{h}+\cdots+(a_{n-3}+a_{n-2}\,a_{n-2})\boldsymbol{G}^{n-2}\boldsymbol{h} \tag{6.117}$$

Equation (6.117) shows that the n dimension column vector $\boldsymbol{G}^n\boldsymbol{h}$ is the same as $\boldsymbol{G}^{n-1}\boldsymbol{h}$, they are all linearly dependent of $n-1$ column vectors $\boldsymbol{h}, \boldsymbol{G}\boldsymbol{h}, \cdots, \boldsymbol{G}^{n-2}\boldsymbol{h}$. In other words, if equation (6.114) holds true then inequality (6.118) must hold true.

$$\operatorname{rank}[\boldsymbol{h}\quad \boldsymbol{G}\boldsymbol{h}\quad \cdots \quad \boldsymbol{G}^{n-1}\boldsymbol{h}\quad \boldsymbol{G}^n\boldsymbol{h}]<n \tag{6.118}$$

That is to say, if equation (6.103) without solution, then equation (6.119)

$$-\boldsymbol{G}^{l+1}\boldsymbol{X}_0=[\boldsymbol{G}\boldsymbol{h}\quad \boldsymbol{G}^2\boldsymbol{h}\quad \cdots \quad \boldsymbol{G}^l\boldsymbol{h}]\begin{bmatrix}\boldsymbol{u}(l-1)\\ \boldsymbol{u}(l-2)\\ \vdots\\ \boldsymbol{u}(0)\end{bmatrix} \tag{6.119}$$

must have no solution. In the same way, for any $m>n$, the following inequality

$$\operatorname{rank}[\boldsymbol{h}\quad \boldsymbol{G}\boldsymbol{h}\quad \cdots \quad \boldsymbol{G}^{m-1}\boldsymbol{h}\quad \boldsymbol{G}^m\boldsymbol{h}]<n \tag{6.120}$$

holds true.

Consequently, the non-zero state $\boldsymbol{X}_0$ would never be transmitted to zero state if it can not be transmitted to zero state through n beats.

6.5.2 Observability

Definition 6.12 Consider the discrete time LTI system described by (6.99), given initial time instant h. If the control input sequence $\boldsymbol{u}(k)$ is known, then a arbitrary non-zero initial state $\boldsymbol{X}(h)=\boldsymbol{X}_0$ can be uniquely determined by the output sequence $\boldsymbol{y}(k)$ which is measured at the finite sampling interval $[h, l]$, where $l>h$. Then the system (state) is said to be **completely observable.** Otherwise, the system (state) is said to be **unobservable.**

Consider the discrete time LTI system described by (6.99) and assume its control input $\boldsymbol{u}(k)=\boldsymbol{0}$, then the system is reduced to

$$\begin{cases}\boldsymbol{X}(k+1)=\boldsymbol{G}\boldsymbol{X}(k)\\ \boldsymbol{y}(k)=\boldsymbol{C}\boldsymbol{X}(k)\end{cases} \tag{6.121}$$

Where $\boldsymbol{G}$、$\boldsymbol{C}$ are system matrix and output matrix, respectively.

Theorem 6.12 Consider the discrete time LTI system described by (6.121), the system is completely observable if and only if matrix $\boldsymbol{Q}_o$ satisfies

$$\operatorname{rank} Q_o=\operatorname{rank}\begin{bmatrix} C \\ CG \\ \vdots \\ CG^{n-2} \\ CG^{n-1} \end{bmatrix}=n \tag{6.122}$$

Where Q_o is called the **observability matrix** of the discrete time LTI system.

Proof. According to the state response of system (6.121), equation (6.123) can be got.

$$y(k)=CX(k)=CG^kX(0) \tag{6.123}$$

Suppose that any non-zero initial state $X(0)=X_0$ can be uniquely determined by the measured output sequence $y(k), k=0,1,2,\cdots,n-1$ through n beats. Rewrite (6.123) in a matrix form as

$$\begin{bmatrix} y(0) \\ y(1) \\ \vdots \\ y(n-2) \\ y(n-1) \end{bmatrix}=\begin{bmatrix} C \\ CG \\ \vdots \\ CG^{n-2} \\ CG^{n-1} \end{bmatrix}X_0 \tag{6.124}$$

This is a non-homogeneous linear equation set with a n-dimension unknown vector X_0, which is determined by $q\times n$ equations. According to linear algebraic theory, X_0 has a unique solution if and only if the $qn\times n$ matrix

$$Q_o=\begin{bmatrix} C \\ CG \\ \vdots \\ CG^{n-2} \\ CG^{n-1} \end{bmatrix} \tag{6.125}$$

is full rank, that is rank $Q_o=n$.

We need to indicate that, different with the controllability, the observability criteria has no requirement about the singularity of the system matrix G. In addition, the observability and its criterion can be derived from the controllability by applying dual principle conveniently.

Example 6.14 Consider the discrete time LTI system described by

$$X(k+1)=\begin{bmatrix} 1 & 0 & -1 \\ 0 & -2 & 1 \\ 3 & 0 & 2 \end{bmatrix}X(k)+\begin{bmatrix} 2 \\ -1 \\ 1 \end{bmatrix}u(k)$$

$$y(k)=[0 \quad 1 \quad 0]X(k)$$

Please try to determine its observability.

Solution The system matrix G and output matrix c are obtained from the state space description, respectively

$$G=\begin{bmatrix} 1 & 0 & -1 \\ 0 & -2 & 1 \\ 3 & 0 & 2 \end{bmatrix}, c=[0 \quad 1 \quad 0]$$

The rank of its observability matrix is

$$\text{rank}\boldsymbol{Q}_o=\text{rank}\begin{bmatrix}\boldsymbol{c}\\ \boldsymbol{cG}\\ \boldsymbol{cG}^2\end{bmatrix}=\text{rank}\begin{bmatrix}0 & 1 & 0\\ 0 & -2 & 1\\ 3 & 4 & 0\end{bmatrix}=3=n$$

Then the system is completely observable.

6.5.3 Condition of Remaining Controllability and Observability by Sampling

A completely controllable and observable continuous system whether remains its controllability and observability after discretization, is an important problem for data-sampled control system and computer control system.

Consider the continuous time LTI system described by

$$\begin{cases}\dot{\boldsymbol{X}}(t)=\boldsymbol{AX}(t)+\boldsymbol{Bu}(t)\\ \boldsymbol{y}(t)=\boldsymbol{CX}(t)\end{cases}\tag{6.126}$$

Suppose the system eigenvalues are $\lambda_1,\lambda_2,\cdots,\lambda_l$, where $\lambda_i\,(i=1,2,\cdots l)$ maybe real or conjugate complex pair, single or multiply eigenvalues, and $l\leqslant n$. The discretization system can be described by

$$\begin{cases}\boldsymbol{X}(k+1)=\boldsymbol{GX}(k)+\boldsymbol{Hu}(k)\\ \boldsymbol{y}(k)=\boldsymbol{CX}(k)\end{cases}\tag{6.127}$$

Where the system matrix $\boldsymbol{G}=\boldsymbol{\Phi}(T)=\mathrm{e}^{\boldsymbol{A}T}$; the input matrix $\boldsymbol{H}=(\int_0^T \mathrm{e}^{\boldsymbol{A}t}\,\mathrm{d}t)\boldsymbol{B}$; T denotes the sampling period.

A typical example about the controllability and observability of the discretization system is given out firstly.

Example 6.15 Consider the controllability and observability of the continuous time system and its discretization system, the continuous system is

$$\begin{cases}\dot{\boldsymbol{X}}=\begin{bmatrix}0 & 1\\ -1 & 0\end{bmatrix}\boldsymbol{X}+\begin{bmatrix}1\\ 0\end{bmatrix}u\\ y=[0\quad 1]X\end{cases}$$

Solution First, determine the controllability and observability of the continuous system, the controllability matrix and observability matrix of the continuous system and the rank of them are

$$\boldsymbol{Q}_c=[\boldsymbol{b}\quad \boldsymbol{Ab}]=\begin{bmatrix}1 & 0\\ 0 & -1\end{bmatrix},\ \text{rank}\boldsymbol{Q}_c=2=n$$

$$\boldsymbol{Q}_o=\begin{bmatrix}\boldsymbol{c}\\ \boldsymbol{cA}\end{bmatrix}=\begin{bmatrix}0 & 1\\ -1 & 0\end{bmatrix},\ \text{rank}\boldsymbol{Q}_o=2=n$$

thus, the continuous system is completely controllable and observable.

In addition, its state transition matrix is

$$\boldsymbol{\Phi}(t)=\mathrm{e}^{\boldsymbol{A}t}=L^{-1}[(s\boldsymbol{I}-\boldsymbol{A})^{-1}]=L^{-1}\left(\begin{bmatrix} s & -1 \\ 1 & s \end{bmatrix}^{-1}\right)$$

$$=L^{-1}\left(\begin{bmatrix} \frac{s}{s^2+1} & \frac{1}{s^2+1} \\ \frac{-1}{s^2+1} & \frac{s}{s^2+1} \end{bmatrix}\right)=\begin{bmatrix} \cos t & \sin t \\ -\sin t & \cos t \end{bmatrix}$$

The discretization system (6.127) is obtained by sampling the continuous system (6.126), its system matrix and input matrix are $\boldsymbol{G}$ and $\boldsymbol{h}$, respectively.

$$\boldsymbol{G}=\boldsymbol{\Phi}(T)=\mathrm{e}^{\boldsymbol{A}T}=\begin{bmatrix} \cos T & \sin T \\ -\sin T & \cos T \end{bmatrix}$$

$$\boldsymbol{h}=\left(\int_0^T \mathrm{e}^{\boldsymbol{A}t}\,\mathrm{d}t\right)\boldsymbol{B}=\int_0^T \begin{bmatrix} \cos t & \sin t \\ -\sin t & \cos t \end{bmatrix}\begin{bmatrix} 1 \\ 0 \end{bmatrix} dt=\int_0^T \begin{bmatrix} \cos t \\ -\sin t \end{bmatrix} \mathrm{d}t=\begin{bmatrix} \sin T \\ \cos T-1 \end{bmatrix}$$

the controllability matrix $\boldsymbol{Q}_c$ and observability matrix $\boldsymbol{Q}_o$ of the discretization system are

$$\boldsymbol{Q}_c=[\boldsymbol{h} \quad \boldsymbol{G}\boldsymbol{h}]=\begin{bmatrix} \sin T & 2\sin T\cos T-\sin T \\ \cos T-1 & \cos^2 T-\sin^2 T-\cos T \end{bmatrix}$$

$$\boldsymbol{Q}_o=\begin{bmatrix} \boldsymbol{c} \\ \boldsymbol{c}\boldsymbol{G} \end{bmatrix}=\begin{bmatrix} 0 & 1 \\ -\sin T & \cos T \end{bmatrix}$$

their determinants are $\det\boldsymbol{Q}_c$ and $\det\boldsymbol{Q}_o$, respectively.

$$\det\boldsymbol{Q}_c=\det\begin{bmatrix} \sin T & 2\sin T\cos T-\sin T \\ \cos T-1 & \cos^2 T-\sin^2 T-\cos T \end{bmatrix}=2\sin T(\cos T-1)$$

$$\det\boldsymbol{Q}_o=\det\begin{bmatrix} 0 & 1 \\ -\sin T & \cos T \end{bmatrix}=\sin T$$

Obviously, whether the controllability matrix and observability matrix are full rank depend on the sampling period. If $T=k\pi$ $(k=1,2,\cdots)$, there exists $\det\boldsymbol{Q}_c=\det\boldsymbol{Q}_o=0$, the two discriminate matrixes are all not full rank, the discretization system is not controllable and observable. If $T\neq k\pi$ $(k=1,2,\cdots)$, there exists $\det\boldsymbol{Q}_c\neq 0$, $\det\boldsymbol{Q}_o\neq 0$, the two discriminate matrixes are all full rank, the discretization system is completely controllable and observable.

It is obvious that whether the discretization system (6.127) remains the controllability and observability of continuous system (6.126) is determined by the sampling period. Condition of the discretization system remaining the controllability and observability of continuous system is given out as follows.

Theorem 6.13 Consider the continuous time LTI system described by (6.126), its discretization system described by (6.127) remains its controllability and observability if and only if, for any eigenvalue of the system matrix $\boldsymbol{A}$ which satisfies

$$\mathrm{Re}(\lambda_i-\lambda_j)=0 \quad (i,j=1,2,\cdots,l) \tag{6.128}$$

such that the sampling period T satisfies

$$T\neq\frac{2k\pi}{\mathrm{Im}(\lambda_i-\lambda_j)} \quad (k=\pm 1,\pm 2,\cdots) \tag{6.129}$$

This conclusion indicates the relations that the sampling period and imaginary parts of eigenvalues whose real parts are equal to each other should satisfy. If the eigenvalues are real, no matter they are single or multiply eigenvalues, there is no restriction on sampling period. Only if the eigenvalues are not real and its real parts are equal, the sampling period must be restricted by (6.129). In Example 6.15, the system matrix $\boldsymbol{A}$ has a pair of conjugate complex eigenvalues $\lambda_{1,2}=\pm j$. According to equation (6.129), in order to remain its controllability and observability, the sampling period should satisfy

$$T\neq\frac{2k\pi}{\mathrm{Im}(\lambda_i-\lambda_j)}=\frac{2k\pi}{2}=k\pi \quad (k=\pm1,\pm2,\cdots)$$

This is the same as the analysis above.

Pay attention to, all the eigenvalues with the same real part should restrict the sampling period by (6.129), this is not limited to the conjugate complex eigenvalues. For example, eigenvalues of some systems are $\lambda_{1,2}=-3\pm j$ and $\lambda_{3,4}=-3\pm 2j$, then the sampling period should be restricted by the following six equations.

$$T\neq\frac{2k\pi}{1-(-1)},T\neq\frac{2k\pi}{1-2},T\neq\frac{2k\pi}{1-(-2)},T\neq\frac{2k\pi}{-1-2},T\neq\frac{2k\pi}{-1-(-2)},T\neq\frac{2k\pi}{2-(-2)}$$

Where $k=\pm1,\pm2,\cdots$.

By eliminating the insignificance and repeated equations, the sampling period T should satisfy:

$$T\neq\frac{k\pi}{2},k\pi,2k\pi,\frac{2k\pi}{3}(k=\pm1,\pm2,\cdots)$$

6.6 Control Synthesis of Discrete Time LTI System

We have indicated that many theories of discrete time system are parallel with continuous system. Analysis problems of discrete time system have been introduced in the preceding sections, in this section we will discuss the design problem of discrete time system. Two basic problems will be discussed mainly, they are poles placement and observer design. Since discrete system and continuous system has internal relations, such that the basic problems have many similar features on description and solving method. Therefore, we only introduce them simply, pay more attention to the existence condition of them and solving method, and the difference from continuous system.

6.6.1 Design of Poles Placement

A open loop control system is shown in Figure 6.8(a). Its state equation is

$$\boldsymbol{X}(k+1)=\boldsymbol{GX}(k)+\boldsymbol{Hu}(k) \tag{6.130}$$

Suppose the magnitude of the control signal $\boldsymbol{u}(k)$ has no limit. If the control input signal $\boldsymbol{u}(k)$ is selected as

$$u(k) = -KX(k) + v(k) \tag{6.131}$$

Where K is a $1\times n$ state feedback gain matrix; $v(k)$ is a reference input of the system, then the system is changed into a closed-loop system as shown in Figure 6.8(b), its state equation is

$$X(k+1) = (G-HK)X(k) + Hv(k) \tag{6.132}$$

By selecting matrix K, such that the eigenvalues of matrix $(G-HK)$ that is the poles of system (6.132) are the expected poles $z_i, i=1,2,\cdots,n$, namely

$$\lambda_i(G-HK) = z_i, \qquad i=1,2,\cdots,n \tag{6.133}$$

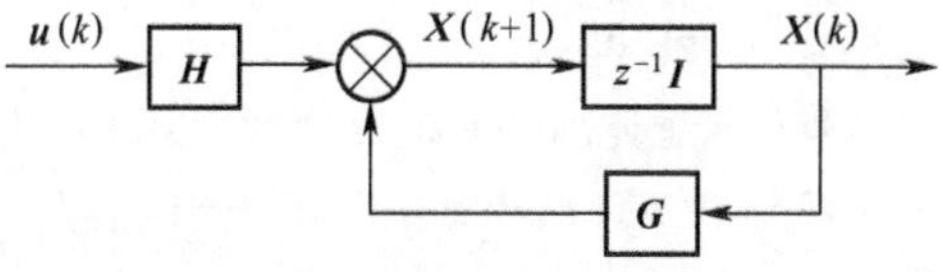

(a) Block Diagram of Open-Loop Control System

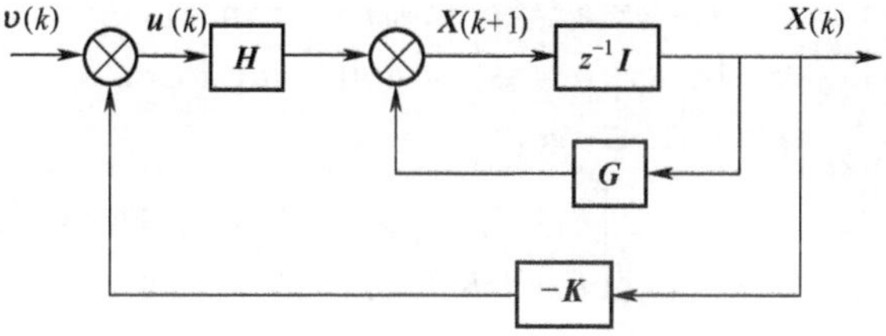

(b) Block Diagram of Closed-Loop Control System

Figure 6.8 Block Diagram of Open-Loop and Closed-Loop Control System

If for any given closed-loop poles $z_i, i=1,2,\cdots,n$, the poles placement problem has solution, then the poles of system (6.130) can be assigned freely by state feedback (6.131).

Theorem 6.14 Poles Placement Theorem

The poles of system (6.130) can be assigned freely if and only if the system state is completely controllable.

Proof. (1) Necessity. If the system is not completely controllable, then the state feedback control (6.131) has no control effect on some eigenvalues of G-HK.

If the state of system (6.130) is not completely controllable, then the rank of controllability matrix Q_c is less than n, namely

$$\text{rank}Q_c = \text{rank}[H \quad GH \quad \cdots \quad G^{n-1}H] = q < n \tag{6.134}$$

This illustrates that the controllability matrix has q linear independent column vectors. Denotes the q linear independent column vectors as $f_1, f_2, \cdots, f_q$, and selects additional $n-q$ linear independent n dimension column vectors $v_{q+1}, v_{q+2}, \cdots, v_n$, such that the rank of matrix P is equal to n, and

$$P = [f_1 \quad f_2 \quad \cdots \quad f_q \quad v_{q+1} \quad v_{q+2} \quad \cdots \quad v_n]$$

Take P as transfer matrix, and define

$$P^{-1}GP = \overline{G}, P^{-1}H = \overline{H}$$

Then, we have

$$\boldsymbol{GP}=\boldsymbol{P}\overline{\boldsymbol{G}}$$

namely

$$\begin{aligned}&[\boldsymbol{G}f_1\quad \boldsymbol{G}f_2\quad \cdots\quad \boldsymbol{G}f_q\quad \boldsymbol{G}v_{q+1}\quad \boldsymbol{G}v_{q+2}\quad \cdots\quad \boldsymbol{G}v_n]\\&=[f_1\quad f_2\quad \cdots\quad f_q\quad v_{q+1}\quad v_{q+2}\quad \cdots\quad v_n]\overline{\boldsymbol{G}}\end{aligned} \tag{6.135}$$

In the same way, we have

$$\boldsymbol{H}=\overline{\boldsymbol{PH}}=[f_1\quad f_2\quad \cdots\quad f_q\quad v_{q+1}\quad v_{q+2}\quad \cdots\quad v_n]\overline{\boldsymbol{H}} \tag{6.136}$$

For the q linear independent column vectors $f_1, f_2, \cdots, f_q$, by applying Cayley-Hamilton Theorem, matrix $\boldsymbol{G}f_1, \boldsymbol{G}f_2, \cdots, \boldsymbol{G}f_q$ can be linearly expressed by the q linear independent column vectors, then we have

$$\begin{aligned}\boldsymbol{G}f_1&=g_{11}f_1+g_{21}f_2+\cdots+g_{q1}f_q\\ \boldsymbol{G}f_2&=g_{12}f_1+g_{22}f_2+\cdots+g_{q2}f_q\\ &\vdots\\ \boldsymbol{G}f_q&=g_{1q}f_1+g_{2q}f_2+\cdots+g_{qq}f_q\end{aligned}$$

Thus, equation (6.135) can be written as the following form.

$$[\boldsymbol{G}f_1\quad \boldsymbol{G}f_2\quad \cdots\quad \boldsymbol{G}f_q\quad \boldsymbol{G}v_{q+1}\quad \boldsymbol{G}v_{q+2}\quad \cdots\quad \boldsymbol{G}v_n]$$

$$=[f_1\quad f_2\quad \cdots\quad f_q\quad v_{q+1}\quad v_{q+2}\quad \cdots\quad v_n]\begin{bmatrix} g_{11} & g_{12} & \cdots & g_{1q} & g_{1q+1} & g_{1q+2} & \cdots & g_{1n}\\ g_{21} & g_{22} & \cdots & g_{2q} & g_{2q+1} & g_{2q+2} & \cdots & g_{2n}\\ \vdots & \vdots & & \vdots & \vdots & \vdots & & \vdots\\ g_{q1} & g_{q2} & \cdots & g_{qq} & g_{qq+1} & g_{qq+2} & \cdots & g_{qn}\\ 0 & 0 & \cdots & 0 & g_{q+1q+1} & g_{q+1q+2} & \cdots & g_{q+1n}\\ \vdots & \vdots & & \vdots & \vdots & \vdots & & \vdots\\ 0 & 0 & \cdots & 0 & g_{nq+1} & g_{nq+2} & \cdots & g_{nn}\end{bmatrix}$$

In order to simplify the expression, define

$$\begin{bmatrix} g_{11} & g_{12} & \cdots & g_{1q}\\ g_{21} & g_{22} & \cdots & g_{2q}\\ \vdots & \vdots & \ddots & \vdots\\ g_{q1} & g_{q2} & \cdots & g_{qq}\end{bmatrix}=\boldsymbol{G}_{11}$$

$$\begin{bmatrix} g_{1q+1} & g_{1q+2} & \cdots & g_{1n}\\ g_{2q+1} & g_{2q+2} & \cdots & g_{2n}\\ \vdots & \vdots & \ddots & \vdots\\ g_{qq+1} & g_{qq+2} & & g_{qn}\end{bmatrix}=\boldsymbol{G}_{12}$$

$$\begin{bmatrix} 0 & 0 & \cdots & 0\\ \vdots & \vdots & & \vdots\\ 0 & 0 & \cdots & 0\end{bmatrix}=\boldsymbol{G}_{21}\text{ is a }(n-q)\times q\text{ zero matrix}$$

$$\begin{bmatrix} g_{q+1q+1} & g_{q+1q+2} & \cdots & g_{q+1n}\\ g_{q+2q+1} & g_{q+2q+2} & \cdots & g_{q+2n}\\ \vdots & \vdots & \ddots & \vdots\\ g_{nq+1} & g_{nq+2} & \cdots & g_{nn}\end{bmatrix}=\boldsymbol{G}_{22}$$

then equation (6.135) can be written as

$$[\boldsymbol{G}f_1 \quad \cdots \quad \boldsymbol{G}f_q \quad \boldsymbol{G}v_{q+1} \quad \cdots \quad \boldsymbol{G}v_n]=[f_1 \quad \cdots \quad f_q \quad v_{q+1} \quad \cdots \quad v_n]\begin{bmatrix}\boldsymbol{G}_{11} & \boldsymbol{G}_{12}\\ \boldsymbol{0} & \boldsymbol{G}_{22}\end{bmatrix} \tag{6.137}$$

Comparing (6.135) with (6.137), (6.138) is obtained.

$$\overline{\boldsymbol{G}}=\begin{bmatrix}\boldsymbol{G}_{11} & \boldsymbol{G}_{12}\\ \boldsymbol{0} & \boldsymbol{G}_{22}\end{bmatrix} \tag{6.138}$$

Then, referring to equation (6.136), and (6.139) is got.

$$\boldsymbol{H}=[f_1 \quad f_2 \quad \cdots \quad f_q \quad v_{q+1} \quad v_{q+2} \quad \cdots \quad v_n]\overline{\boldsymbol{H}} \tag{6.139}$$

Referring to equation (6.134), matrix $\boldsymbol{H}$ can be linearly expressed by the q linear independent column vectors $f_1, f_2, \cdots, f_q$, then we can obtain

$$\boldsymbol{H}=h_{11}f_1+h_{21}f_2+\cdots+h_{q1}f_q$$

and equation (6.139) is written to the following form

$$\boldsymbol{H}=h_{11}f_1+h_{21}f_2+\cdots+h_{q1}f_q$$

$$=[f_1 \quad f_2 \quad \cdots \quad f_q \quad v_{q+1} \quad v_{q+2} \quad \cdots v_n]\begin{bmatrix}h_{11}\\ h_{21}\\ \vdots\\ h_{q1}\\ 0\\ \vdots\\ 0\end{bmatrix}$$

Therefore

$$\overline{\boldsymbol{H}}=\begin{bmatrix}\boldsymbol{H}_{11}\\ \boldsymbol{0}\end{bmatrix} \tag{6.140}$$

Where

$$\boldsymbol{H}_{11}=\begin{bmatrix}h_{11}\\ h_{21}\\ \vdots\\ h_{q1}\end{bmatrix}$$

The closed-loop system described by (6.131) will be analyzed know. Its characteristic equation is

$$|z\boldsymbol{I}-\boldsymbol{G}+\boldsymbol{HK}|=0$$

Define $\widetilde{\boldsymbol{K}}=\boldsymbol{KP}$, (6.141) is obtained by blocking matrix $\widetilde{\boldsymbol{K}}$

$$\widetilde{\boldsymbol{K}}=[\boldsymbol{K}_{11} \quad \boldsymbol{K}_{12}] \tag{6.141}$$

Where $\boldsymbol{K}_{11}$ is a $1\times q$ matrix; $\boldsymbol{K}_{12}$ is a $1\times(n-q)$ matrix. Then the $1\times n$ matrix $\boldsymbol{K}$ can be expressed in the following form.

$$\boldsymbol{K}=\widetilde{\boldsymbol{K}}\boldsymbol{P}^{-1}=[\boldsymbol{K}_{11} \quad \boldsymbol{K}_{12}]\boldsymbol{P}^{-1}$$

Then the characteristic equation of the closed-loop system (6.131) can be expressed in

(6.142).

$$
\begin{aligned}
|z\boldsymbol{I}-\boldsymbol{G}+\boldsymbol{HK}| &= |\boldsymbol{P}^{-1}||z\boldsymbol{I}-\boldsymbol{G}+\boldsymbol{HK}||\boldsymbol{P}| \\
&= |\boldsymbol{P}^{-1}(z\boldsymbol{I}-\boldsymbol{G}+\boldsymbol{HK})\boldsymbol{P}| = |z\boldsymbol{I}-\boldsymbol{P}^{-1}\boldsymbol{GP}+\boldsymbol{P}^{-1}\boldsymbol{HKP}| \\
&= |z\boldsymbol{I}-\overline{\boldsymbol{G}}+\overline{\boldsymbol{H}}\tilde{\boldsymbol{K}}|
\end{aligned}
\tag{6.142}
$$

Substituting (6.138)、(6.140) and (6.141) into (6.142), (6.143) is obtained as

$$
\begin{aligned}
|z\boldsymbol{I}-\overline{\boldsymbol{G}}+\overline{\boldsymbol{H}}\tilde{\boldsymbol{K}}| &= \left| z\begin{bmatrix} \boldsymbol{I}_q & \boldsymbol{0} \\ \boldsymbol{0} & \boldsymbol{I}_{n-q} \end{bmatrix} - \begin{bmatrix} \boldsymbol{G}_{11} & \boldsymbol{G}_{12} \\ \boldsymbol{0} & \boldsymbol{G}_{22} \end{bmatrix} + \begin{bmatrix} \boldsymbol{H}_{11} \\ \boldsymbol{0} \end{bmatrix} [\boldsymbol{K}_{11} \quad \boldsymbol{K}_{12}] \right| \\
&= \begin{vmatrix} z\boldsymbol{I}_q - \boldsymbol{G}_{11} + \boldsymbol{H}_{11}\boldsymbol{K}_{11} & -\boldsymbol{G}_{12} + \boldsymbol{H}_{11}\boldsymbol{K}_{12} \\ \boldsymbol{0} & z\boldsymbol{I}_{n-q} - \boldsymbol{G}_{22} \end{vmatrix} \\
&= |z\boldsymbol{I}_q - \boldsymbol{G}_{11} + \boldsymbol{H}_{11}\boldsymbol{K}_{11}||z\boldsymbol{I}_{n-q} - \boldsymbol{G}_{22}|
\end{aligned}
\tag{6.143}
$$

Equation (6.143) shows that, matrix $\boldsymbol{K}$ has control effect on the q eigenvalues of $\boldsymbol{G}_{11}+\boldsymbol{H}_{11}\boldsymbol{K}_{11}$, but has no control effect on the $n-q$ eigenvalues of $\boldsymbol{G}_{22}$. That is to say, $n-q$ eigenvalues of $\boldsymbol{G}-\boldsymbol{HK}$ are in dependent of the state feedback gain matrix $\boldsymbol{K}$.

Thus, the system state completely controllable is the necessary condition of all the eigenvalues of matrix $\boldsymbol{G}-\boldsymbol{HK}$ can be assigned arbitrarily.

(2) Sufficiency. If the system state is completely controllable, then there exists a matrix $\boldsymbol{K}$ such that all the eigenvalues of $\boldsymbol{G}-\boldsymbol{HK}$ are the expected closed-loop poles, namely assigns the closed-loop poles to the expected locations.

The expected eigenvalues of matrix $\boldsymbol{G}-\boldsymbol{HK}$ are $z_i, i=1,2,\cdots,n$; the complex eigenvalues are often in the form of conjugate complex pair. The characteristic equation of the primary system is

$$
|z\boldsymbol{I}-\boldsymbol{G}| = z^n + a_1 z^{n-1} + a_2 z^{n-2} + \cdots + a_{n-1}z + a_n = 0
$$

Define $\boldsymbol{T}$ as the transfer matrix

$$
\boldsymbol{T}=\boldsymbol{MW}
$$

Where

$$
\boldsymbol{M}=[\boldsymbol{H} \quad \boldsymbol{GH} \quad \cdots \quad \boldsymbol{G}^{n-1}\boldsymbol{H}]
\tag{6.144}
$$

Its rank is n, in addition

$$
\boldsymbol{W}=\begin{bmatrix} a_{n-1} & a_{n-2} & \cdots & a_1 & 1 \\ a_{n-2} & a_{n-3} & \cdots & 1 & 0 \\ \vdots & \vdots & \iddots & \vdots & \vdots \\ a_1 & 1 & \cdots & 0 & 0 \\ 1 & 0 & \cdots & 0 & 0 \end{bmatrix}
$$

By referring to the controllable canonical form, we have

$$
\boldsymbol{T}^{-1}\boldsymbol{GT}=\overline{\boldsymbol{G}}=\begin{bmatrix} 0 & 1 & 0 & \cdots & 0 \\ 0 & 0 & 1 & \cdots & 0 \\ \vdots & \vdots & \vdots & & \vdots \\ 0 & 0 & 0 & \cdots & 1 \\ -a_n & -a_{n-1} & -a_{n-2} & \cdots & -a_1 \end{bmatrix}
$$

$$T^{-1}H=\bar{H}=\begin{bmatrix}0\\0\\\vdots\\0\\1\end{bmatrix}$$

By defining

$$\bar{K}=KT=[k_n^* \quad k_{n-1}^* \quad \cdots \quad k_2^* \quad k_1^*] \tag{6.145}$$

we have

$$\bar{H}\bar{K}=\begin{bmatrix}0\\0\\\vdots\\0\\1\end{bmatrix}[k_n^* \quad k_{n-1}^* \quad \cdots k_2^* \quad k_1^*]=\begin{bmatrix}0 & 0 & \cdots & 0 & 0\\0 & 0 & \cdots & 0 & 0\\\vdots & \vdots & & \vdots & \vdots\\0 & 0 & \cdots & 0 & 0\\k_n^* & k_{n-1}^* & \cdots & k_2^* & k_1^*\end{bmatrix}$$

The characteristic eigenvalues of system (6.131) can be written as the following form.

$$|zI-G+HK|=|zI-\bar{G}+\bar{H}\bar{K}| \tag{6.146}$$

Substituting $\bar{G}$、$\bar{H}$ and $\bar{K}$ into (6.146), (6.147) is gained.

$$\begin{aligned}|zI-G+HK|&=|zI-\bar{G}+\bar{H}\bar{K}|\\&=\begin{bmatrix}z & -1 & 0 & \cdots & 0\\0 & z & -1 & \cdots & 0\\\vdots & & \ddots & & \vdots\\0 & \cdots & 0 & z & -1\\a_n+k_n^* & a_{n-1}+k_{n-1}^* & \cdots & a_2+k_2^* & a_1+k_1^*\end{bmatrix}\\&=z^n+(a_1+k_1^*)z^{n-1}+\cdots+(a_{n-1}+k_{n-1}^*)z+(a_n+k_n^*)=0\end{aligned} \tag{6.147}$$

Otherwise, the characteristic equation (6.148) of closed-loop system (6.131) determined by the expected eigenvalues is

$$(z-z_1)(z-z_2)\cdots(z-z_n)=z^n+a_1^*z^{n-1}+\cdots+a_{n-1}^*z+a_n^*=0 \tag{6.148}$$

Comparing (6.147) with (6.148), such that the coefficients of $z^i, i=0,1,\cdots,n-1$ are equal, the following equations are gained as

$$\begin{aligned}a_1^*&=a_1+k_1^*\\a_2^*&=a_2+k_2^*\\&\vdots\\a_n^*&=a_n+k_n^*\end{aligned}$$

Therefore, from equation (6.145) we can get (6.149).

$$\begin{aligned}K&=\bar{K}T^{-1}=[k_n^* \quad k_{n-1}^* \quad \cdots \quad k_2^* \quad k_1^*]T^{-1}\\&=[a_n^*-a_n \quad a_{n-1}^*-a_{n-1} \quad \cdots \quad a_2^*-a_2 \quad a_1^*-a_1]\end{aligned} \tag{6.149}$$

Where a_i^* and a_i are the coefficients of the known closed-loop and open-loop characteristic polynomial respectively; T is a known matrix.

Consequently, if system (6.130) is completely controllable, the state feedback gain matrix $\boldsymbol{K}$ can be determined by the original poles and the expected poles.

Therefore, the poles can be assigned arbitrarily if and only if the system state is completely controllable.

Consider the discrete time system described by (6.130), assume its state is completely controllable, as long as the expected characteristic equation is determined. Then, the corresponding state feedback gain matrix $\boldsymbol{K}$ can be calculated as the following steps.

Step1: Determine the characteristic equation of the original system.

$$|z\boldsymbol{I}-\boldsymbol{G}|=z^n+a_1z^{n-1}+a_2z^{n-2}+\cdots+a_{n-1}z+a_n=0$$

Step2: Determine the expected characteristic equation of the closed-loop system.

$$\begin{aligned}|z\boldsymbol{I}-\boldsymbol{G}+\boldsymbol{HK}|&=(z-z_1)(z-z_2)\cdots(z-z_n)\\&=z^n+a_1^*z^{n-1}+\cdots+a_{n-1}^*z+a_n^*=0\end{aligned}$$

Step3: Determine the coefficient matrix $\overline{\boldsymbol{K}}$.

$$\begin{aligned}\overline{\boldsymbol{K}}&=[k_n^* \quad k_{n-1}^* \quad \cdots \quad k_2^* \quad k_1^*]\\&=[a_n^*-a_n \quad a_{n-1}^*-a_{n-1} \quad \cdots \quad a_2^*-a_2 \quad a_1^*-a_1]\end{aligned}$$

Step4: Determine the state feedback gain matrix $\boldsymbol{K}$.

$$\boldsymbol{K}=\overline{\boldsymbol{K}}\boldsymbol{T}^{-1}$$

Where

$$\boldsymbol{T}=\boldsymbol{MW},\boldsymbol{M}=[\boldsymbol{H} \quad \boldsymbol{GH} \quad \cdots \quad \boldsymbol{G}^{n-1}\boldsymbol{H}]$$

$$\boldsymbol{W}=\begin{bmatrix}a_{n-1} & a_{n-2} & \cdots & a_1 & 1\\ a_{n-2} & a_{n-3} & \cdots & 1 & 0\\ \vdots & \vdots & \iddots & \vdots & \vdots\\ a_1 & 1 & \cdots & 0 & 0\\ 1 & 0 & \cdots & 0 & 0\end{bmatrix}$$

If the system state equation is the controllable canonical form already, the transfer matrix $\boldsymbol{T}$ is unit matrix $\boldsymbol{I}$, the calculation of state feedback gain matrix $\boldsymbol{K}$ is simplified, and $\boldsymbol{K}=\overline{\boldsymbol{K}}\boldsymbol{I}^{-1}=\overline{\boldsymbol{K}}$.

Example 6.16 Consider the discrete time LTI system described by

$$\boldsymbol{X}(k+1)=\begin{bmatrix}0 & 1\\ -0.16 & -1\end{bmatrix}\boldsymbol{X}(k)+\begin{bmatrix}0\\1\end{bmatrix}\boldsymbol{u}(k)$$

Calculate a suitable state feedback gain vector $\boldsymbol{k}$ such that the expected poles of closed-loop system are

$$z_1=0.5+\mathrm{j}0.5, \quad z_2=-0.5+\mathrm{j}0.5$$

Solution The state equation yields the system matrix and input matrix

$$\boldsymbol{G}=\begin{bmatrix}0 & 1\\ -0.16 & -1\end{bmatrix},\boldsymbol{h}=\begin{bmatrix}0\\1\end{bmatrix}$$

Detecting the rank of controllability matrix firstly

$$\operatorname{rank}[\boldsymbol{h} \quad \boldsymbol{Gh}]=\operatorname{rank}\begin{bmatrix}0 & 1\\ 1 & -1\end{bmatrix}=2=n$$

Consequently, the system is completely controllable. Then the poles can be assigned freely.

The characteristic equation of the original system is

$$|z\boldsymbol{I}-\boldsymbol{G}|=\begin{vmatrix} z & -1 \\ 0.16 & z+1 \end{vmatrix}=z^2+z+0.16=0$$

Therefore

$$a_1=1, a_2=0.16$$

The expected characteristic equation determined by the closed-loop poles is

$$|z\boldsymbol{I}-\boldsymbol{G}+\boldsymbol{hk}|=(z-0.5-\mathrm{j}0.5)(z+0.5-\mathrm{j}0.5)==z^2-z+0.5=0$$

Hence

$$a_1^*=-1, a_2^*=0.5$$

Then, the coefficient vector $\bar{\boldsymbol{k}}$ is

$$\bar{\boldsymbol{k}}=[k_2^* \quad k_1^*]=[a_2^*-a_2 \quad a_1^*-a_1]=[0.34 \quad -2]$$

and its trans for mation matrix $\boldsymbol{T}$ is

$$\boldsymbol{T}=\boldsymbol{MW}=[\boldsymbol{h} \quad \boldsymbol{Gh}]\begin{bmatrix} a_1 & 1 \\ 1 & 0 \end{bmatrix}=\begin{bmatrix} 0 & 1 \\ 1 & -1 \end{bmatrix}\begin{bmatrix} 1 & 1 \\ 1 & 0 \end{bmatrix}=\begin{bmatrix} 1 & 0 \\ 0 & 1 \end{bmatrix}$$

Thus
$$\boldsymbol{k}=\bar{\boldsymbol{k}}\boldsymbol{T}^{-1}=\bar{\boldsymbol{k}}=[0.34 \quad -2]$$

In an other side, since the original system is controllable canonical form, then the transfer matrix $\boldsymbol{T}$ is the unit matrix $\boldsymbol{I}$, so $\boldsymbol{k}=\bar{\boldsymbol{k}}=[0.34 \quad -2]$.

In spite of discrete time system is the same as continuous time system about the poles placement problem in many aspects, such as problem description、solving condition and algorithm and so on. But there exists a very important point that is different in actual design. For continuous time LTI system, the condition of asymptotically stable is that all the eigenvalues (poles) have negative real part. While for discrete time LTI system, the condition is that all the modulus values of its eigenvalues (poles) are less than one. Therefore, in the poles placement problem of discrete time system, all the modulus values of the expected closed-loop poles should be less than one.

6.6.2 State Observer

In the discussion above of applying state feedback method to realize poles placement, we have assumed that all the state can be used directly to feedback. In fact, since the state is immeasurable, it is very difficult to do that. Thus state observer must be designed for reconstructing the system state by applying the measurable output $\boldsymbol{y}(k)$ and known quantity $\boldsymbol{u}(k)$. This process is also called **state estimation.** Thus the real feedback variables are the reconstructed state or estimated state $\hat{\boldsymbol{X}}(k)$, not the real state $\boldsymbol{X}(k)$. That is to say $\boldsymbol{u}(k)=-\boldsymbol{K}\hat{\boldsymbol{X}}(k)+\boldsymbol{v}(k)$, as shown in Figure 6.9 , a control system with state observer.

Consider the discrete time LTI system described by

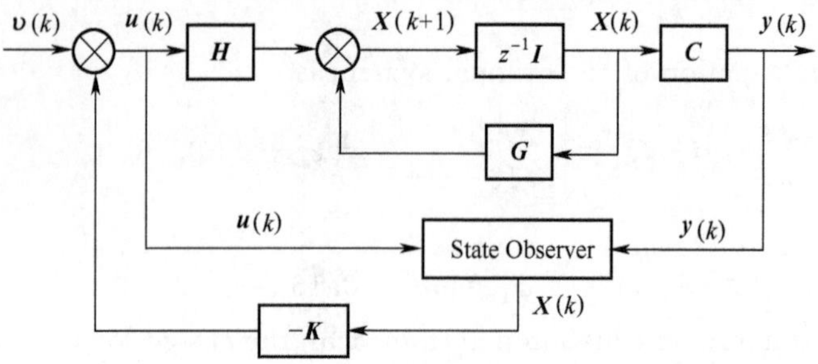

Figure 6.9 Control System with State Observer

$$\begin{cases}\boldsymbol{X}(k+1)=\boldsymbol{GX}(k)+\boldsymbol{Hu}(k)\\ \boldsymbol{y}(k)=\boldsymbol{CX}(k)\end{cases} \tag{6.150}$$

A full-order state observer of discrete time LTI system described by (6.150) is the discrete time dynamic system with n dimension vectors $\boldsymbol{y}(k)$ and $\boldsymbol{u}(k)$ as its input vectors. No matter what initial values of the dynamic system and system (6.150) are, the output vectors $\hat{\boldsymbol{X}}(k)$ of the dynamic system and the state $\boldsymbol{X}(k)$ of the original system has the following relationship.

$$\lim_{k\to\infty}\hat{\boldsymbol{X}}(k)=\lim_{k\to\infty}\boldsymbol{X}(k) \tag{6.151}$$

Similar to the continuous system, the state observer of system (6.150) is selected as the following form.

$$\hat{\boldsymbol{X}}(k+1)=\boldsymbol{G}\hat{\boldsymbol{X}}(k)+\boldsymbol{Hu}(k)+\boldsymbol{L}[\boldsymbol{C}\hat{\boldsymbol{X}}(k)-\boldsymbol{y}(k)] \tag{6.152}$$

Where $\boldsymbol{L}$ is a $n\times m$ real matrix. Denote the observation error as

$$\boldsymbol{e}(k)=\boldsymbol{X}(k)-\hat{\boldsymbol{X}}(k) \tag{6.153}$$

it is also called **reconstruction error** or **estimation error**. Equation (6.151) is equivalent to (6.154).

$$\lim_{k\to\infty}\boldsymbol{e}(k)=\boldsymbol{0} \tag{6.154}$$

Subtracting the state equation of system(6.150)by(6.152)and substituting(6.153) into it, then(6.155)is obtained as

$$\boldsymbol{e}(k+1)=(\boldsymbol{G}+\boldsymbol{LC})\boldsymbol{e}(k) \tag{6.155}$$

It can be seen that in order to hold equation (6.151) true, its sufficient and necessary condition is system (6.155) should be asymptotically stable. That is to say, all the eigenvalues of $\boldsymbol{G}+\boldsymbol{LC}$ must make their modulus values less than one. Based on the discussion above, the design of state observer described by (6.152) can be concluded as the following algebraic problem.

The known quantities are $\boldsymbol{G}\in\mathbf{R}^{n\times n}$, $\boldsymbol{C}\in\mathbf{R}^{q\times n}$, try to determine matrix $\boldsymbol{L}\in\mathbf{R}^{n\times q}$, such that inequality (6.156)

$$|\lambda_i(\boldsymbol{G}+\boldsymbol{LC})|<1, i=1,2,\cdots,n \tag{6.156}$$

holds true.

Theorem 6.15 The discrete time LTI system (6.150) has full order state observer described by (6.152) if and only if system (6.150) or matrix pair $(\boldsymbol{G},\boldsymbol{C})$ is observable.

The practical design problem of observer (6.152) can be realized by applying the dual relationship of observer design and state feedback stabilization. In fact, let

$$G'=G^{\mathrm{T}}, H'=C^{\mathrm{T}}, K=L^{\mathrm{T}} \tag{6.157}$$

Then

$$G+LC=[G'+H'K]^{\mathrm{T}} \tag{6.158}$$

Consequently

$$\lambda_i(G+LC)=\lambda_i(G'+H'K), i=1,2,\cdots,n \tag{6.159}$$

Then, the solving problem is transferred to a state feedback stabilization problem of the discrete time system (6.160).

$$X(k+1)=G'X(k)+H'u(k) \tag{6.160}$$

That is to say, the observer can be designed by applying poles placements method. In addition, the property of separation principle is also suitable for designing the discrete time control system with observer, as shown in Figure 6.9. Namely, the state feedback control law and observer can be carried out independently. Here, we will not explain it detailed.

Example 6.17 Consider the discrete time LTI system described by

$$X(k+1)=\begin{bmatrix} 0 & 1 \\ -0.16 & -1 \end{bmatrix}X(k)+\begin{bmatrix} 0 \\ 1 \end{bmatrix}u(k)$$

$$y(k)=[0 \quad 1]X(k)$$

Please design a full-order observer, its expected characteristic values are

$$z_1=0.5+\mathrm{j}0.5, z_2=-0.5+\mathrm{j}0.5$$

Solution The state equation yields the system matrix, input matrix and output matrix

$$G=\begin{bmatrix} 0 & 1 \\ -0.16 & -1 \end{bmatrix}, h=\begin{bmatrix} 0 \\ 1 \end{bmatrix}, c=[0 \quad 1]$$

According to Theorem 6.15, detecting the rank of observability matrix firstly, its rank is

$$\mathrm{rank}\begin{bmatrix} c \\ cG \end{bmatrix}=\mathrm{rank}\begin{bmatrix} 0 & 1 \\ -0.16 & -1 \end{bmatrix}=2=n$$

Consequently, the system is observable. Then a full order state observer described by (6.152) exists.

Next, we will design the observer gain vector l by applying the poles placement method.

Let

$$G'=G^{\mathrm{T}}, h'=c^{\mathrm{T}}$$

The expected characteristic equation determined by the closed-loop poles is

$$|zI-G'+h'k|=(z-0.5-\mathrm{j}0.5)(z+0.5-\mathrm{j}0.5)=z^2-z+0.5=0$$

Hence $$a_1^*=-1, a_2^*=0.5$$

The characteristic equation of the original system is

$$|zI-G'|=\begin{vmatrix} z & -1 \\ 0.16 & z+1 \end{vmatrix}=z^2+z+0.16=0$$

Therefore

$$a_1=1, a_2=0.16$$

Then, the coefficient vector $\bar{k}$ is

$$\bar{k}=[k_2^* \quad k_1^*]=[a_2^*-a_2 \quad a_1^*-a_1]=[0.34 \quad -2]$$

It is obvious that the original system or (G', h') is in controllable canonical form. Then the trans for mation matrix T is the unit matrix I, so $k=\bar{k}=[0.34 \quad -2]$.

Then, by equation (6.157)

$$l=k^{\mathrm{T}}=\begin{bmatrix} 0.34 \\ -2 \end{bmatrix}$$

Problems

6.1 Consider the discrete time system, its difference equation is

$$y(k+3)+3y(k+2)+5y(k+1)+y(k)=u(k+1)+2u(k)$$

(1) Determine the state space description of the discrete time system.

(2) Draw the simulation diagram.

6.2 The following discrete time systems are described by their impulse transfer function as follows.

(1) $G(z)=\dfrac{6(z+1)}{z(z+2)(z+4)}$ (2) $G(z)=\dfrac{2z+1}{z^2+6z+5}$

Determine the state space description of them.

6.3 Consider the discrete time LTI system described by

$$\begin{cases} X(k+1)=\begin{bmatrix} 0 & 1 \\ 1 & 3 \end{bmatrix} X(k)+\begin{bmatrix} 0 \\ 1 \end{bmatrix} u(k) \\ y(k)=[1 \quad 2] X(k) \end{cases}$$

Determine the impulse transfer function matrix of the discrete system.

6.4 Consider the discrete time LTI system, its state equation is

$$X(k+1)=\begin{bmatrix} 0 & 1 \\ -0.1 & -0.7 \end{bmatrix} X(k)+\begin{bmatrix} 0 \\ 1 \end{bmatrix} u(k)$$

Calculate the state transition matrix of the system.

6.5 Consider the continuous time system, its state equation is

$$\begin{bmatrix} \dot{x}_1 \\ \dot{x}_2 \end{bmatrix}=\begin{bmatrix} 0 & 1 \\ 0 & 0 \end{bmatrix}\begin{bmatrix} x_1 \\ x_2 \end{bmatrix}+\begin{bmatrix} 0 \\ 1 \end{bmatrix} u$$

Determine the discretization model of the continuous time, T denotes the sampling period, $T=2$s.

6.6 Consider the continuous time system, its state equation is

$$\dot{X}=\begin{bmatrix}0 & 1\\0 & -2\end{bmatrix}X+\begin{bmatrix}0\\1\end{bmatrix}u, t\geqslant 0$$

(1) Suppose the sampling period is $T=0.1$s, determine the discretization model of the continuous system.

(2) Suppose the initial state is $X(0)=\begin{bmatrix}1\\-1\end{bmatrix}$, the input signal is unit step, i. e. $u(k)=1$, determine the state response and output response at every sampling points.

6.7 Determine the stability of the following discrete time LTI system

$$\begin{cases}x_1(k+1)=-x_1(k)-2x_2(k)\\x_2(k+1)=x_1(k)-4x_2(k)\end{cases}$$

(1) Determine the stability of the system.

(2) If the system is stable, then determine a Lyapunov function of the system.

6.8 Determine the controllability and observability of the following systems.

(1) $X(k+1)=\begin{bmatrix}0 & 1\\0 & -3\end{bmatrix}X(k)+\begin{bmatrix}0\\1\end{bmatrix}u(k), y(k)=[0\quad 2]X(k)$

(2) $X(k+1)=\begin{bmatrix}0 & 1 & 0\\0 & 0 & 1\\0 & -2 & -3\end{bmatrix}X(k)+\begin{bmatrix}0\\1\\1\end{bmatrix}u(k), y(k)=[1\quad 0\quad 1]X(k)$

6.9 Consider the discrete time LTI system described by

$$X(k+1)=\begin{bmatrix}a & b\\c & d\end{bmatrix}X(k)+\begin{bmatrix}1\\1\end{bmatrix}u(k)$$

$$y(k)=[1\quad 0]X(k)$$

Please determine the conditions of a, b, c and d such that the system state is completely controllable and observable.

6.10 The state equation of the discrete time LTI system is

$$X(k+1)=\begin{bmatrix}0 & 1\\-0.16 & -1\end{bmatrix}X(k)+\begin{bmatrix}1\\0.5\end{bmatrix}u(k), \text{and } X(0)=\begin{bmatrix}1\\-1\end{bmatrix}$$

(1) Determine the controllability of the system.

(2) If it is controllable, determine a control input sequence $u(0)$ and $u(1)$, such that $X(2)=[-1\quad 2]^{\mathrm{T}}$.

6.11 Consider the continuous time system described by

$$\dot{X}=\begin{bmatrix}0 & 1\\-25 & -6\end{bmatrix}X+\begin{bmatrix}0\\1\end{bmatrix}u$$

$$y=[3\quad 1]X$$

The system is completely controllable and observable, please determine the condition of the discretization system remaining its controllability and observability. Suppose the sampling period is T.

6.12 Consider the discrete time LTI system described by

$$X(k+1)=\begin{bmatrix}0 & 1\\ -2 & -3\end{bmatrix}X(k)+\begin{bmatrix}0\\ 1\end{bmatrix}u(k)$$

$$y(k)=[1 \quad 2]X(k)$$

(1) Design a state feedback control law that assigns the set of closed-loop eigenvalues as $\{-3, -6\}$.

(2) Try to draw the state simulation diagram of the closed-loop system after state feedback.

6.13 Consider the discrete time LTI system described by

$$X(k+1)=\begin{bmatrix}0 & 0 & 0\\ 1 & -6 & 0\\ 0 & 1 & -12\end{bmatrix}X(k)+\begin{bmatrix}1\\ 0\\ 0\end{bmatrix}u(k)$$

$$y(k)=[1 \quad 0 \quad 0]X(k)$$

Please design a full-order observer, its expected characteristic values are $z_1=-2$, $z_{2,3}=-1\pm j$.

Index

A

adjoint matrix 伴随矩阵

asymptotically stable i. s. L 李亚普诺夫意义下渐近稳定

B

block diagonal matrix 分块对角矩阵

block diagram 方框图

C

canonical decomposition 规范分解

Cayley-Hamilton Theorem 凯莱-哈密尔顿定理

characteristic equation 特征方程

characteristic polynomial 特征多项式

closed-loop system 闭环系统

coefficient matrix 系数矩阵

column vector 列向量

companion matrix 友矩阵

completely controllable 完全能控

completely reachable 完全能达

complete solution of the state equation 状态方程的全解

complex conjugate pairs 共轭复数对

composite system 组合系统

controllability 能控性

controllability matrix 能控性矩阵

controllable canonical decomposition 能控规范分解

controllable canonical form 能控规范型

controllable canonical form realization 能控规范型实现

control matrix 控制矩阵

curl condition 旋度条件

D

data-sampled system 数据采样系统

definiteness 定号性

desired or expected closed-loop poles 期望的闭环极点

determinant 行列式

diagonal canonical form 对角线规范型

diagonal canonical form criteria 对角线规范型判据

diagonal matrix 对角阵

dimension 维数

discrete-time control system 离散时间控制系统

discrete-time LTI system 离散时间线性定常系统

discrete-time model 离散时间模型

discretization 离散化

duality principle 对偶原理

dual system 对偶系统

dynamic system 动态系统

E

eigenvalue 特征值

eigenvector 特征向量

equilibrium point 平衡点

equivalent 等价

Euclidean norm 欧几里得范数

F

forward matrix 前馈矩阵

full column rank 列满秩

full-order observer 全维观测器

full row rank 行满秩

G

generalized eigenvector 广义特征向量
global asymptotically stability 大范围渐进稳定
global uniformly asymptotically stable 大范围一致渐近稳定
Gramian criteria 格莱姆判据
Gramian matrix 格莱姆矩阵

H

homogeneous equation 齐次方程

I

iff (if and only if) 当且仅当
impulse response matrix 脉冲响应矩阵
indefinite 不定
initial condition 初始条件
input matrix 输入矩阵
invertible 可逆的
I/O description 输入输出描述
irreducible realization 不可简约实现

J

Jacobi matrix 雅可比矩阵
Jordan block 约当块
Jordan canonical form 约当规范型
Jordan canonical form criteria 约当规范型判据
Jordan matrix 约当阵

L

Laplace transform 拉普拉斯变换
leading principal minor determinant 顺序主子式
linear independently 线性无关
linearization 线性化
linearly dependant 线性相关
linear time invariant (LTI) system 线性定常系统
linear time varying (LTV) system 线性时变系统
linear transformation 线性变换
LTI homogeneous system 线性定常自治系统
Lyapunov equation 李亚普诺夫方程
Lyapunov function 李亚普诺夫函数
Lyapunov stability theory 李亚普诺夫稳定性理论

M

matrix exponential function 矩阵指数函数
MIMO (multiple input multiple output) 多输入多输出
minimum beat control 最小拍控制
minimum beat observation 最小拍观测
minimal realization 最小实现
minimum set of variables 最小变量集

N

necessary and sufficient condition 充分必要条件
negative definite 负定
negative semi-definite 负半定
nonlinear system 非线性系统
nonsingular transformation 非奇异变换
non-zero initial condition 非零初始条件
norm 范数

O

observability 能观(测)性
observability matrix 能观性矩阵
observable 能观(测)的
observable canonical decomposition 能观规范分解
observable canonical form 能观规范型
observer gain matrix 观测器增益矩阵
order 阶次
output equation 输出方程
output feedback 输出反馈
output matrix 输出矩阵

P

PBH rank criteria PBH 秩判据

poles 极点
poles placement 极点配置
pole-zero cancellation 零极点对消
positive definite 正定
positive semi-definite 正半定
power series 幂级数
principal minor determinant 主子式
proper rational function matrix 正则函数矩阵
proper rational system 正则系统

Q

quadratic form function 二次型函数

R

rank criteria 秩判据
reachability 可达性
reachable 可达的
realization problem 实现问题
reduced-order observer 降维观测器

S

sampling period 采样周期
scalar function 标量函数
separation principle 分离性原理
similarity transformation 相似变换
simulation diagram 仿真图
SISO (single input single output) 单输入单输出
stability 稳定性
stable i. s. L 李亚普诺夫意义下稳定
state 状态
state equation 状态方程
state estimation 状态估计
state feedback control law 状态反馈控制律
state feedback matrix 状态反馈矩阵
state observer 状态观测器
state reconstruction 状态重构
state response 状态响应
state space 状态空间
state space description 状态空间描述
state solution 状态解
state transformation 状态变换
state transition matrix 状态转移矩阵
state trajectory 状态轨迹
state variable 状态变量
state vector 状态向量
strictly proper rational function matrix 严格正则函数矩阵
strictly proper rational system 严格正则系统
Sylvester criterion 赛尔维斯特准则
system matrix 系统矩阵

T

Taylor series 泰勒级数
time response 时间响应
transfer function matrix 传递函数矩阵
transpose 转置

U

uncontrollable 不能控
uniformly asymptotically stable 一致渐近稳定
uniformly stable 一致稳定
unobservable 不能观测
unreachable 不能达
unstable i. s. L 李亚普诺夫意义下不稳定

V

Vandermonde matrix 范德蒙矩阵
variable gradient method 变量梯度法

Z

zero initial condition 零初始条件
zero initial state time response 零状态时间响应
zero input response 零输入响应
zero input state time response 零输入状态时间响应
zero-order hold 零阶保持器
zeros 零点
z transform z 变换

References

[1] Steven J Leon. Linear Algebra with Application [M]. 6th ed. 北京：机械工业出版社，2004.

[2] 杨明，刘先忠．矩阵论[M]. 2 版．武汉：华中科技大学出版社，2005.

[3] John Dorsey. Continuous and Discrete Control systems[M]. 北京：电子工业出版社，2002.

[4] Gene F Franklin，J David Powell，Abbas Emami-Naeini. Feedback Control of Dynamic System [M]. 4th ed. 北京：高等教育出版社，2003.

[5] Benjamin C Kuo，Farid Golnaraghi. Automatic Control Systems [M]. 8th ed. 北京：高等教育出版社，2003.

[6] Katsuhiko Ogata. Discrete-Time Control systems[M]. 2nd ed. 北京：机械工业出版社，2004.

[7] Richard C Dorf，Robert H Bishop. Modern Control Systems [M]. 11th ed. 北京：电子工业出版社，2005.

[8] 李素玲，胡健．自动控制原理[M]. 西安：西安电子科技大学出版社，2007.

[9] 郑大钟．线性系统理论[M]. 2 版．北京：清华大学出版社，2004.

[10] 段广仁．线性系统理论[M]. 2 版．哈尔滨：哈尔滨工业大学出版社，2004.

[11] Jing Yuanwei，Zhou Yucheng，Jiang Nan. Modern Cotrol Theory[M]. 北京：冶金工业出版社，2006.

[12] 陆军，王晓陵．线性系统理论[M]. 哈尔滨：哈尔滨工程大学出版社，2006.

[13] 张嗣瀛，高力群．现代控制理论[M]. 北京：清华大学出版社，2006.

[14] 谢克明．现代控制理论[M]. 北京：清华大学出版社，2007.

[15] 施颂椒，陈学中，杜秀华．现代控制理论基础[M]. 2 版．北京：高等教育出版社，2009.

[16] 李道根．Modern Control Theory[M]. 哈尔滨：哈尔滨工业大学出版社，2009.

[17] 卢泽生，吴振顺，孙雅洲．控制理论及其应用[M]. 北京：高等教育出版社，2009.

[18] 赵光宙．现代控制理论[M]. 北京：机械工业出版社，2010.

[19] 许委胜，朱劲，王中杰．Foundation of Modern Control Theory[M]. 上海：同济大学出版社，2011.